D1481871

BINDING
CONSTANTS

BINDING
CONSTANTS

The Measurement
of Molecular Complex Stability

KENNETH A. CONNORS
School of Pharmacy
University of Wisconsin-Madison

A WILEY-INTERSCIENCE PUBLICATION
JOHN WILEY & SONS
New York . Chichester . Brisbane . Toronto . Singapore

Copyright © 1987 by John Wiley & Sons, Inc.

All rights reserved. Published simultaneously in Canada.

Reproduction or translation of any part of this work
beyond that permitted by Section 107 or 108 of the
1976 United States Copyright Act without the permission
of the copyright owner is unlawful. Requests for
permission or further information should be addressed to
the Permissions Department, John Wiley & Sons, Inc.

Library of Congress Cataloging-in-Publication Data

Connors, Kenneth A. (Kenneth Antonio), 1932-
    Binding constants.

    "A Wiley-Interscience publication."
    Includes index.
    1. Chemical equilibrium.   2. Chemical bonds.
I. Title.
QD503.C67   1987   541.2′24   86-29016
ISBN 0-471-83083-6

10 9 8 7 6 5 4 3 2

To my mother,
Adeline Gioia Connors Lombardo

PREFACE

Equilibrium processes in which noncovalent interactions take place occur widely in chemical and biochemical systems, and the measurement of their equilibrium constants is an important step in describing and understanding them. Among these processes are metal ion–ligand coordination equilibria; acid–base reactions; molecular complex formation involving hydrogen bonding, charge transfer, dispersion forces, and the hydrophobic effect; ion pairing; self-association, including micellization; enzyme–substrate complexation; and the binding of small molecules to many types of macromolecules. The study of each of these phenomena tends to be a specialty, with its own methods and literature. An experimental technique developed in one field may be useful to workers in another field, yet communication among fields is often not very effective. One goal of this book is to go beyond the limits of application defined by the type of reaction and bring together descriptions of methods from many fields as an aid to their more widespread application.

There is another way that I think this book can be helpful, especially to readers without much experience in measuring binding constants. Some of the original journal literature may be difficult to comprehend as a guide to experimental work, for several reasons, including its variable nomenclature and symbolism, historical shifts of emphasis, individual styles, and especially its conciseness. By making use of a fairly consistent symbolism and by more expansively developing quantitative relationships, I have tried to minimize these difficulties, and in this way I hope the book will help the reader approach the journal literature for fuller understanding. Derivations are outlined or given in detail for all equations critical to the determination or interpretation of binding constants. However, the original literature usually contains much more detail than does my treatment, and readers may find it instructive to make use of my description together with the original.

In my citations of the literature I have tried to give credit to the innovators and important contributors, and I have also attempted to lead the interested reader into the literature of the field by citing reviews and significant applications. It is very probable that I have made some misjudgments and have overlooked some notable contributions. I have pointed out where there are controversies in interpretation, and sometimes have made my own contribution to the argument. I have presumed that the reader is familiar with the basic chemistry of each of the measurement techniques whose application to binding is treated here.

It must be evident that my own experience cannot include all of the methods discussed in this book. My coworkers and I have made use of some of these techniques, although we have not used many others. To some extent this personal experience has influenced the emphasis of treatment in the book, but I hope that the book is sufficiently balanced to serve as a useful guide to experimentalists in many fields.

KENNETH A. CONNORS

Madison, Wisconsin
January, 1987

CONTENTS

PART I

GENERAL MATTERS

1

MOLECULAR ASSOCIATION

1.1. INTRODUCTION

Nomenclature

This book is about molecular complexes, species formed by the association of two or more interactant molecules. We use this terminology to describe the interactants:

The *substrate* S is that interactant whose physical or chemical properties are experimentally observed.

The *ligand* L is that interactant whose concentration is the independent variable.

Thus the substrate is, from the observer's point of view, that which is acted on. Among the useful properties are light absorption, the nuclear magnetic resonance phenomenon, solubility, conductance, and chemical reactivity. It is by a change in the observed property as a function of the ligand concentration that the complex formation is recognized and studied. The ligand, in this usage, is merely the "other interactant," with the word carrying no structural or electronic meaning. [The word *ligand*, from the Latin *ligare*, "to bind," was first used in the German literature to describe species bound to a central inorganic ion or atom (1).]

Evidently, the terms *substrate* and *ligand* are operational and imply nothing about the nature of the interaction or the properties of the complex; this mechanistic neutrality is an advantage of the terminology. In principle, it is arbitrary which interactant is made to serve as the substrate and which as the ligand, but in practice the choice is often forced on us by the suitability of properties for observation and by the solubilities of the interactants. In studying self-association, a single interactant is both substrate and ligand.

We define a *molecular complex* as a noncovalently bound species of definite substrate-to-ligand stoichiometry that is formed in a facile equilibrium process in solution. There is no restriction as to charge type. Complexes may also exist in the solid state, but our concern is with solution equilibria. Obviously, there is room for interpretation in this definition. For one thing, the distinction between covalent and noncovalent binding is not clearcut. The term *facile* is vague and may be interpreted to mean only that the rate of attainment of equilibrium is much more rapid than the rates of the measurement processes, so that the system is at equilibrium. Actually, many of the experimental techniques to be described can be applied, with attention to specific requirements of the case, to slow covalent equilibria as well as to fast noncovalent equilibria.

The terms *complex formation, complexation, molecular association,* and *binding* are synonymous. Complex and molecular complex mean about the same thing; the adjective is prescriptive rather than descriptive and serves to rule out some species to which the much-used word *complex* has been applied, such as the σ-complexes that appear as intermediates in aromatic substitution reactions. Some authors would not consider coordinate covalent complexes of metal ions to be molecular complexes, and complexes of small oppositely charged ions are commonly called ion pairs rather than molecular complexes.

We shall always express stoichiometric ratios in the order *substrate : ligand*, so that 1:2 stoichiometry denotes SL_2, 2:1 means S_2L, and so on.

Examples of Complexes

Table 1.1 describes some systems that show complex formation within the sense of our definition. The equilibrium constant K_{11} is a stability constant for the formation of the 1:1 complex SL, namely, $K_{11} =$

Table 1.1. Examples of Molecular Complexes

Substrate	Ligand	Solvent	K_{11}/M^{-1} [a]	$\Delta G^\circ/\text{Kcal mol}^{-1}$ [a]	References
Sodium ion	Chlorate ion	H_2O	3.2	−0.7	2
Iodine	Hexamethylbenzene	CCl_4	1.35	−0.2	3
Tetracyanoethylene	Hexamethylbenzene	CH_2Cl_2	17	−1.7	4
7,7,8,8-Tetracyanoquinodimethan	Pyrene	CH_2Cl_2	0.94	0.0	5
Salicylic acid	Caffeine	H_2O	44	−2.3	6
Hydrocortisone	Benzoate ion	H_2O	2.9	−0.6	7
Methyl trans-cinnamate	Imidazole	H_2O	1.0	0.0	8
Naphthalene	Theophylline	H_2O	64	−2.5	9
p-Hydroxybenzoic acid	α-Cyclodextrin	H_2O	1130	−4.2	10
Caffeine	Caffeine	H_2O	19	−1.7	8
Phenol	Dimethylformamide	C_6H_6	442	−3.6	11

[a] Temperatures range from 22 to 30°C.

5

[SL]/[S][L], all concentrations being expressed in molar units. Note that K_{11} for the S ≡ caffeine, L ≡ caffeine system is a dimerization constant.

The stabilities of these complexes, expressed as K_{11} or as ΔG°, are typical. Although many equilibrium constants smaller than $1\,M^{-1}$ have been reported, there are difficulties in interpreting such small effects as due to complex formation; this problem is treated in Section 2.6. There is no unambiguous upper limit to the stability of a molecular complex, for this is closely related to the problem of distinguishing between non-covalent and covalent bonding. Complexes of transition metal ions with basic ligands may have extremely large stability constants; for example, $\log K_{11} = 18.8$ for the complex of Cu(II) and EDTA tetraanion (12), but this represents a type of bonding different from the examples in Table 1.1. The range 1 to $10^4\,M^{-1}$, corresponding to a ΔG° range (molar concentration basis, 25 °C) of 0 to $-5.5\,\mathrm{kcal\,mol^{-1}}$ or $-23\,\mathrm{kJ\,mol^{-1}}$, will include most noncovalently bound complexes of small molecules. There are more stable complexes than these, however, in some macromolecular systems. Enzyme–inhibitor and antigen–antibody complexes of great stability can form when there exist very favorable mutual structural and electronic features at the binding sites of the two interactants.

Plan of the Book

This book is organized into two parts. Part I (Chapters 1 to 3) includes matters of general validity, applicable to essentially all studies and methods; Part II deals with the experimental methods for measuring stability constants. Concepts, methods, and equations in Chapters 2 and 3 are often referred to in later chapters, and attention is sometimes drawn to specific applications of general developments in these chapters. To workers new to this field, Chapter 2 should be especially helpful as a survey of many important and useful ideas for the design of experiments and the interpretation of results. The appendixes include specialized topics that serve as background to discussions in Chapters 2 and 3.

1.2. BINDING FORCES

The forces responsible for complex formation are not the subject of this book, but one of the reasons to measure binding constants is to learn something about these forces. Moreover, much of the literature on stability constants makes use of concepts and terms dealing with the intermolecular forces, so it is necessary to consider these. This brief treatment is designed to provide the background necessary for understanding the literature.

Potential Energy Functions

Although we commonly speak of intermolecular forces, it is the intermolecular energies that appear in quantum mechanical and statistical mechanical treatments. Let $V(r)$ be the potential energy of interaction of two particles and $F(r)$ be the force of interaction, where r is the interparticle distance of separation. Then these quantities are related by

$$V(r) = \int_r^\infty F(r)\, dr \tag{1.1}$$

$$F(r) = -\frac{dV(r)}{dr} \tag{1.2}$$

Thus, for example, if $V(r)$ were proportional to r^{-n}, $F(r)$ would be proportional to $r^{-(n+1)}$. It is conventional to take as the zero of potential energy the state in which the particles are infinitely separated. A negative $V(r)$ is attractive, a positive value is repulsive. We are interested in the dependence of $V(r)$ on r. The following treatment is drawn largely from Hirschfelder et al. (13). Consider the two-particle system of S and L. The (noncovalent) attractive forces are of three classes: electrostatic, induction, and dispersion.

The *electrostatic* interactions consist of interactions between the multipole moments of polar molecules; these moments are charges (C), dipole moments (μ), and quadrupole moments (Q). Except for charge–charge interactions, the electrostatic potential energies depend on the mutual orientation of the interacting moments. However, the average potential energy $\bar{V}(r)$, which is the angle-dependent potential energy averaged over the angles, weighted by Boltzmann factors, is dependent only on intermolecular distance. The results follows, with the subscript symbols indicating the type of interaction.

$$\bar{V}_{C,C} = +\frac{C_S C_L}{r} \tag{1.3}$$

$$\bar{V}_{C,\mu} = -\frac{1}{3kT}\frac{C_S^2 \mu_L^2}{r^4} \tag{1.4}$$

$$\bar{V}_{C,Q} = -\frac{1}{20kT}\frac{C_S^2 Q_L^2}{r^6} \tag{1.5}$$

$$\bar{V}_{\mu,\mu} = -\frac{2}{3kT}\frac{\mu_S^2 \mu_L^2}{r^6} \tag{1.6}$$

$$\bar{V}_{\mu,Q} = -\frac{1}{kT}\frac{\mu_S^2 Q_L^2}{r^8} \tag{1.7}$$

$$\bar{V}_{Q,Q} = -\frac{7}{40kT}\frac{Q_S^2 Q_L^2}{r^{10}} \tag{1.8}$$

In these equations k is the Boltzmann constant and T the absolute temperature. Equation (1.3) is Coulomb's law; the charges are to be accompanied by their signs. Because of the relatively high-order reciprocal dependence seen in Eqs. (1.7) and (1.8), these quadrupolar contributions will usually be negligible. For neutral polar molecules the most important contributor to the electrostatic potential energy is the dipole-dipole interaction, Eq. (1.6), which has the r^{-6} dependence.

The *induction* or polarization forces arise from the effect of a moment in a polar molecule inducing a charge separation in an adjacent molecule. The average potential energy functions are

$$\bar{V}_{C,\text{ind}\mu} = -\frac{C_S^2 \alpha_L}{2r^4} \tag{1.9}$$

$$\bar{V}_{\mu,\text{ind}\mu} = -\frac{\mu_S^2 \alpha_L}{r^6} \tag{1.10}$$

where α_L is the polarizability of L. The r^{-6} dependence in Eq. (1.10) is to be noted. A quadrupole-induced dipole contribution is also possible (14).

The *dispersion* (London) force is a quantum mechanical phenomenon. At any instant the electronic distribution in molecule S may result in an instantaneous dipole moment, even if S is a spherical nonpolar molecule. This instantaneous dipole induces a moment in L, which interacts with the moment in S. For nonpolar spheres the "induced dipole-induced dipole" dispersion energy function is

$$V_{\text{disp}} = -\frac{3}{4}\left[\frac{\epsilon_S \epsilon_L}{\epsilon_S + \epsilon_L}\right]\frac{\alpha_S \alpha_L}{r^6} \tag{1.11}$$

where ϵ_S, ϵ_L are characteristic energies. There are less important terms contributing to V_{disp} from higher induced moments. For asymmetric molecules the dispersion contribution is angle dependent, but the distance dependence remains as in Eq. (1.11). Conjugated and aromatic molecules show more complicated behavior (15). The London dispersion force is also called the van der Waals force (though some authors use "van der Waals forces" as a general term to include all the long-range noncovalent interactions).

As the two molecules approach each other, they are subject to these long-range r^{-n} attractive forces. Eventually, they will experience a repulsive force as r becomes very small, the origin of this force being electron–electron repulsion (ultimately due to the Pauli exclusion principle). The repulsive interaction is approximately an exponential function of the intermolecular distance, and still more approximately is often taken as a reciprocal power of distance. The net potential energy function is a combination of the repulsive and attractive components, and it has the shape seen in Figure 1.1. This curve has been calculated with Eq. (1.12), which is called the Lennard-Jones 6-12 potential.

$$V(r) = 4V_{min}\left[\left(\frac{r_0}{r}\right)^{12} - \left(\frac{r_0}{r}\right)^{6}\right] \tag{1.12}$$

In Eq. (1.12), $-V_{min}$ is the value of $V(r)$ at the minimum in the potential well, that is, where $r = r_e$, the equilibrium intermolecular distance; r_0 is the value of r when $V(r) = 0$. The term in r^{-12} is the repulsive term. Note that the attractive term has the r^{-6} dependence seen in several of the earlier potential functions. Equation (1.12) gives a reasonable description of the properties of real nonpolar gases. V_{min} is typically of the order $5 \, kcal \, mol^{-1}$ or less. (Interaction leading to covalent bond formation can be described with a qualitatively similar picture, but the potential well is of the order $100 \, kcal \, mol^{-1}$.) At the potential minimum, the net force is zero according to Eq. (1.2).

Chemical Interactions

The fundamental (in a chemical sense) forces of interaction consist of electron–electron (exchange) and internuclear repulsive forces, the short-range covalent bonding attractive forces, and the long-range noncovalent attractive forces. This list includes all the important possibilities. Yet chemists are accustomed to invoking certain specifically *chemical* types of binding, and we should consider how these are related to the basic forces.

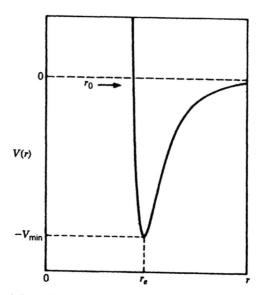

Figure 1.1. Plot of Eq. (1.12), the Lennard-Jones 6–12 potential energy function. The curve was calculated with the parameters for nitrogen: $V_{min}/k = 95$ K, $r_0 = 3.70$ Å, $r_e = 4.15$ Å.

del Re (16) has pointed out that we still lack a unifying concept of binding and lists a wide variety of chemical bonds, of which we are particularly interested in charge-transfer and hydrogen bonds.

The *charge-transfer* concept was introduced by Mulliken (17) as a description of complex formation in which a new electronic absorption band, attributable to neither the substrate nor the ligand, is observable. The benzene–iodine system studied by Benesi and Hildebrand (18) is typical. Consider an interactant D capable of acting as an electron donor and an interactant A that can function as an electron acceptor. The ground state of the 1:1 molecular complex formed by D and A is formally described by the wave function Ψ_N:

$$\Psi_N = a\Psi_0(D, A) + b\Psi_1(D^+\!\!-\!A^-) \qquad (1.13)$$

In this representation (D, A) is often called the "no-bond" form, but Ψ_0 is defined to include all of the classical noncovalent interactions. $(D^+\!\!-\!A^-)$ is the "dative" form, and Ψ_1 represents a covalent contribution to the complex; the physical picture is that an electron is transferred from D to A. [Equation (1.13) strictly should be extended to include a contribution from $(D^-—A^+)$, but this will often be negligible.] The molecular complex is called a charge-transfer (CT) complex or an electron donor–acceptor (EDA) complex. The parameters a and b are related by the normalization $a^2 + b^2 + 2abS_{01} = 1$, where S_{01} is the overlap integral between Ψ_0 and Ψ_1. For weak complexes S_{01} may be negligible, but in general the weights or fractions of the no-bond (F_0) and dative (F_1) structures contributing to the complex are (19) as follows:

$$F_0 = a^2 + abS_{01} ; \qquad F_1 = b^2 + abS_{01}$$

The wave function for the excited state of the complex is

$$\Psi_V = a^*\Psi_1(D^+—A^-) - b^*\Psi_0(D, A) \qquad (1.14)$$

where $a^* \approx a$ and $b^* \approx b$. Normally, $a^2 \gg b^2$, so the no-bond structure makes the principal contribution to the ground state, whereas the dative structure makes the main contribution to the excited state $(a^{*2} \gg b^{*2})$.

This theory of charge-transfer complexes is an extension of the more familiar quantum mechanical description of molecules like HCl, whose wave function has contributions from a nonionic covalent structure H—Cl and an ionic dative structure H^+Cl^-. The theory can be generalized to other types of electron transfer reactions such as acid–base reactions, where an electron pair is transferred; in this application the base is the electron donor, the acid is the acceptor. In the context of molecular complexes, perhaps the most novel feature of charge-transfer theory is the idea that nonlocalized covalent binding is possible. That is, the theory does not require that the electron transfer occur between specific atoms, but admits the possibility of charge transfer between delocalized orbitals, as in aromatic compounds.

There has been much discussion of the importance of the dative structure in determining the stability of the complex. This has been assessed theoretically and experimentally. For example, if neither D nor A has a permanent dipole moment, whereas the complex has a dipole moment, it is inferred that the dative structure is responsible. The charge-transfer contribution to the complex stability can be appreciable in unusual cases. Calculations on the trimethylamine–sulfur dioxide complex in the gas phase (20) give a total interaction energy of $-14.8 \, \text{kcal mol}^{-1}$, this being composed of the following contributions:

electrostatic, -31.8; induction, -4.9; charge transfer, -14.1; exchange repulsion, $+36.0$. For most complexes, however, the charge-transfer contribution to the complex stability appears to be minor or negligible, and these complexes may be considered to be essentially noncovalently bound, despite their possession of intense electronic absorption spectra. Reference (21) cites several sources on this question.

The second class of specific interaction to be considered is *hydrogen bonding*. Pimentel and McClellan (22) have given a useful definition: a hydrogen bond exists when (1) there is evidence of a bond, and (2) this bond involves a hydrogen atom already bonded to another atom. Thus the bond formation process can be written

$$A—H + B \rightleftharpoons A—H \cdots B$$

with the nature of the hydrogen bond, denoted by dots, being of paramount interest.

As with charge-transfer complexes, the key question concerns the extent of covalent bonding contributing to the strength of the hydrogen bond. We have pointed out that acid–base reactions can be treated within the context of charge-transfer theory, and so hydrogen bonding can be discussed in such terms. Calculations of the contributions of the several interaction energies to hydrogen-bonded complexes suggest that the electrostatic component is usually dominant, with a small but significant contribution from charge transfer (21e, f). There is an indication that very strong hydrogen bonding is primarily electrostatic and that very weak hydrogen bonding has a larger (relative) charge-transfer component. For example, Morukuma (21e) obtained for the strong complex $H_3N \cdots HF$, electrostatic (ES), $-25.6 \, \text{kcal mol}^{-1}$, CT, $-4.1 \, \text{kcal mol}^{-1}$; for the weaker complex $H_2O \cdots HOH$, ES, $-10.5 \, \text{kcal mol}^{-1}$, CT, $-2.4 \, \text{kcal mol}^{-1}$. (The total interaction energy, of course, also has induction and exchange repulsion components.)

The Solvent

Thus far in our description of binding forces the solvent has not been explicitly included, and most of the discussion has been applicable to the gas phase rather than the solution phase. We now introduce the solvent. It is obvious, but easy to forget, that the solvent is itself a molecular system rather than a continuous medium and that solvent molecules are subject to the same intermolecular forces already described for interactants (solutes). Writing M for a molecule of the solvent (medium), the coming together of interactants S and L to form a complex is represented approximately as

$$SM + LM \rightleftharpoons SL + MM$$

where SM and LM represent solvated species and MM describes solvent–solvent interaction. If the system molecular properties do not lead to strong solute–solvent or solvent–solvent interactions, then the complex formation will reflect predominantly the solute–solute (SL) interactions. If, however, solvation and solvent–solvent interactions are extensive, these may dominate the process, and SL intermolecular interactions may be masked and perhaps relatively unimportant.

The overall free energy change for this process is conveniently written as the sum in Eq. (1.15).

$$\Delta G = \Delta G_{MM} + \Delta G_{MS} + \Delta G_{SS} \tag{1.15}$$

where ΔG_{MM} is the contribution arising from medium–medium interactions; ΔG_{MS} includes all medium–solute interactions, that is, all solvation phenomena; and ΔG_{SS} includes all solute–solute contributions. These three free energy terms owe their existence to the basic modes of interaction that have already been discussed; ΔG_{MM}, ΔG_{MS}, and ΔG_{SS} can be thought of as the "driving forces" for the complexation.

The ΔG_{SS} term can only make a stabilizing contribution to ΔG at equilibrium. The ΔG_{MS} term includes a substrate–medium, ligand–medium, and complex–medium interactions, so it can be either stabilizing or destabilizing, but in most circumstances it will be destabilizing, because SM and LM interactions compete with SL interactions.

The ΔG_{MM} term is the most important one in the present context. If MM interactions are very weak (nonpolar solvents subject only to dispersion interactions), then solvation effects will also be small, and the ΔG_{SS} term will dominate ΔG. In polar solvents, however, most notably in water, the MM interactions can be very strong, and then ΔG_{MM} may be practically the only contributor to ΔG, since ΔG_{MS} and ΔG_{SS} may roughly cancel each other. ΔG_{MM} constitutes the *hydrophobic effect* in aqueous solutions (the solvophobic effect in general). Jencks (23) has given a very useful review of interactions in aqueous solutions, and Tanford's book (24) is a treatment of the hydrophobic effect.

The hydrophobic effect is manifested by the very low solubility of nonpolar substances in water and by the tendency of nonpolar solutes to form aggregates in aqueous solution. Both phenomena are explained on the basis of very strong water–water interactions, resulting in "squeezing out" of the nonpolar solutes, which themselves do not interact strongly with water. Two points of view have developed to account for this effect. One focuses attention on the structure of water in the vicinity of a nonpolar solute. When a nonpolar solute dissolves in water, the strong intermolecular hydrogen bonding characteristic of water must be modified. No H-bonds from water to the solute can form, but a new network of water–water H-bonds develops close to the solute. There may actually be an increase in the number of H-bonds relative to those in the absence of the solute; the solute is said to be "structure making." However, the possible orientations are more restricted in the presence of the solute, so the entropy change is unfavorable.

Now consider two such dissolved molecules brought into mutual contact. The structured solvent that originally accompanied each of them must reorganize about this new entity; some of the formerly structured solvent is released to become bulk water. Thus there may be a net loss of hydrogen bonding (resulting in a positive enthalpy change for the process), but the entropy change will be positive. This picture describes the "classical" view of aggregation via the hydrophobic effect, a process driven by a favorable entropy change.

Some workers have used this result as a diagnostic tool for (or against) a hydrophobic contribution to binding, but the situation is not clearcut, and it appears that hydrophobic association may be either entropy driven or enthalpy driven (23, 25). Moreover, moderately polar solutes also may bind with hydrophobic assistance, and the water structuring may be different; such solutes may be "structure breakers."

The second approach to the hydrophobic effect is the cavity model. The high surface tension of a polar solvent like water is a strong driving force for the minimization of solvent surface area. A dissolved solvent molecule is considered to be contained within a cavity of definite surface area. When two solute molecules come together, there is a reduction in cavity surface area as the two cavities containing the separated solute species coalesce into a single cavity containing the complex. The driving force for association is the reduction in surface free energy (which is the product of surface tension and surface area). Theory shows that, to a first approximation, the solvophobic contribution to the free energy of binding is proportional to the product of solvent surface tension and the decrease in cavity surface area (26). There is experimental support for these surface tension and surface area dependencies (9, 27). Refinement of these relationships encounters the problem that the macroscopic concepts (surface area, surface tension) may not be valid on the molecular scale (28).

Returning to Eq. (1.15), we note that ΔG_{MM}, ΔG_{MS}, and ΔG_{SS} may not be independent. If, for example, we adopt the cavity model viewpoint, we find that the ΔG_{MM} (hydrophobic) contribution depends on the change in surface area on complex formation, that is, on the area of overlap between S and L in the complex. But this may be controlled by specific substrate–ligand interactions, so that ΔG_{MM} and ΔG_{SS} are coupled. This has been suggested for some α-cyclodextrin inclusion complexes (29). In another example of such interdependence, it has been proposed that hydrophobic association promotes charge-transfer interaction in the ion pair of tetraphenylborate–Rhodamine 6G (30).

Conclusion

An understanding of the intermolecular forces involved in complexation is, from a chemical point of view, the deepest kind of understanding that we seek. This will result from structural, thermodynamic, and kinetic experimental results interpreted with quantum mechanical and statistical mechanical theory. It is therefore an oversimplification or a trivialization to claim, as a consequence of a few experimental observations, that the binding forces responsible for a particular complex are known. Of course, it is reasonable to draw inferences that are consistent with available knowledge and to make tentative and productive suggestions; but it is not much of a contribution simply to name the possibilities, of which there are not very many, as we have seen. One purpose of the preceding review has been to acquaint the experimentalist with some of the difficulties.

1.3. ALTERNATIVE INTERPRETATIONS

When a measured property of a solution of solutes S and L is not equal to the sum of the values of the property of separate solutions of S and L, each at the same concentration as in their mixture, then it is inferred that some sort of interaction has taken place between them. There are two approaches to the investigation of such behavior. One of these, which will be called the *physical* point of view, is to ascribe the nonadditivity to nonideal behavior and to describe it quantitatively in terms of an activity coefficient, excess thermodynamic function, or virial coefficient. This is an empirical interim treatment, because the ultimate goal is to be able to interpret the nonideality in molecular terms.

The second approach is the *chemical* point of view, in which the nonadditive behavior is ascribed to the occurrence of a chemical reaction, with the formation of one or more new species, which we call molecular complexes. The quantitative description of the system now involves the evaluation of equilibrium constants. As a first level of approximation it is assumed that the nonideality is completely ascribable to the complex formation, and all species act ideally, but closer analysis may require adjustments in the form of additional complexes or the evaluation of activity coefficients of the species.

It is universally believed that complexes can exist, and the purpose of this book is to describe means by which their equilibrium constants may be determined. Nevertheless, we must consider anew for each system encountered whether the physical or the chemical point of view is the more appropriate one. This is a classic problem in the study of ionic solutions, but it is a general problem. An apparent approach is to try to detect the postulated complex experimentally. However suggestive this may be, any observed effects can equally well be discussed in terms of the physical or chemical interpretations. The observation of a very specific effect, such as a charge-transfer spectral transition, does not alter this position, because, first, proximity of solute species is prerequisite to interactions that could be interpreted either as general deviations from ideality or specific complexation, and, second, such proximity is to be expected on purely statistical grounds, with an increase in concentration resulting in a greater rate of bimolecular collisions. (This statistical aspect is discussed in Section 2.6.)

Nor is the magnitude of the nonadditivity an unambiguous diagnostic. For example, Edsall and Wyman (31) point out that $0.1\ M$ NaCl results in a hundredfold increase in solubility of β-lactoglobulin relative to the solubility in water, but this is not considered to be a consequence of complex formation.

The postulate of complex formation does carry with it a stringent restriction that can be tested experimentally, namely, the requirement that the mass-action law be followed. It may on this account be possible to rule out complex formation if it proves impossible to evaluate constant equilibrium quotients for all chemically reasonable stoichiometric models; however, the possibility of nonideal behavior combined with complexation must also be considered.

We should also consider the contrary case in which an equilibrium constant does give a satisfactory description of the solution behavior and ask what restrictions this implies in a physical description of the system. This can be treated analytically. For the physical description of nonideal behavior in a solution of S and L we write

$$a_S^p = \gamma_S^p c_S^p \;; \qquad a_L^p = \gamma_L^p c_L^p \tag{1.16}$$

where the superscript identifies the physical interpretation, a, γ, and c representing activity, activity coefficient, and concentration. For the chemical interpretation let us invoke the formation of $1:1$ complex SL, with the assumption that all species behave ideally. Thus

$$a_S^c = c_S^c \;; \qquad a_L^c = c_L^c \;; \qquad a_{SL}^c = c_{SL}^c \tag{1.17}$$

and

$$K = \frac{c_{SL}^c}{c_S^c c_L^c} \tag{1.18}$$

The mass balance expressions on S are

$$S_t^p = S_t^c \;; \qquad S_t^p = c_S^p \;; \qquad S_t^c = c_S^c + c_{SL}^c \tag{1.19}$$

where S_t represents total (formal) concentration of S. Combining these gives

$$c_S^c = \frac{c_S^p}{1 + Kc_L^c} \tag{1.20}$$

Now suppose that K evaluated experimentally is satisfactorily constant. In the physical interpretation the apparent change in activity of S is equivalent, within the chemical association hypothesis, to a reduction in the concentration of S caused by its conversion to another species. Thus the equivalence of the two viewpoints is expressed in the equality $a_S^p = c_S^c$.

Combining this with Eqs. (1.16) and (1.2) gives

$$\gamma_S^p = \frac{1}{1 + Kc_L^c} \tag{1.21}$$

According to Eq. (1.21), for the conditions adopted here the physical and chemical viewpoints give consistent interpretations if K is constant and if γ_S^p depends on c_L^c as shown. Presumably a significant discrepancy would rule out one of the interpretations, but in practice this would not be decisive. For example, Eq. (1.21) is inconsistent with $\gamma_S^p > 1$, but if, as a next level of approximation, an activity coefficient γ_S^c is defined, Eq. (1.22) is obtained:

$$\gamma_S^p = \frac{\gamma_S^c}{1 + K^c c_L^c} \tag{1.22}$$

where K^c is the equilibrium concentration quotient. Hammett (32) has given a similar treatment, describing our "physical" and "chemical" interpretations as "primitive" and "sophisticated," respectively.

An example of the combination of these viewpoints is provided by the work of Aveyard et al. (33) on solutions of n-alkanols in n-octane. They first measured activity coefficients by vapor pressure osmometry and by infrared absorption, finding good agreement. Then they interpreted the activity coefficients in terms of a self-association model, finding that at high concentration a tetramer forms, whereas at low concentration it is necessary to include the formation of a dimer or trimer.

REFERENCES

1. G. B. Kauffman, W. H. Brock, K. A. Jensen, and C. K. Jørgensen, *J. Chem. Educ.*, **60**, 509 (1983).
2. C. W. Davies, *Ion Association*, Butterworth's, Washington, D.C., 1962, p. 169.
3. L. J. Andrews and R. M. Keefer, *J. Am. Chem. Soc.*, **74**, 4500 (1952).
4. R. E. Merrifield and W. D. Phillips, *J. Am. Chem. Soc.*, **80**, 2778 (1958).
5. L. R. Melby, R. J. Harder, W. R. Hertler, W. Mahler, R. E. Benson, and W. E. Mochel, *J. Am. Chem. Soc.*, **84**, 3374 (1962).

6. T. Higuchi and D. A. Zuck, *J. Am. Pharm. Assoc., Sci. Ed.*, **42**, 138 (1953).
7. T. Higuchi and A. Drubulis, *J. Pharm. Sci.*, **50**, 905 (1961).
8. J. A. Mollica and K. A. Connors, *J. Am. Chem. Soc.*, **89**, 308 (1967).
9. J. L. Cohen and K. A. Connors, *J. Pharm. Sci.*, **59**, 1271 (1970).
10. K. A. Connors, S.-F. Lin, and A. B. Wong, *J. Pharm. Sci.*, **71**, 217 (1982).
11. H. P. Lundgren and C. H. Binkley, *J. Polymer Sci.*, **14**, 139 (1954).
12. G. Schwarzenbach, R. Gut, and G. Anderegg, *Helv. Chim. Acta*, **37**, 937 (1954).
13. J. O. Hirschfelder, C. F. Curtiss, and R. B. Bird, *Molecular Theory of Gases and Liquids*, Wiley, New York, 1954, Ch. 1.
14. M. W. Hanna, *J. Am. Chem. Soc.*, **90**, 285 (1968).
15. J. O. Hirschfelder, C. F. Curtiss, and R. B. Bird, *Molecular Theory of Gases and Liquids*, Wiley, New York, 1954, pp. 968–983.
16. G. del Re, *Int. J. Quantum Chem.*, **19**, 981 (1981).
17. R. S. Mulliken, *J. Am. Chem. Soc.*, **74**, 811 (1952).
18. H. A. Benesi and J. H. Hildebrand, *J. Am. Chem. Soc.*, **71**, 2703 (1949).
19. R. S. Mulliken and W. B. Person, *Molecular Complexes*, Wiley-Interscience, New York, 1969, Ch. 1.
20. J. E. Douglas and P. A. Kollman, *J. Am. Chem. Soc.*, **100**, 5226 (1978).
21. (a) E. M. Kosower, *Prog. Phys. Org. Chem.*, **3**, 81 (1965); (b) P. Claverie, *Molecular Associations in Biology*, B. Pullman, Ed., Academic, New York, 1968, p. 115; (c) M.-J. Mantione, *Molecular Associations in Biology*, B. Pullman, Ed., Academic, New York, 1968, p. 411; (d) R. S. Mulliken and W. B. Person, *J. Am. Chem. Soc.*, **91**, 3409 (1969); (e) K. Morokuma, *Acc. Chem. Res.*, **10**, 294 (1977); (f) P. A. Kollman, *Acc. Chem. Res.*, **10**, 365 (1977).
22. G. C. Pimentel and A. L. McClellan, *The Hydrogen Bond*, Freeman, San Francisco, 1960, p. 195.
23. W. P. Jencks, *Catalysis in Chemistry and Enzymology*, McGraw-Hill, New York, 1969, Ch. 5, 6–9.
24. C. Tanford, *The Hydrophobic Effect: Formation of Micelles and Biological Membranes*, 2nd ed., Wiley-Interscience, New York, 198C.
25. (a) M. H. Abraham, *J. Am. Chem. Soc.*, **104**, 2085 (1982); (b) D. Mirejovsky and E. M. Arnett, *J. Am. Chem. Soc.*, **105**, 1112 (1983).
26. (a) O. Sinanoğlu and S. Abdulnur, *Fed. Proc.*, **24**, Suppl. 15, S-12 (1965); (b) O. Sinanoğlu, *Molecular Associations in Biology*, B. Pullman, Ed., Academic, New York, 1968, p. 427.
27. K. A. Connors and S. Sun, *J. Am. Chem. Soc.*, **93**, 7239 (1971).
28. C. Tanford, *Proc. Nat. Acad. Sci.*, US, **76**, 4175 (1979).
29. K. A. Connors and D. D. Pendergast, *J. Am. Chem. Soc.*, **106**, 7607 (1984).
30. K. Hirano and M. Tokuhara, *Bull. Chem. Soc. Jap.*, **57**, 2031 (1984).

31. J. T. Edsall and J. Wyman, *Biophysical Chemistry*, Vol. I, Academic, New York, 1958, p. 593.

32. L. P. Hammett, *Physical Organic Chemistry*, 2nd ed., McGraw-Hill, New York, 1970, pp. 16–19.

33. R. Aveyard, B. J. Briscoe, and J. Chapman, *J. Chem. Soc. Faraday Trans.*, *I*, **69**, 1772 (1973).

2

BINDING CONSTANTS

2.1. THE EQUILIBRIUM CONSTANT

Definitions

The three simplest complex stoichiometries are SL, SL_2, and S_2L. It is chemically reasonable to assume that every complex is formed in a bimolecular process, so these three complexes are related by the following equilibria:

$$S + L \rightleftharpoons SL$$

$$SL + L \rightleftharpoons SL_2$$

$$S + SL \rightleftharpoons S_2L$$

and we define the *stepwise binding constants*

$$K_{11} = \frac{[SL]}{[S][L]} \tag{2.1}$$

$$K_{12} = \frac{[SL_2]}{[SL][L]} \tag{2.2}$$

$$K_{21} = \frac{[S_2L]}{[S][SL]} \tag{2.3}$$

where brackets signify equilibrium molar concentrations. Of course, these constants should be defined strictly in terms of activities, but, in practice, concentration quotients are nearly always reported; we comment on this later.

Evidently, any complex S_mL_n can be constructed in this manner; thus S_2L_3 is produced from SL and SL_2, and so on. It is also possible to write the formation of higher complexes directly from the substrate and ligand, as in

$$mS + nL \rightleftharpoons S_mL_n$$

and to define the *overall binding constant*

$$\beta_{mn} = \frac{[S_mL_n]}{[S]^m[L]^n} \tag{2.4}$$

Thus, for example, if $m = 1$, $n = 2$, from Eqs. (2.1), (2.2), and (2.4) we get $\beta_{12} = K_{11}K_{12}$. Since products of stepwise constants often appear in mathematical expressions, this kind of substitution can be very convenient.

These K_{mn} quantities have been defined for the formation of the complex S_mL_n, and they are referred to as binding, stability, formation, or association constants. The reciprocal quantity is a dissociation or instability constant.

The units of a stepwise stability constant are (concentration)$^{-1}$, and since the common practice is to use the molar concentration scale, K_{mn} usually has the units M^{-1}. When the mole fraction scale is used, the constant is dimensionless, but some authors specify (mf)$^{-1}$ as its unit in order to clarify the meaning of the number. Sometimes the experiment is carried out by expressing the concentrations of S and SL in molar units and of L as a mole fraction; then K_{11} has the (mf)$^{-1}$ unit.

The constants K_{mn} can be further described as stoichiometric binding constants. The existence of a complex for which m or n is an integer greater than unity means that the ligand or substrate possesses more than one binding site. Consider, as the simplest such example, a system in which the substrate has two binding sites and the ligand has a single binding site. We use the symbol XY to denote the substrate with binding sites X and Y. Now the 1:1 complex can be formed by the interaction of ligand L at either of the sites X or Y, giving the isomeric 1:1 complexes

X'Y and XY', where the primed site indicates the location of the bound ligand. The 1:2 complex X'Y' is formed by the addition of a second ligand to either of the 1:1 complexes. These equilibria are shown in this scheme:

$$
\begin{array}{ccc}
 & \mathrm{X'Y} & \\
{\scriptstyle K_{X'Y}}\nearrow & & \nwarrow{\scriptstyle K^{*}_{X'Y'}} \\
\mathrm{XY} & & \mathrm{X'Y'} \\
{\scriptstyle K_{XY'}}\searrow & & \swarrow{\scriptstyle K^{**}_{X'Y'}} \\
 & \mathrm{XY'} &
\end{array}
$$

The equilibrium constants, $K_{X'Y}$, $K_{XY'}$, $K^{*}_{X'Y'}$, and $K^{**}_{X'Y'}$, whose definitions can be inferred from the reaction scheme, are called *microscopic binding constants*.

The microscopic and stoichiometric binding constants are related. As we shall see in Section 2.6, any experimental method for measuring K_{11} actually gives the sum of the isomeric 1:1 binding constants, or, for the scheme being considered,

$$K_{11} = K_{X'Y} + K_{XY'} \tag{2.5}$$

The 1:2 complex is formed from the 1:1 complex in either form, or

$$K_{12} = \frac{[X'Y']}{([X'Y] + [XY'])[L]}$$

which, upon substitution, gives

$$K_{12} = \frac{a_{XY}K_{X'Y}K_{XY'}}{K_{11}} \tag{2.6}$$

where $a_{XY} = K^{*}_{X'Y'}/K_{XY'} = K^{**}_{X'Y'}/K_{X'Y}$. The quantity a_{XY} is an interaction parameter that measures the extent of interaction between sites X and Y in 1:2 complex formation. In fact, since a_{XY} is a ratio of equilibrium constants, it must itself be an equilibrium constant; a_{XY} is the equilibrium constant for this disproportionation:

$$X'Y + XY' \rightleftharpoons X'Y' + XY$$

Thus if the sites are independent, $a_{XY} = 1$, but there is in general no restriction on its value. The existence of a 1:2 complex implies the existence (though not necessarily the detectability) of both 1:1 complexes.

This connection between stoichiometric and microscopic binding constants has been developed in great detail for protonation equilibria, for which the hydronium ion plays the role of ligand, and the reciprocal of the conventional acid dissociation constant corresponds to our stepwise stoichiometric binding constant (1). The simple case leading to Eqs. (2.5) and (2.6) has been applied to complexation equilibria (2).

Determination of Stoichiometry

It is obvious that before we can define and evaluate a binding constant we must know the stoichiometric coefficients m and n in S_mL_n. This information is obtainable in several ways, which are discussed at appropriate points in the book. Here we examine the techniques called the continuous variation and mole ratio methods. Some of the other approaches are by the analysis of solid complexes, the interpretation of solubility phase diagrams (Chapter 8), the observation of isosbestic points and generalized variants (Chapter 4), the dependence of experimental stability constants on system parameters such as wavelength (Chapter 4), the direct measurement of complex molecular weight (3), the comparison of stability constants evaluated by several techniques (4), and, most generally useful, the fitting of data to assumed stoichiometric models (Sections 2.4, 2.5, 2.6, and Chapter 3).

The possible stoichiometric relationships are determined by the identities of the substrate and ligand, but the detectable stoichiometries may be, to some extent, under the control of the experimenter. Suppose, for example, that 1:1, 1:2, and 2:1 complexes could form. If the system is designed so that the total ligand concentration is very much larger than the total substrate concentration, the formation of SL_2 will tend to be favored relative to S_2L, the ratio of their concentrations being $[SL_2]/[S_2L] = K_{12}[L]/K_{21}[S]$.

The *method of continuous variations*, often called Job's method (5), is sometimes capable of yielding the ratio n/m. The experimental procedure consists of preparing a series of solutions of substrate and ligand subject to the condition that the sum of the total substrate and ligand concentrations is constant. Some property whose value changes when the substrate forms a complex is measured in each solution, and an extremum in this property is related to n/m. To develop this quantitatively we consider the formation of the single complex S_mL_n, with overall stability constant β_{mn} [Eq. (2.4)]. The actual manner of preparing the solutions is irrelevant, but for concreteness suppose separate solutions of S and L, each of the same concentration c mol/L, are mixed by taking v_S mL of substrate solution and v_L mL of ligand solution, these quantities being related by the condition $v_S + v_L = v$, where v is a constant. The solutions may be diluted to a total constant volume V. The total substrate concentration in one of these final solutions is $S_t = v_S c/V$, and similarly $L_t = v_L c/V$. It follows that $S_t + L_t = vc/V$, a constant. Define the dimensionless quantity x:

$$x = \frac{L_t}{S_t + L_t} = \frac{L_t V}{vc} \tag{2.7}$$

(x is often called a mole fraction, but this is an incorrect use of the term.) The mass balance equations for this system are

$$S_t = [S] + m[S_mL_n] \tag{2.8a}$$

$$L_t = [L] + n[S_mL_n] \tag{2.8b}$$

Combining these with the preceding expressions and using Eq. (2.4) gives the equations for the method of continuous variations, where we write $c_f = vc/V$.

$$c_f(1 - x) = [S] + m[S_mL_n] \tag{2.9a}$$

$$c_f x = [L] + n[S_mL_n] \tag{2.9b}$$

$$\beta_{mn}[S]^m[L]^n = [S_mL_n] \tag{2.9c}$$

Combining these gives

$$\beta_{mn}\{c_f(1 - x) - m[S_mL_n]\}^m\{c_f x - n[S_mL_n]\}^n = [S_mL_n] \tag{2.10}$$

The simplest procedure is to take the logarithm of Eq. (2.10), differentiate with respect to x, and set $d[S_m L_n]/dx = 0$, giving as the result

$$\frac{n}{m} = \frac{x_{max}}{1 - x_{max}} \tag{2.11}$$

Most applications of the method have used absorption spectroscopy as the experimental tool, a wavelength being chosen at which a large absorbance change is observed on complexation; from the absorbance of each solution containing S and L is subtracted the absorbance that would have been observed in the absence of complexation, and this absorbance difference is plotted against x ($0 \le x \le 1$) to find x_{max} (or x_{min}). Note that the method gives only the ratio n/m.

Figure 2.1 shows hypothetical continuous variation curves for a 1:1 complex, calculated with Eq. (2.10). The ordinate gives the concentration [SL]; in an experimental situation a property of the solution would be plotted on this axis. Clearly, the sharpness of the maximum, and therefore the accuracy with which x_{max} can be located, depends on the magnitude of the binding constant. Moreover, Ingham (6) has analyzed the case $m = 1$ (complex SL_n) for $n = 1$ to 4, finding that x_{max} for given n ($n > 1$) depends on the product $\beta_{1n} c_f$; the larger this quantity, the closer x_{max} will be to the theoretical value.

A considerable literature has developed on this method, and only a few references will be cited. Likussar and Boltz (7) show calculated continuous variation curves for many simulated systems; these plots reveal asymmetries and inflection points when $n \neq m$, and these can be useful diagnostic features. Most treatments and applications of continuous variations have assumed that only a single complex is present. This may be a poor assumption, and Vosburgh and Cooper (8) studied the effect of more than one complex on the appearance of the continuous variations plot. They found that when only one complex is present, the value of x_{max} is independent of the wavelength at which the solutions were studied, but with more than one complex x_{max} may depend on wavelength. This dependence may be diagnostically helpful. Atkinson (9) has given an interesting geometric interpretation of continuous variations, with a list of references on the subject.

In the *mole ratio method* the total substrate concentration S_t is held constant and the total ligand concentration L_t is varied. A plot of a suitable property of the system against the ratio L_t/S_t is examined for

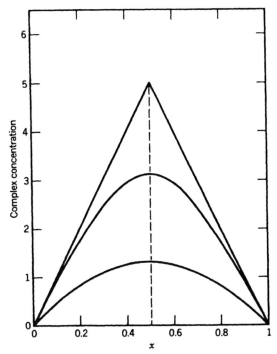

Figure 2.1. Continuous variation curves plotted for hypothetical systems. These were calculated with Eq. (2.10), taking $m = n = 1$, $c_t = 0.01$, and (top to bottom), $\beta_{11} = \infty$, 10^3, 10^2.

discontinuities or abrupt changes of slope corresponding to the stoichiometric ratio. Suppose such a system had the value $\beta_{mn} = \infty$. Then for each m moles of S in the solution, addition of n moles of L will result in the formation of exactly one mole of $S_m L_n$. The concentration of complex will increase until $L_t/S_t = n/m$; beyond this value no more complex can form because all the substrate will have been reacted. Thus a break in the plot will be seen at this point. As the stability constant decreases, of course, the extent of complex dissociation increases, so the break becomes a gently rounded curve, and attempts to locate the discontinuity involve extrapolation of only approximately linear segments over considerable ranges. For elaborations of the method see Meyer and Ayres (10), Chriswell and Schilt (11), and Beltran-Porter et al. (12).

Since the extent of curvature in both the continuous variations and the mole ratio methods depends on the stability constant, evidently it may be possible to extract an estimate of the constant from the same data used to obtain the stoichiometric ratio. Many authors have described ways to do this. The continuous variations experiment is not designed for such service, but the mole ratio experimental design, as we shall see in Section 2.5, is identical with that used to obtain the binding isotherm. Determination of the stability constant will therefore be considered later. Nearly all the theoretical development of the continuous variations and mole ratio techniques has been made in support of studies of metal ion complexes (which are often of high stability); much of this work appears to have been done in virtual isolation from the extensive literature on the measurement of stability constants of organic complexes.

We do not pursue these methods further because of their limitation that they work best when they are needed least, that is, when only a single complex is present. Nevertheless, supporting evidence can be valuable, even if just to strengthen a conclusion that only a 1:1 complex is present. In using a continuous variations plot for this purpose, it should be recalled that the likely additional species are the 2:1 and 1:2 complexes and that any experimentally significant deviation of x_{max} from $\frac{1}{2}$ may be evidence for a second complex.

Reference States

We start with the thermodynamic relationship

$$\mu_i = \mu_i^\circ + RT \ln a_i \qquad (2.12)$$

where μ_i is the *chemical potential* (partial molar free energy) of substance i and a_i is its *activity*. The quantity μ_i° is the *standard chemical potential*; this is evidently the value of μ_i when the activity is unity. The *standard state* is that state of a system in which the activity of substance i is unity.

The activity is an empirical variable related to concentration by Eq. (2.13):

$$a_i = \gamma_i c_i \qquad (2.13)$$

In this relationship γ_i is called the *activity coefficient* of i; γ_i is in general a function of solution composition. Combining Eqs. (2.12) and (2.13),

$$\mu_i = \mu_i^\circ + RT \ln c_i + RT \ln \gamma_i \qquad (2.14)$$

We define the *reference state* as that state of the system in which the activity coefficient of substance i is unity. Thus the standard potential μ_i° is determined when the standard state is defined, and the free energy difference $(\mu_i - \mu_i^\circ)$ is determined when the reference state is defined. In Section 2.2 we treat standard states. Different concentration scales lead to correspondingly different activity coefficients.

For the generalized reaction

$$aA + bB + \cdots \rightleftharpoons mM + nN + \cdots$$

the thermodynamic equilibrium constant is defined as

$$K = \frac{a_M^m a_N^n \cdots}{a_A^a a_B^b \cdots} \qquad (2.15)$$

where it is now understood that the a_i refer to the system at equilibrium. It is a fundamental thermodynamic result that

$$\Delta\mu^\circ = -RT \ln K \qquad (2.16)$$

$\Delta\mu^\circ$ (which is often written ΔG°) is the change in free energy, per mole, accompanying the conversion of reactants into products when all substances are in their standard states.

Equation (2.15) can also be written

$$K = \frac{c_M^m c_N^n \cdots}{c_A^a c_B^b \cdots} \cdot \frac{\gamma_M^m \gamma_N^n \cdots}{\gamma_A^a \gamma_B^b \cdots}$$

$$K = K_c K_\gamma \qquad (2.17)$$

where K_c is the concentration equilibrium constant or concentration quotient. Evidently, when all substances are in their reference states, $K = K_c$. (In these equations c represents any concentration scale and γ the corresponding activity coefficient.)

The activity coefficient is an empirical measure of nonideal behavior, which is a consequence of interparticle interactions. As can be seen from Eq. (2.14), an activity coefficient can be interpreted (in the form $RT \ln \gamma$) as a contribution to the free energy, and it has as much significance in this sense as does a concentration (13). Activity coefficients can be measured, and in some instances calculated theoretically (as with the Debye–Hückel theory of ionic interactions).

The goal of an experimental study of an equilibrium constant is to obtain a quantity that is (at constant temperature and pressure) a true constant. The quantity K is such a constant because the activities appearing in Eq. (2.15) are defined to make it so; it is really a tautology to say that K is a constant. We measure K_c, the concentration constant. Thus in order to obtain K we need to know K_γ. We have several possible ways to proceed.

We must first define the reference state, which is reasonably taken to be the same for all solutes. Often the infinitely dilute solution is taken as the reference state. (For most purposes 10^{-4} M in each component is infinitely dilute.) K_c is then measured as a function of solution composition. If the equilibrium is an ionic association in aqueous solution, K_c is measured as a function of ionic strength in solutions as dilute as possible. Then K is obtained from K_c by estimating activity coefficients with the Debye–Hückel equation (if the ionic strengths are low enough) or with empirical extensions of it, such as Davies' equation (14), for higher ionic strengths. This approach is widely used for the determination of thermodynamic acid dissociation constants and association constants for ion pairs in aqueous solution.

A second method is to extrapolate the K_c values from the experimental solvent range to the reference state. The difficulty with this method is that there is seldom a theoretical basis for the extrapolation, though in aqueous solution at very low ionic strengths the square root of the ionic strength is known to be an appropriate extrapolation variable.

A third approach is to define the experimental solvent as the reference state. The solvent may itself be a more or less complicated mixture, but to the extent possible its composition is held constant throughout the measurements needed to find K_c. Since the solvent is the reference state, $K_\gamma = 1$ by definition, and $K = K_c$. (Of course, the activity coefficients in the infinitely dilute solution may be far from unity when referred to this reference state.) The situation is not as simple as this, however, because the solutions required to measure K_c inevitably alter the composition of the solvent. For example, substrate and ligand are added, and the ligand concentration is varied. The best that can be hoped for is that K_γ remains substantially constant and not greatly different from unity over the range of compositions studied. If the solvent is water, it is advisable to hold the ionic strength constant. In reporting the results it is essential that the reference state be defined; this requires no more than the statement that K_c values were measured and reported, with a full description of solvent composition and range of variables studied (substrate and ligand concentrations in particular). It is important to give the concentration units of K_c.

Most complex binding constants are reported on the basis outlined in the preceding paragraph. This is quite adequate for many purposes, especially considering the typical experimental uncertainties of K_c values. In comparing K_c values from different laboratories or by different techniques, however, it becomes essential to pay attention to their reference states, that is, to the solvent compositions and experimental variables. We must also be prepared to find that K_c values are not truly constant over a range of experimental conditions. When this happens we are presented with a situation much like that outlined in Section 1.3, namely, is the inconstancy to be ascribed to a changing activity coefficient term, the stoichiometric complex model being the correct one; or is it to be ascribed to the inadequacy of the assumed stoichiometric model, the activity coefficient term being constant? Some independent evidence may be obtainable in the form of activity coefficient measurements, but these must be in solutions that do not contain both the substrate and the ligand. There may be a strong chemical basis for postulating the existence of a second complex, but whether or not its binding constant can be reliably estimated will be related to its magnitude because of the complicating possibility of the variable activity coefficient term.

This device of defining an experimental nondilute solution as the reference state is formally quite acceptable, but physically its meaning is different from the assignment of the infinitely dilute solution as the reference state. Deviations of solution behavior from ideality are caused by interionic and intermolecular interactions among solute particles, and these will be vanishingly small only in the infinitely dilute solution, which is the physical basis for preferring this reference state. The nondilute reference state stratagem provides activity coefficients of (approximately) unity in the experimental solvent, but these are not to be interpreted as meaning that interparticle interactions are absent, if the experimental solvent contains solute particles in the nondilute range.

2.2. CONCENTRATION SCALES AND THEIR USES

Concentration Scales

Several concentration scales are in common use in solution chemistry: mole fraction (x), molality (m), and molarity (c). The mole ratio (r) is defined $r_i = n_i/n_1$, where n_i is the number of moles of solute i and n_1 is the number of moles of solvent. The relationship between the mole fraction x_i and the molarity c_i of a solute i is given by

$$x_i = \frac{c_i}{1000\rho} \left(\frac{\Sigma \, n_j M_j}{\Sigma \, n_j} \right) \tag{2.18}$$

where ρ is the density of the solution, $\Sigma \, n_j$ is the sum of the number of moles, and M_j is the molecular weight of constituent j; thus $\Sigma \, n_j M_j$ is the total mass of the solution. The mole fraction and molality are related by

$$x_i = \frac{m_i}{1000} \left(\frac{n_1 M_1}{\Sigma \, n_j} \right) \tag{2.19}$$

where M_1 is the molecular weight of the solvent. The mole ratio and molality are connected by $m_i = 1000 r_i/M_1 = M^* r_i$, where $M^* = 1000/M_1$ is the number of moles of solvent in 1 kg of solvent.

It is to be noticed that the mole fraction, molality, and molarity scales are not directly proportional. In very dilute solutions of solute i, however, Eqs. (2.18) and (2.19) become

$$x_i = \frac{c_i M_1}{1000\rho_1} \qquad (x_i \lll 1) \tag{2.20}$$

$$x_i = \frac{m_i M_1}{1000} \qquad (x_i \lll 1) \tag{2.21}$$

where ρ_1 is the density of the solvent; in this special situation the concentration scales are directly proportional to each other.

The molarity is called by some writers the volume concentration or simply the concentration; its value is temperature dependent, whereas the values of the other scales are not dependent upon the temperature.

Standard States

The equilibrium constant is defined [Eq. (2.15)] in terms of activities, and the activities are undefined until a standard state has been specified [Eq. (2.13)]. In very dilute solution (with respect to the solutes) activity coefficients approach unity, the activity becomes equal to the concentration, and the equilibrium constant can be calculated in terms of concentrations [Eq. (2.17)]. Thus in dilute solution the selection of a standard state is equivalent to selection of a concentration scale. For the solvent the mole fraction scale is usually used, whereas solute concentrations are usually expressed in molarities, although molalities are sometimes used, and for liquid solutes the mole fraction scale may be used.

In nondilute (nonideal) solutions it is necessary to invoke Eq. (2.13) and to employ the activity of the solute. We still in effect choose the standard state when we select the concentration scale, but the interpretation is different; now $\mu°$ is the value of μ at unit activity, not at unit concentration, and these are different conditions. The standard state is then described as a hypothetical state of unit concentration but possessing some of the properties of the infinitely dilute solution. Denbigh (15) suggests that $\mu°$ simply be regarded as a quantity that is independent of solution composition.

We now write Eq. (2.12) for a given solute/system in terms of the three common concentration scales, using subscripts to indicate the standard state definition.

$$\mu = \mu_x° + RT \ln \gamma_x x \qquad (2.22a)$$

$$\mu = \mu_c° + RT \ln \gamma_c c \qquad (2.22b)$$

$$\mu = \mu_m° + RT \ln \gamma_m m \qquad (2.22c)$$

The chemical potential μ has a definite value independent of the concentration scale or standard state, but the standard potentials and the activities have different numerical values in the three equations. The three activity coefficients have different numerical values; Glasstone gives relationships among them (16). Some authors reserve the symbol γ for the molal activity coefficient, using y for the molar activity coefficient and f for the mole fraction activity coefficient (this last also being called the rational activity coefficient).

Since the x, m, and c scales are not proportional to each other over the full concentration range, the several activity coefficients are not proportional to each other, except in dilute solution. Although there is no compelling thermodynamic reason to prefer one concentration scale to another, ideal solution behavior is defined in terms of solute mole fraction, for partial pressure is proportional to x rather than to m or c. We shall return to this point later. It is possible for a solute to exhibit ideal behavior on the mole fraction scale ($\gamma_x = 1$) while at the same time behaving nonideally on the molal and molar scales (17).

Ben-Naim (18) has proposed a standard state definition on the molar scale in which deviations from ideality are assigned to the standard potential rather than to the activity term [Eq. (4.6) of ref. (18)].

Standard State Transformations

Since activities depend on the choice of standard state, and the thermodynamic equilibrium constant is defined in terms of activities, the numerical value of the equilibrium constant may depend on the standard states. But we have seen that the standard state is ordinarily specified by choosing a concentration scale. This means that an equilibrium constant may have different values when calculated on the basis of different concentration scales, and it is useful to be able to convert an equilibrium constant directly from one concentration scale to another. Let us obtain a relationship between the concentration equilibrium constants on the molar (K_c) and mole fraction (K_x) scales, for these are most frequently encountered. These constants are expressed

$$K_c = \prod^{q+\Delta q} c_p \bigg/ \prod^q c_r$$

$$K_x = \prod^{q+\Delta q} x_p \bigg/ \prod^q x_r$$

where subscripts r and p refer to reactants and products and the reaction converts q particles of reactants into $q + \Delta q$ of products. Consider first the very dilute solution range, where K_c and K_x approximate closely to the thermodynamic constants. Then taking the logarithms of both equations, eliminating molarities by means of Eq. (2.20), and rearranging gives

$$K_c = K_x(M^*\rho_1)^{\Delta q} \tag{2.23}$$

where $M^* = 1000/M_1$. In this equation Δq is established by the definition of the equilibrium constant and not by the way the reaction happens to have been written. For all stepwise complex stability constants $\Delta q = -1$.

In a like manner the molality and mole fraction constants are related, in dilute solution, by Eq. (2.24).

$$K_m = K_x(M^*)^{\Delta q} \tag{2.24}$$

Pimentel and McClellan (19) give conversion formulas for gas-phase equilibrium constants whose units are atm^{-1}.

For strong molecular complexes the measurement of binding constants can be accomplished in solutions dilute with respect to the solutes, and Eqs. (2.23) and (2.24) are therefore applicable. Weak complexes, however, require studies over a wider concentration range. In this case it may be found that analysis of the same set of data expressed on the c and x scales yields K_c and K_x values that are not consistent with the dilute solution limit, Eq. (2.23). It becomes necessary to use Eq. (2.18), which can be written in terms of molar volumes:

$$ x_i = \frac{c_i}{1000} \left(\frac{\Sigma\, n_j V_j}{\Sigma\, n_j} \right) $$

For constituent i in a solution containing high concentrations of solute B in solvent S this becomes

$$ x_i = \frac{c_i}{1000} \left[V_S + x_B (V_B - V_S) \right] $$

Thus for the stepwise complex formation reaction

$$ K_x = \frac{1000 K_c}{V_S + x_B (V_B - V_S)} \tag{2.25} $$

Equation (2.25) relates K_x to K_c when x_B is appreciable. This equation also shows that K_x and K_c cannot both be constants over a wide range of solution composition. [If necessary the molar volumes in Eq. (2.25) can be replaced by partial molar volumes.]

Choice of Concentration Scale

We must now face the problem of the meaning of thermodynamic quantities whose values can be altered by changing the concentration scale on which they are evaluated. Which of the possible values is most pertinent to an interpretation of the process? Gurney (20) made the first analysis of this problem.

We start by considering the very dilute solution range where the system behaves ideally on all concentration bases. Then Eq. (2.22) becomes

$$\mu = \mu_x^\circ + RT \ln x \qquad\qquad (2.26a)$$

$$\mu = \mu_c^\circ + RT \ln c \qquad\qquad (2.26b)$$

$$\mu = \mu_m^\circ + RT \ln m \qquad\qquad (2.26c)$$

Now μ is a definite quantity independent of the concentration scale, so the relationship between any two standard potentials can be found from these equations, for example,

$$\mu_x^\circ = \mu_c^\circ + RT \ln M^* \rho_1$$

where Eq. (2.20) has been used.

Consider the stepwise complex formation reaction

$$A + B \rightleftharpoons C$$

For the overall change in molar free energy for this process we obtain, by means of Eqs. (2.26),

$$\Delta\mu = \Delta\mu_x^\circ + RT \ln \frac{x_C}{x_A x_B} \qquad\qquad (2.27a)$$

$$\Delta\mu = \Delta\mu_c^\circ + RT \ln \frac{c_C}{c_A c_B} \qquad\qquad (2.27b)$$

$$\Delta\mu = \Delta\mu_m^\circ + RT \ln \frac{m_C}{m_A m_B} \qquad\qquad (2.27c)$$

where the concentrations do not necessarily refer to the equilibrium condition. Using Eq. (2.16) these become

$$\Delta\mu = -RT \ln K_x + RT \ln L_x \qquad\qquad (2.28a)$$

$$\Delta\mu = -RT \ln K_c + RT \ln L_c \qquad\qquad (2.28b)$$

$$\Delta\mu = -RT \ln K_m + RT \ln L_m \qquad\qquad (2.28c)$$

All of these have the form

$$\Delta\mu = [\text{composition-independent term}] + [\text{composition-dependent term}]$$

Since at equilibrium $\Delta\mu = 0$, evidently at equilibrium these terms are equal in magnitude and opposite in sign. In studying chemical equilibrium we wish to identify quantities characteristic of the chemical process itself rather than of the particular solution; we want therefore to study the composition-independent component. It is for this reason that we focus attention on the equilibrium constant and the standard free energy change. But the equilibrium constants based on all three concentration scales appear to meet this requirement. Thus we wish to learn if any one of these quantities is superior to the others.

The solution of interest contains A, B, and C particles in solvent S. According to thermodynamics the total molar free energy of mixing is the free energy change on forming this solution at specified concentrations from the pure components, exclusive of any contributions from intermolecular forces,

$$\frac{\Delta G_{mix}}{RT} = n_A \ln x_A + n_B \ln x_B + n_C \ln x_C + n_S \ln x_S \qquad (2.29)$$

where n_i is the number of moles of i. Let a small number ∂n_C of moles of C be introduced into the solution, the numbers of moles of other constituents being held constant. Then the change in free energy is found by partial differentiation of Eq. (2.29) term by term:

$$\frac{\partial(\Delta G_{mix}/RT)}{\partial n_C} = -x_A - x_B + (1 - x_C) + \ln x_C - x_S = \ln x_C \qquad (2.30)$$

This contribution is therefore composition dependent. Since $x_i < 1$, an increase in x_i (positive value of ∂n_i) results in a negative contribution to the free energy. This is purely a mixing effect (a configurational contribution in statistical mechanical terms). If a species contribution decreases, the corresponding contribution to the free energy is given by $-\ln x_i$.

Now we apply this result to the equilibrium process. We increase the number of C particles, which must result in a decrease in the numbers of A and B particles. The molar free energy of mixing produced by this process is therefore

$$RT(\ln x_C - \ln x_A - \ln x_B) = RT \ln \frac{x_C}{x_A x_B}$$

This quantity is therefore the composition-dependent mixing contribution to the reaction free energy change. When the x_i correspond to the equilibrium concentrations, this quantity, taken with opposite sign, is the composition-independent quantity that we seek. We therefore find that the equilibrium constant K_x is the preferred expression of the equilibrium state of the system.

Gurney (20) calls $-RT \ln K_x$ the *unitary* part and $RT \ln L_x$ the *cratic* part of the free energy change. Note that this argument has been developed for the case of the extremely dilute solution. In real nonideal solutions all concentrations must be replaced by activities; then $RT \ln L_x$ is called the *communal* part, and it consists of the cratic part plus an activity coefficient term.

Writing ΔG_{un}° for the unitary standard free energy change, we now have

$$\Delta G_{un}^\circ = -RT \ln K_x = -RT \ln K_c + \Delta q\, RT \ln (M^* \rho_1) \qquad (2.31)$$

It can be shown (20) that the cratic part of the free energy change makes no contribution to the enthalpy change; we demonstrate this graphically. The standard enthalpy change ΔH° is determined from the slope of a van't Hoff plot of $\ln K$ versus $1/T$. Changing the K value from one concentration scale to another has the effect of displacing this line vertically, which does not change its slope. Thus the same value of ΔH° is obtained no matter which standard state is chosen. The entropy change is, in effect, obtained as the extrapolated intercept of this plot; hence ΔS° will be affected by the concentration scale. Writing $\Delta G^\circ = \Delta H^\circ - T\,\Delta S^\circ$ for both the x and c scales, eliminating ΔH° between these because it is the same on both scales, and combining the result with Eq. (2.31) gives

$$\Delta S_{un}^\circ = \Delta S_c^\circ - \Delta q\, R \ln (M^* \rho_1) \qquad (2.32)$$

where ΔS_{un}° has been identified with ΔS_x°.

Suppose, as an example, the complex binding constant ($\Delta q = -1$) is studied in dilute aqueous solution at 25°C. Then Eqs. (2.31) and (2.32) become

$$\Delta G_{un}^\circ / \text{cal mol}^{-1} = -1364.4 \log K_c - 2379.5$$

$$\Delta S_{un}^\circ / \text{e.u.} = \Delta S_c^\circ + 8.0$$

This means that ΔS_{un}° and ΔS_c° may even have different signs; thus it was found (21) that for the 1:1 complex of methyl *trans*-cinnamate with

8-chlorotheophylline anion $\Delta S^{\circ}_{un} = +3$ e.u., whereas $\Delta S^{\circ}_{c} = -5$ e.u. Since the sign of ΔS° is often a critical factor in mechanistic interpretations, the choice of concentration scale is important. Note, by the way, that M^* can be calculated for mixed solvents; for example, for 12.50% (w/w) aceto-nitrile in water,

$$M^* = \frac{125}{41.054} + \frac{875}{18.016} = 51.608$$

The treatment thus far suggests that the mole fraction concentration scale yields equilibrium constants, standard free energy changes, and standard entropy changes that are the best numerical values for interpretation in molecular terms because the purely configurational (mixing) contribution is not included in these numbers. We note that of all possible standard states the mole fraction state is unique in that it is the only one that is not a mixture. Jencks (22) and Tanford (23) have discussed unitary quantities in relation to specific chemical processes.

Opinion is not unanimous on the unique suitability of the mole fraction scale for the interpretation of thermodynamic quantities. Since thermodynamics itself cannot provide an unambiguous basis for a choice, it is necessary to invoke molecular models for this purpose. Ben-Naim (18, 24) has employed a statistical mechanical argument that leads him to advocate the molar concentration scale for the interpretation of transfer free energies, that is, free energy changes for the transfer (in dilute solutions) of a solute from one solvent to another. An extremely over-simplified sketch of this treatment follows; it is emphasized that the following presentation is not sufficiently realistic to be useful, but it serves to show the origin of the concentration preference. We consider a system of N particles in a volume V. The quantum mechanical description of the particle in a box leads to the translational partition function of an individual particle (25),

$$q = \frac{V}{\Lambda^3} \qquad (2.33)$$

where $\Lambda = h/(2\pi m k T)^{1/2}$, h being Planck's constant and m the mass of the particle. With the assumption of ideal behavior (no intermolecular forces), the ensemble partition function for N indistinguishable particles is

$$Q = \frac{q^N}{N!}$$

The chemical potential (per particle) for this ensemble is defined

$$\mu = -kT\left(\frac{\partial \ln Q}{\partial N}\right)_{V,T}$$

which leads to

$$\mu = kT \ln \frac{N}{q}$$

or, using Eq. (2.33),

$$\mu = kT \ln \frac{N\Lambda^3}{V}$$

Defining the density $\rho = N/V$ gives

$$\mu = kT \ln \Lambda^3 + kT \ln \rho$$

which may be written

$$\mu = \mu° + kT \ln \rho \tag{2.34}$$

where $\mu°$ has the character of a standard potential, and the density is related to molar concentration through Avogadro's number: $\rho = N_A c$. Intermolecular forces can be incorporated via a potential energy function in the partition function. This approach has been worked out in great detail by Ben-Naim (18, 24). We do not pursue it further here, but point out that it is a controversial issue (26). The use of a volume concentration (actually a volume fraction) as the most natural concentration scale has precedent in the Flory–Huggins statistical mechanical theory of polymer solutions (27), in which a solute molecule is orders of magnitude larger than a solvent molecule, so that relative volume is a more pertinent characteristic than is relative number.

Because of the ambiguous guidance provided by theory, some workers have sought a resolution in experiment. We have seen that equilibrium "constants" evaluated on the various concentration scales of mole fraction (number/number), molarity (number/volume), and molality (number/mass) cannot all be constants over a wide range of composition, as shown by Eq. (2.25). Many studies have sought to establish which of these quantities is most nearly constant (28). The experimental problem is a difficult one and the results have been inconclusive in any general sense, with each of the scales being preferred in some system.

We conclude this section with some advice and with a thought experiment. The advice to the experimentalist is to give a crystal-clear description of the nature of all equilibrium constants, standard free energy changes, and standard entropy changes when reporting these quantities. This includes their units, obviously, but for $\Delta G°$ and $\Delta S°$ it means also the standard states; that is, were these calculated from K_x, K_c, or K_m? Note, by the way, that it is not possible to take the logarithm of a unit or dimension, so an expression like $\Delta G_c° = -RT \ln K_c$ really implicitly means $\Delta G_c° = -RT \ln (K_c/M^{-1})$, for K_c/M^{-1} is a pure number. This device can always be used to generate a number without dimensions. Its use as a proper way to present data is shown in Table 1.1.

This is the thought experiment: Rather than asking which concentration scale generates an invariant equilibrium constant, let us ask which scale produces a standard entropy change having the correct sign. We consider the complex formation $A + B = C$, with $\Delta q = -1$. For our first case suppose we have identified a system for which a valid statistical mechanical model predicts $\Delta S° < 0$. This would probably involve an inert solvent so that there are no solvation–desolvation effects. For each molecule of C formed there will be a loss of three translational and three rotational degrees of freedom, and there will be some restriction of vibrational modes in bound A and B, partly offset by new vibrational modes in C. Now suppose the system is studied experimentally and both $\Delta S_x°$ and $\Delta S_c°$ are evaluated. (The discussion could include other scales as desired.) If both of these quantities are negative, no decision can be made, but if one is negative and one is positive, the scale giving the positive result can be eliminated. Actually, we do not have to do the experiment, because we know how $\Delta S_x°$ and $\Delta S_c°$ are related [Eq. (2.32)]; for $\Delta q = -1$ and any realistic solvent $\Delta S_x°$ will be more positive than $\Delta S_c°$. Thus the mole fraction scale is ruled out, whereas the molar scale is admitted (though not proved to be the only appropriate one).

Next we consider a system whose standard entropy change is positive according to a statistical mechanical description. Presumably, solvation and solvent-structure effects could produce such a result; water comes to mind. A repetition of the preceding argument shows that the contrary conclusion would be reached; $\Delta S_c°$ would be ruled out whereas $\Delta S_x°$ would be an admissible scale.

Although these are hypothetical situations, they are not unrealistic, and the argument strongly suggests that there is no single best concentration scale for the evaluation of thermodynamic quantities, even at fixed Δq. There is a hint that the molar scale may be preferable in inert solvents, and the mole fraction scale in aqueous solutions, for complex formation equilibria; but the question is still an open one.

2.3. THE GENERAL PROBLEM

The problem of determining binding constants for molecular complex formation can be expressed as follows. We have a solution containing substrate S in the presence of ligand L, and we measure a property Y of the solution as a function of L_t, the total ligand concentration. The system is described by Eqs. (2.35) and (2.36).

$$Y_i = f([L]_i, \eta_j) \tag{2.35}$$

$$L_{ti} = g([L]_i, \lambda_k) \tag{2.36}$$

In these equations Y_i is the value of the measured property (the dependent variable) at total ligand concentration L_{ti}; $[L]_i$ is the free (unbound) ligand concentration corresponding to L_{ti}; and the η_j and λ_k are parameters. The forms of the functions and the nature of the parameters depend on the system being studied and on the assumed stoichiometric model. The parameters include the desired stability constant(s). Equation (2.35) is one form of the binding isotherm.

The parameters are estimated from the data set (the Y_i as a function of L_{ti}) by solving Eq. (2.35) algebraically, graphically, or by least-squares regression analysis. In order to carry out this analysis, however, the free ligand concentration $[L]_i$ is required. (There are some methods in which $[L]_i$ is measured.) Equation (2.36) relates $[L]_i$ to the known total ligand concentration, but without knowledge of the parameters λ_k it is not possible to solve Eq. (2.36). There are several ways to proceed.

1. Assume $[L]_i = L_{ti}$. This assumption is widely used. It is a good approximation when $L_t \ggg S_t$ (S_t is the total substrate concentration), or when the extent of complexation is very small, but it clearly is not generally acceptable.

2. If possible, solve Eq. (2.36) explicitly for $[L]_i$ in terms of L_{ti} and the

parameters λ_k. With preliminary estimates of the parameters (perhaps obtained by means of procedure 1 above), calculate the free ligand concentration and use this in Eq. (2.35) to obtain improved parameter estimates. Repeat this process, iterating as required until the parameter estimates converge. This method, or equivalent versions, has been widely applied to spectroscopic studies of complex formation (29). It can be very effective, but is only applicable when Eq. (2.36) can be explicitly solved, which usually means only for 1:1 stoichiometry.

3. It may be possible to combine Eqs. (2.35) and (2.36) in a useful way. For example, in the solubility method for 1:1 stoichiometry $[L]_i$ can be eliminated between these equations, and then Y_i is obtained as a function of L_{ti} (Chapter 8). In the potentiometric method combination of these equations gives $[L]_i$ as a function of L_{ti} and Y_i, which simplifies the functional relationship (the algebraic complexity being inherent in the Y_i themselves), though at the cost of introducing additional experimental uncertainty (Chapter 7).

4. There is a general graphical solution. Although Eq. (2.36) may not be solvable explicitly for $[L]_i$, it is solved for L_{ti}. Hence, substitute arbitrarily chosen values of $[L]_i$ into Eq. (2.36) (covering the experimentally studied range), and with initial estimates of the parameters calculate the corresponding L_{ti}. Now plot the $[L]_i$, L_{ti} pairs. From this plot the $[L]_i$ corresponding to each experimental L_{ti} can be obtained by interpolation. Iteration is then carried out. As described, this method has the disadvantage of requiring a manual interpolation step, which interrupts the iteration computation, but this could probably be overcome.

5. Equation (2.36) can be expanded in a Taylor's series, which is truncated to give an expression that can be solved explicitly for $[L]_i$ (30). The Taylor's series expansion of Eq. (2.36) is

$$L_{ti} = g(L_{ti}) + g'(L_{ti})([L]_i - L_{ti}) + \frac{g''(L_{ti})}{2}([L]_i - L_{ti})^2 + \cdots$$

$$(2.37)$$

where $g(L_{ti})$ is Eq. (2.36), $g'(L_{ti})$ is $dL_{ti}/d[L]_i$, and $g''(L_{ti})$ is $d^2L_{ti}/d[L]_i^2$, all evaluated at $[L]_i = L_{ti}$.

The procedure will be illustrated with the very common problem consisting of the spectrophotometric study of a 1:1 complex. The equations corresponding to Eqs. (2.35) and (2.36) are, from Chapter 4,

$$\frac{\Delta A}{b} = \frac{K_{11} S_t \, \Delta\epsilon [\text{L}]_i}{1 + K_{11}[\text{L}]_i} \tag{2.38}$$

$$\text{L}_{ti} = [\text{L}]_i + \frac{K_{11} S_t [\text{L}]_i}{1 + K_{11}[\text{L}]_i} \tag{2.39}$$

where $\Delta A/b$ is the change in absorbance per centimeter when the total ligand concentration is changed from zero to L_{ti}, K_{11} is the stability constant, S_t is the total substrate concentration, and $\Delta\epsilon$ is the difference in molar absorptivities of the complexed and uncomplexed substrate. The derivative $g'(\text{L}_{ti})$ is, from Eq. (2.39),

$$g'(\text{L}_{ti}) = 1 + \frac{K_{11} S_t}{1 + 2K_{11}\text{L}_{ti} - K_{11}^2 \text{L}_{ti}^2} \tag{2.40}$$

Let us truncate Eq. (2.37) at the linear term. Then substituting (2.40) into (2.37) and solving for $[\text{L}]_i$ gives

$$[\text{L}]_i = \text{L}_{ti} - \frac{1}{g'(\text{L}_{ti})}\left[\frac{K_{11} S_t \text{L}_{ti}}{1 + K_{11}\text{L}_{ti}}\right] \tag{2.41}$$

In this example it is possible to solve Eq. (2.39) algebraically, and comparison of this solution for $[\text{L}]_i$ with the Taylor's series approximation, Eq. (2.41), shows that the approximation is very satisfactory for typical simulated systems; Table 2.1 shows such a comparison.

This example involves approximation of a quadratic with a linear function when only a 1:1 complex is present. A more complicated system is exemplified by the solubility study of a system containing both 1:1 and 1:2 complexes. As is shown in Chapter 8, the equation corresponding to Eq. (2.36) is

$$\text{L}_t = (1 + K_{11} s_0)[\text{L}] + 2K_{11} K_{12} s_0 [\text{L}]^2 \tag{2.42}$$

where s_0 is the (constant) free substrate concentration and we have dropped the index i for convenience. The functions needed in the Taylor's series expansion are

Table 2.1. Comparison of Exact Solution and Taylor's Series Approximation for the 1:1 Complexing Example[a]

	$[L]_i/M$	
L_{ti}/M	Exact[b]	Approximate[c]
0.005	0.004 564	0.004 563
0.01	0.009 161	0.009 160
0.02	0.018 44	0.018 44
0.04	0.037 28	0.037 28
0.06	0.056 39	0.056 39
0.08	0.075 69	0.075 69
0.10	0.095 12	0.095 12
0.20	0.193 41	0.193 41
0.50	0.491 7	0.491 7
1.00	0.990 9	0.990 9

[a] Using $K_{11} = 10.0 \ M^{-1}$, $S_t = 0.010 \ M$.
[b] With Eq. (2.39).
[c] With Eq. (2.41).

$$g(L_t) = (1 + K_{11}s_0)L_t + 2K_{11}K_{12}s_0L_t^2$$

$$g'(L_t) = (1 + K_{11}s_0) + 4K_{11}K_{12}s_0L_t$$

$$g''(L_t) = 4K_{11}K_{12}s_0$$

Applying the linear truncation gives

$$[L] = \frac{L_t + 2K_{11}K_{12}s_0L_t^2}{1 + K_{11}s_0 + 4K_{11}K_{12}s_0L_t} \qquad (2.43)$$

Table 2.2 gives some results of Eq. (2.43) for a hypothetical system having $K_{11} = 400 \ M^{-1}$, $K_{12} = 100 \ M^{-1}$, $s_0 = 1 \times 10^{-4} \ M$. [These values are similar to those found (2) for the dimethylterephthalate-α-cyclodextrin system.] Such calculations are useful in showing the level of approximation achieved by the linear truncation, and they can be carried out whenever the method is to be employed, using reasonable estimates of the system parameters. If the linear approximation is not satisfactory, the quadratic term in Eq. (2.37) can be retained. In the example shown in Table 2.2 use of the quadratic term in the Taylor's series expansion simply generates the exact equation, (2.42).

In Chapter 7 additional examples of this method are shown.

Table 2.2. Taylor's Series Approximation for a 1:1 and 1:2 Complexing System[a]

	[L]/M	
L_t/M[b]	Actual	Calculated[c]
0.0112	0.01	0.0100
0.0240	0.02	0.0201
0.0384	0.03	0.0303
0.0544	0.04	0.0409
0.0720	0.05	0.0518
0.0912	0.06	0.0631
0.1120	0.07	0.0750
0.1344	0.08	0.0874
0.1584	0.09	0.1005
0.1840	0.10	0.1142

[a] $K_{11} = 400\ M^{-1}$; $K_{12} = 100\ M^{-1}$; $s_0 = 1 \times 10^{-4}\ M$.
[b] With Eq. (2.42) from [L] values in second column.
[c] With Eq. (2.43).

2.4. MODELS

A model is a hypothesis. It is formulated on the basis of prior knowledge of the system, it provides guidance in designing experiments, and it is tested against experimental data. If not enough information is available to allow the construction of a reasonable model, the first step may be to carry out exploratory studies to help define the problem. In the present context a model is a statement of the possible complex stoichiometries, perhaps with auxiliary statements that amplify the description.

Measures of Binding

Let m be the (maximum) number of binding sites on ligand L, with h the number of occupied sites; thus $h = 1, 2, 3, \ldots, m$; similarly, n is the maximum number of binding sites on substrate S, with $i = 0, 1, 2, \ldots, n$ being the corresponding running index. Then the basic statement of a stoichiometric model involves the values of m and n.

We need to express the extent to which binding takes place as a function of system variables. In Part II this is done in terms of chemical and physical properties of the system, but here we develop more generalized quantities. We start by defining the mass balance expressions for S and L, where S_t and [S] represent total and free (unbound) substate concentrations, respectively, and similarly for ligand, all concentrations being on the molarity scale:

$$S_t = [S] + \sum_{\substack{h=1 \\ i=1}}^{m,n} h[S_h L_i] \tag{2.44}$$

$$L_t = [L] + \sum_{\substack{h=1 \\ i=1}}^{m,n} i[S_h L_i] \tag{2.45}$$

For example, if $m = 1$, $n = 2$, then $S_t = [S] + [SL] + [SL_2]$ and $L_t = [L] + [SL] + 2[SL_2]$. Equations (2.44) and (2.45) are not completely general, because they do not include provision for dimers, trimers, and so on, of S and L, but these can be added as necessary.

Now as one measure of extent of binding we define $\bar{i}$, the average number of ligand molecules bound per molecule of substrate, or

$$\bar{i} = \frac{\Sigma \, (\text{L bound to S})}{\Sigma \, (\text{all S})} = \frac{L_t - [L]}{S_t} \tag{2.46}$$

We shall see shortly why $\bar{i}$ is properly interpreted as an average of i. Combination of Eqs. (2.44) to (2.46) gives a general expression for $\bar{i}$. We seldom have need of such generality, however, and can limit our expressions to suit the appropriate model. The most widely applicable limitation of this type is to take $m = 1$, a reasonable choice for many equilibria, including protonations, metal ion complexation, and macromolecule–small molecule binding. If we set $m = 1$, we find for $\bar{i}$:

$$\bar{i} = \frac{\sum_{i=1}^{n} i[SL_i]}{S_t}$$

We now eliminate all complex concentrations by substitution from the stability constant expressions, most conveniently in terms of the overall constants

$$\beta_{hi} = \frac{[S_h L_i]}{[S]^h [L]^i}$$

which gives

$$\bar{i} = \frac{\sum_{i=1}^{n} i\beta_{1i}[L]^i}{1 + \sum_{i=1}^{n} \beta_{1i}[L]^i} \tag{2.47}$$

Again taking the $[m = 1; n = 2]$ model as an example, we have

$$\bar{i} = \frac{[SL] + 2[SL_2]}{[S] + [SL] + [SL_2]} = \frac{\beta_{11}[L] + 2\beta_{12}[L]^2}{1 + \beta_{11}[L] + \beta_{12}[L]^2} \tag{2.48}$$

where $\beta_{11} = K_{11}$ and $\beta_{12} = K_{11}K_{12}$, these constants having been defined in Section 2.1.

Next we define the fraction of substrate f_{hi} in the form of species $S_h L_i$:

$$f_{hi} = \frac{h[S_h L_i]}{S_t} \tag{2.49}$$

so that

$$\sum_{\substack{h=1 \\ i=0}}^{m,n} f_{hi} = 1 \tag{2.50}$$

and $f_{10} = [S]/S_t$, the fraction of free (uncomplexed) substrate. Again choosing the important special case $m = 1$, and expanding Eq. (2.49),

$$f_{1i} = \frac{\beta_{1i}[L]^i}{1 + \sum_{i=1}^{n} \beta_{1i}[L]^i} \tag{2.51}$$

Comparing Eqs. (2.47) and (2.51) gives a succinct expression for $\bar{i}$:

$$\bar{i} = \sum_{i=1}^{n} i f_{1i} \tag{2.52}$$

This shows why $\bar{i}$ is an average value of the index i. In the literature the quantity $\bar{i}$ is variously written $\bar{n}$, $\bar{\nu}$, or r, depending on the writer and the field.

The quantity that is the most appropriate measure of the extent of binding depends on the model and the experimental measurements made on the system. Often $\bar{i}$ is the best measure, but others can be useful. We define $\theta = \bar{i}/n$ and $F_S = 1 - f_{10}$. From Eqs. (2.50) and (2.52), these quantities are

$$\theta = \frac{1}{n} \sum_{i=1}^{n} i f_{1i} \tag{2.53}$$

$$F_S = \sum_{i=1}^{n} f_{1i} \tag{2.54}$$

Both θ and $F_{\tilde{S}}$ are fractions but they have quite different meanings. $F_{\tilde{S}}$ is the fraction of substrate *molecules* to which one or more ligand molecules are bound, whereas θ is the fraction of substrate *sites* to which ligand molecules are bound. θ is called the degree of saturation. $F_{\tilde{S}}$ is a sorting function that distinguishes between molecules of substrate S, and molecules of "not-substrate" $\tilde{S}$, so $F_S = f_{10}$ and $F_S + F_{\tilde{S}} = 1$. Notice that the range of θ and $F_{\tilde{S}}$ is 0 to 1, whereas the range of $\bar{i}$ is 0 to n. In the special case $n = 1$, $\bar{i}$, θ, and $F_{\tilde{S}}$ become identical.

Continuing to use the $[m = 1; n = 2]$ model in illustration, we find

$$f_{10} = \frac{1}{1 + K_{11}[L] + K_{11}K_{12}[L]^2} \tag{2.55a}$$

$$f_{11} = \frac{K_{11}[L]}{1 + K_{11}[L] + K_{11}K_{12}[L]^2} \tag{2.55b}$$

$$f_{12} = \frac{K_{11}K_{12}[L]^2}{1 + K_{11}[L] + K_{11}K_{12}[L]^2} \tag{2.55c}$$

and $f_{10} + f_{11} + f_{12} = 1$. From Eqs. (2.52), (2.53), and (2.54),

$$\bar{i} = f_{11} + 2f_{12} \tag{2.56a}$$

$$\theta = \frac{f_{11}}{2} + f_{12} \tag{2.56b}$$

$$F_{\tilde{S}} = f_{11} + f_{12} \tag{2.56c}$$

The development of a model into a useful and testable form involves converting the basic model into a functional relationship called the *binding isotherm*. Equation (2.35) is a generalized isotherm, and Eqs. (2.52)–(2.54) are explicit expressions of isotherms. (The terminology implies that the binding constants are temperature dependent.) The experimentally observed quantity may be incorporated into the equation. Note that the isotherm [Eq. (2.35)] is, in effect, developed by operating on the mass balance expression for S_t, whereas the supporting relationship Eq. (2.36) is obtained by operating on the mass balance expression for L_t.

Examples of Models

$[m = 1; \ n = 1]$ For this simplest of models we find $\bar{i} = \theta = F_{\bar{s}} = f_{11}$, where $f_{11} = [SL]/S_t$. Substitution from the expression $K_{11} = [SL]/[S][L]$ gives for the binding isotherm

$$f_{11} = \frac{K_{11}[L]}{1 + K_{11}[L]} \tag{2.57}$$

Equation (2.57) is of fundamental importance in the study of equilibria of all types, and it is treated in detail in Section 2.5 and Chapter 3. The functional dependence of f_{11} on $[L]$ is the characteristic feature; observe in particular that f_{11} depends on free ligand concentration, not on the ratio of ligand to substrate.

Note that with but a single binding site on both S and L, only a single $1:1$ complex is formed.

$[m = 1; \ n = 2]$ We have used this model as an example in the earlier development. The stepwise binding constants are defined in Eqs. (2.1) and (2.2), the fractions f_{10}, f_{11}, and f_{12} are given in Eq. (2.55), and expressions for $\bar{i}$, θ, and $F_{\bar{s}}$ in terms of the fractions are given in Eq. (2.56). Equation (2.48) gives $\bar{i}$ as a function of $[L]$.

Since there are two binding sites on S, there are two possible isomeric $1:1$ complexes. This means that the $1:2$ complex SL_2 can be formed by two routes, each with its own microscopic binding constants. This situation was discussed in Section 2.1, where K_{11} and K_{12} were related to the microscopic constants $K_{X'Y}$ and $K_{XY'}$ by

$$K_{11} = K_{X'Y} + K_{XY'} \tag{2.58}$$

$$K_{12} = \frac{a_{XY}K_{X'Y}K_{XY'}}{K_{11}} \tag{2.59}$$

where a_{XY} is a parameter describing interaction between ligands bound to the two sites. Evidently, the three quantities a_{XY}, $K_{X'Y}$, $K_{XY'}$ cannot be obtained from experimental measurements of K_{11} and K_{12} by means of these two equations. In the special circumstance that the two binding sites on the substrate are identical, however, so that $K_{X'Y} = K_{XY'}$, then (letting the substrate be denoted XX), Eqs. (2.58) and (2.59) become

$$K_{11} = 2K_{X'X} \tag{2.60}$$

$$K_{12} = \frac{a_{XX}K_{11}}{4} \tag{2.61}$$

For such a system the microscopic binding constant $K_{X'X}$ and the interaction parameter a_{XX} can be determined. Note that when the sites are identical and when there is no interaction between the sites ($a_{XX} = 1$), $K_{11} = 4K_{12}$. When the sites are not identical and when there is no interaction between the sites, Eq. (2.50) gives $K_{11}K_{12} = K_{X'Y}K_{XY'}$. Thus this model includes some interesting special cases. Hill (31) gives a statistical mechanical description of this model.

[$m = 2$; $n = 1$] Since $m \neq 1$, we cannot use expressions such as Eqs. (2.47) and (2.51), but must return to the general definitions. The mass balance expressions are

$$S_t = [S] + [SL] + 2[S_2L]$$

$$L_t = [L] + [SL] + [S_2L]$$

Substituting into Eq. (2.46) gives

$$\bar{i} = \frac{\beta_{11}[L] + \beta_{12}[S][L]}{1 + \beta_{11}[L] + 2\beta_{21}[S][L]}$$

Similarly, since $f_{10} = [S]/S_t$, for $F_{\bar{S}}$ we get

$$F_{\bar{S}} = \frac{\beta_{11}[L] + 2\beta_{21}[S][L]}{1 + \beta_{11}[L] + 2\beta_{21}[S][L]}$$

The interesting feature is the presence of the substrate concentration in the isotherm expressions. This will always happen for $m > 1$, and it provides a means for diagnosing this condition, since if $m > 1$, then $\bar{i}$ and $F_{\bar{S}}$ will be functions of the substrate concentration.

[$m = 1$; n **identical, distinguishable, independent sites, without interactions between bound ligands**] This model is not very realistic, but it leads to a simple result that is widely used, so we consider it in detail. We first note that there will be $n!/(n - i)!i!$ isomeric complexes of formula SL_i, since this is the number of different ways i ligand molecules can be distributed among n sites. (Recall that $0! = 1$.)

Let k be the microscopic binding site constant. This is the constant for binding to a single site, whereas we need the stepwise binding constants K_{1i}, which describe the binding of the ith L molecule. K_{1i} will be the consequence of two factors besides k, namely, the number of unoccupied sites on $SL_{(i-1)}$, a larger value of which will favor association; and the number of occupied sites on SL_i, a larger value of which will favor dissociation. This argument gives

$$K_{1i} = \frac{\text{No. unoccupied sites on } SL_{(i-1)}}{\text{No. occupied sites on } SL_i} k$$

$$= \frac{n-(i-1)}{i} k$$

$$= \frac{n-i+1}{i} k \tag{2.62}$$

We first give the standard (and not obvious) derivation of the binding isotherm (32). The overall binding constant is

$$\beta_{1i} = \prod_{j=1}^{i} K_{1j} = k^i \prod_{j=1}^{i} \frac{n-j+1}{j}$$

This product is seen to be equivalent to the binomial coefficient:

$$\prod_{j=1}^{i} \frac{n-j+1}{j} = \frac{n(n-1)(n-2)\cdots(n-i+1)}{1\cdot 2\cdot 3\cdots i} = \frac{n!}{i!(n-i)!}$$

Therefore

$$\beta_{1i} = \frac{n!}{i!(n-i)!} k^i \tag{2.63}$$

This result is inserted into Eq. (2.47).

$$\bar{i} = \frac{\displaystyle\sum_{i=1}^{n} i \frac{n!}{i!(n-i)!} x^i}{1 + \displaystyle\sum_{i=1}^{n} \frac{n!}{i!(n-i)!} x^i} \tag{2.64}$$

where $x = k[L]$. Now we note that the denominator is the binomial expansion of $(1+x)^n$.

$$(1 + x)^n = 1 + \sum_{i=1}^{n} \frac{n!}{i!(n-i)!} \, x^i \tag{2.65}$$

Differentiate both sides of Eq. (2.65) with respect to x, and then multiply through by x.

$$nx(1 + x)^{n-1} = \sum_{i=1}^{n} i \, \frac{n!}{i!(n-i)!} \, x^i \tag{2.66}$$

Substitute from Eqs. (2.65) and (2.66) into (2.64), obtaining

$$\bar{i} = \frac{nk[L]}{1 + k[L]} \tag{2.67}$$

According to this remarkable result the binding isotherm has the same functional form as does the simple 1:1 model. Introducing the degree of saturation $\theta = \bar{i}/n$ gives an alternative form:

$$\frac{\theta}{1 - \theta} = k[L] \tag{2.68}$$

We next give a kinetic derivation. The rate of association is written as a second-order rate equation

$$v_a = k_a(n - \bar{i})[L]$$

where $(n - \bar{i})$ is proportional to the concentration of unoccupied sites. Similarly, a rate of dissociation is written

$$v_d = k_d \bar{i}$$

At equilibrium these rates are equal, leading immediately to Eq. (2.68) by equating k and k_a/k_d. This argument is often used in the context of adsorption to a solid surface from a gas or solution phase, and then Eq. (2.68) is called the Langmuir adsorption isotherm.

A statistical mechanical argument (33) leads to a molecular interpretation of the constant k. Strictly speaking, this is for adsorption from the gas phase, but the argument is also applicable to solutions. Let q be the partition function of a single bound ligand molecule, where q will have contributions from vibrational motions in its bound state as determined by the potential energy function for binding. The system of n sites and $\bar{i}$ bound molecules has the partition function Q:

$$Q = \frac{n!}{\bar{i}!(n-\bar{i})!} \, q^{\bar{i}}$$

where the origin of the binomial coefficient has already been discussed. In logarithmic form,

$$\ln Q = n \ln n - \bar{i} \ln \bar{i} - (n-\bar{i}) \ln (n-\bar{i}) + \bar{i} \ln q$$

where we have used Stirling's approximation, $\ln y! = y \ln y - y$. The chemical potential per particle of ligand is defined

$$\mu = -kT \left(\frac{\partial \ln Q}{\partial \bar{i}} \right)_{n,T}$$

which leads to

$$\mu = kT \ln \frac{\theta}{(1-\theta)q}$$

where $\theta = \bar{i}/n$. The thermodynamic expression for the ligand chemical potential is

$$\mu = \mu^{\circ} + kT \ln [\mathrm{L}]$$

Setting these equal gives

$$\frac{\theta}{1-\theta} = q e^{\mu^{\circ}/kT}[\mathrm{L}] \tag{2.69}$$

Comparison of Eqs. (2.68) and (2.69) shows that $k = q \exp(\mu^{\circ}/kT)$.

Some extensions of this model can be made (32). Suppose that the substrate possesses two sets of sites, having numbers n_{I} and n_{II} and microscopic binding constants k_{I} and k_{II}. Then because of the independence of sites assumption in this model, we have $\bar{i} = \bar{i}_{\mathrm{I}} + \bar{i}_{\mathrm{II}}$, or

$$\bar{i} = \frac{n_{\mathrm{I}} k_{\mathrm{I}}[\mathrm{L}]}{1 + k_{\mathrm{I}}[\mathrm{L}]} + \frac{n_{\mathrm{II}} k_{\mathrm{II}}[\mathrm{L}]}{1 + k_{\mathrm{II}}[\mathrm{L}]} \tag{2.70}$$

This can be generalized. Clearly in this situation the simple functional relationship between $\bar{i}$ and $[\mathrm{L}]$ that is seen in Eq. (2.67) has been lost.

As another variation consider that the substrate possesses n identical sites, but that the system contains two competing ligands, A and B. Now the number of sites available to A is decreased from n to $(n - \bar{i}_B)$, and similarly the number of sites accessible to B is $(n - \bar{i}_A)$. Thus Eq. (2.67) leads to

$$\bar{i}_A = \frac{(n - \bar{i}_B)k_A[A]}{1 + k_A[A]}$$

$$\bar{i}_B = \frac{(n - \bar{i}_A)k_B[B]}{1 + k_B[B]}$$

Combining these equations:

$$\bar{i}_A = \frac{nk_A[A]}{1 + k_A[A] + k_B[B]} \tag{2.71}$$

$$\bar{i}_B = \frac{nk_B[B]}{1 + k_A[A] + k_B[B]} \tag{2.72}$$

Equation (2.71) can also be written as (34)

$$\bar{i}_A = \frac{nk'_A[A]}{1 + k'_A[A]} \tag{2.73}$$

where $k'_A = k_A/(1 + k_B[B])$; thus if [B] is held constant, Eq. (2.73) has the same functional form as Eq. (2.67).

[$m = 1$; n different, dependent sites, with interactions between bound ligands] In this quite general and realistic model a wide range of behavior is included. It is possible for binding of one ligand molecule to alter the affinity in a subsequent binding step. Steric and electrostatic effects by the ligand particles are often responsible for such behavior, but it is also possible for binding to induce a conformational change in the substrate, altering the nature of the remaining unoccupied sites.

A general thermodynamic description of the model can be given in terms of the stoichiometric stepwise binding constants K_{1i}. Then Eqs. (2.47) for $\bar{i}$ and Eq. (2.51) for f_{1i} are applicable, with

$$\beta_{1i} = \prod_{i=1}^{n} K_{1i} \tag{2.74}$$

and the isotherm can be written in terms of $\bar{i}$, θ, or $F_{\bar{S}}$ [Eqs. (2.52) to

(2.54)]. We have already considered the $[m = 1; n = 2]$ model; this is a special case, but it illustrates most of the features of the general model. If the stepwise stability constants can all be measured, the system is completely described thermodynamically. Nevertheless, we often seek knowledge on the molecular level, in the form of microscopic site binding constants. For the $[m = 1; n = 2]$ case we found that there are four microscopic constants, of which three are independent, but there are only two stoichiometric constants, so it is not possible to obtain the microscopic constants from the stoichiometric constants [see discussion of Eqs. (2.5), (2.6), (2.58), and (2.59)]. This is a general problem; Klotz and Hunston (35) obtained the results in Table 2.3. Thus without additional relationships or information (such as equivalence or independence of sites) the microscopic binding constants are not accessible.

Methods of analyzing experimental data to obtain the stoichiometric binding constants are treated in Sections 2.5 and 2.6 and Chapter 3, but it is obvious that the value of n is a critical piece of information. If the structure of the substrate is known, it may be possible to specify n or at least to place a limit on it, but with macromolecular substrates n may be unknown. The data analysis will yield an estimate of n, but any contributing information on stoichiometric ratios will be helpful.

[Self-Association]   In this model of complexing a solute interacts with itself to form dimers, trimers, . . . , m-mers. The stepwise binding constants are defined by these equilibria, where we call the interactant the substrate:

$$2S \underset{}{\overset{K_2}{\rightleftharpoons}} S_2$$

$$S + S_2 \underset{}{\overset{K_3}{\rightleftharpoons}} S_3$$

$$\vdots$$

$$S + S_{(h-1)} \underset{}{\overset{K_h}{\rightleftharpoons}} S_h$$

For generality we write $S \rightleftharpoons S$ and define $K_1 = 1$. Then the overall binding constant β_h is

$$\beta_h = \prod_{j=1}^{h} K_j$$

We present the model of Mukerjee and Ghosh (36). Let S_t be, as in the earlier use of this symbol, the total concentration of substrate in equivalents of S per liter:

Table 2.3. Numbers of Binding Constants for a Substrate with n Binding Sites

n	No. of Stoichiometric Constants	Total No. of Microscopic Constants	No. of Independent Microscopic Constants
2	2	4	3
3	3	12	7
4	4	32	15
6	6	192	63
8	8	1024	255
12	12	24576	4095

$$S_t = [S] + 2[S_2] + 3[S_3] + \cdots = \sum_{h=1}^{m} h\beta_h[S]^h \qquad (2.75)$$

where $[S]$ represents the monomer concentration and m is the maximum value of h. Now define σ_t as the total concentration of substrate in moles of species per liter, or

$$\sigma_t = [S] + [S_2] + [S_3] + \cdots = \sum_{h=1}^{m} \beta_h[S]^h \qquad (2.76)$$

Mukerjee and Ghosh define the average number of monomers in all multimers by Eq. (2.77).

$$\bar{\nu} = \frac{S_t - [S]}{\sigma_t - [S]} \qquad (2.77)$$

Combining Eqs. (2.75)–(2.77),

$$\bar{\nu} = \frac{\sum_{h=1}^{m} h\beta_h[S]^h - [S]}{\sum_{h=1}^{m} \beta_h[S]^h - [S]} = \frac{\sum_{h=2}^{m} h\beta_h[S]^h}{\sum_{h=2}^{m} \beta_h[S]^h} \qquad (2.78)$$

Equation (2.78) is the isotherm, which is unusual in that the range of $\bar{\nu}$ is from 2 to m, for nonzero β_2. Dividing out the quotient in Eq. (2.78) gives a series in $[S]$.

$$\bar{\nu} = 2 + K_3[S] + K_2K_3(2K_4 - K_3)[S]^2 + \cdots \qquad (2.79)$$

An alternative measure of binding can be defined by analogy with a substrate–ligand system. Taking Eq. (2.46) as a guide, we write

$$\bar{h} = \frac{\Sigma \,(\text{S bound to S, therefore acting as ligand})}{\Sigma \,(\text{all S acting as substrate})}$$

$$= \frac{S_t - [S]}{\sigma_t} \tag{2.80}$$

$$= \frac{\Sigma_{h=2}^{m} \, h\beta_h [S]^h}{\Sigma_{h=1}^{m} \, \beta_h [S]^h}$$

$$= \frac{\Sigma_{h=2}^{m} \, h\beta_h [S]^{h-1}}{1 + \Sigma_{h=2}^{m} \, \beta_h [S]^{h-1}} \tag{2.81}$$

which is similar in form to Eq. (2.47). The limits of $\bar{h}$ are 0 and m. Dividing out this quotient gives

$$\bar{h} = 2K_2 [S] + K_2 (3K_3 - 2K_2)[S]^2 + \cdots \tag{2.82}$$

The experimental study of self-association is more difficult than studying substrate–ligand interactions because there is one fewer degree of freedom (the ligand concentration).

[Micelle formation] Micelle formation is a form of self-association among amphiphiles, molecules possessing both a polar "head" group and a nonpolar "tail." Micelles have high aggregation numbers, 30–100 being typical, and the distribution of h in S_h about the maximum value is relatively narrow. We can model micelle formation as in the model of self-association. Define f_h as the fraction of amphiphile in the form of micelles of formula S_h, so that $f_h = h[S_h]/S_t$, or

$$f_h = \frac{h\beta_h [S]^h}{S_t}$$

The standard free energy change per mole of micelle S_h is $-RT \ln \beta_h$, or per mole of monomeric amphiphile contained in the micelle

$$\Delta\mu^{\circ}_{\text{mono}} = -\frac{RT}{h} \ln \beta_h$$

Combining these expressions

$$f_h = \frac{h[S]^h}{S_t} \exp\left(-h\,\Delta\mu^{\circ}_{\text{mono}}/RT\right) \tag{2.83}$$

Equation (2.83) is an expression for the dependence of f_h on h. The preexponential term is an increasing function of h. The exponential term may increase or decrease with h, depending on the dependence of $\Delta\mu^{\circ}_{mono}$ on h. At low values of h, the hydrophobic effect will be dominant, making $\Delta\mu^{\circ}_{mono}$ negative (attractive), and therefore increasing f_h. At very high h, however, the increasing proximity of the polar head groups will introduce an opposing repulsive interaction, and the exponential term will act to decrease f_h. Thus a maximum is anticipated in the distribution function. Tanford (37) has used empirical measures of the attractive and repulsive interaction contributions to describe micelle formation with Eq. (2.83). Mukerjee (38) has discussed models of stepwise self-association and micelle formation.

2.5. THE BINDING ISOTHERM

In Section 2.4 we developed several general expressions for binding isotherms and obtained analytical descriptions for some models. We next treat the simplest of these models, namely, 1:1 complex formation. This is also the most important isotherm, first because it is often applicable to real systems, and second because many other systems can be viewed as extensions of, or deviations from, this simple case. It is usually appropriate to test this model against the data before adopting more complicated assumptions.

Graphical Presentation

In its simplest form the isotherm for 1:1 binding is

$$f_{11} = \frac{K_{11}[L]}{1 + K_{11}[L]} \tag{2.84}$$

and the same functional relationship is shown by the isotherm for binding to n identical sites without interactions:

$$\bar{i} = \frac{nk[L]}{1 + k[L]} \tag{2.85}$$

These equations and symbols were introduced in Section 2.4. Often the isotherm will be expressed in terms of experimental variables and parameters [see Eq. (2.38) for an example], so for generality we write the function as Eq. (2.86):

$$y = \frac{dx}{f + ex} \qquad (2.86)$$

where y is the dependent variable, x is the independent variable (free ligand concentration), and d, e, f are constants or parameters. The properties of this function are developed in Appendix A. Evidently, it is always possible to express Eq. (2.86) so that one of the parameters is unity.

A graphical display of the isotherm can be a powerful means of assisting qualitative interpretations and obtaining quantitative estimates. To demonstrate these plotting forms we use the simulated data of Table A.2, which are calculated for Eq. (2.86) with $d = e = f = 1$. In Figure 2.2 is shown the *direct plot* of y against x; this plot is often called the binding curve. It is shown in Appendix A that this curve is (if carried out to $x = \infty$) one-half of one branch of a rectangular hyperbola, the vertex coinciding with the origin in the experimental x, y frame. As x is made very small so that $ex \lll f$, then y approaches $(d/f)x$, a linear function; but it is noted in Appendix A that the curvature of the hyperbola is greatest at the origin. Observe that the points shown in Figure 2.2 have been selected with equal increments in x. It will be evident from the figure that unless x is made extremely large it will be difficult or impossible to estimate the asymptotic limit of y.

Figure 2.3 is the *semilogarithmic plot* of y against $\log x$. This is called the Bjerrum formation function, and it is most familiar in the context of acid–base equilibria, when it is equivalent to a titration curve. Although it may not be possible to estimate y_{max} accurately, it still may be possible to find the value of x corresponding to $y_{max}/2$, for this marks the inflection point in the curve. An advantage of the semilog plot is its capability of presenting data covering several orders of magnitude in x.

Equation (2.86) can be put in the form

$$\frac{y}{(d/f) - (e/f)y} = x \qquad (2.87)$$

If $(d/f) = (e/f)$, this becomes

$$\log \frac{y}{1 - y} = \log x - \log \frac{d}{f} \qquad (2.88)$$

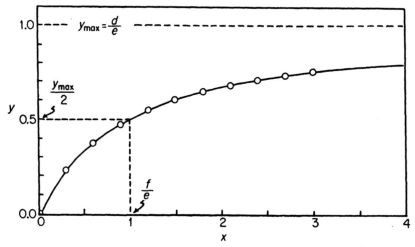

Figure 2.2. Direct binding curve plot of points in Table A.2. The parameters d, e, f are defined by Eq. (2.86).

For example, Eq. (2.84) gives

$$\log \frac{f_{11}}{1 - f_{11}} = \log [L] - \log K_{11} \qquad (2.89)$$

Thus a *log–log plot* of the left side of the equation against $\log x$ should give a straight line with slope = 1 (Figure 2.4). This is called the (A.V.) Hill plot (39). The value of the slope at the point $\log [y/(1 - y)] = 0$ is the Hill coefficient. For the simple isotherm considered here obviously the Hill plot is linear with coefficient unity. Deviations from this behavior are diagnostic of more complicated phenomena, and Hill plots have been widely applied as aids in the study of enzyme inhibition and cooperative binding; the references cited in (40) lead to this literature, and in Section 2.6 cooperativity is discussed. Equation (2.89) shows that for different systems obeying this equation the curve shapes in the semilog and log–log plots will be identical, but will be positioned along the $\log [L]$ axis according to their $\log K_{11}$ values.

There are three nonlogarithmic linear plotting forms of the rectangular hyperbola. From Eq. (2.86) we get

$$\frac{1}{y} = \frac{f}{d} \cdot \frac{1}{x} + \frac{e}{d} \qquad (2.90)$$

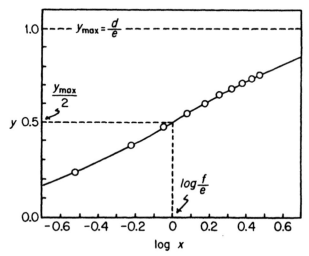

Figure 2.3. Semilogarithmic plot of binding data.

A plot of $1/y$ against $1/x$ is linear; see Figure 2.5. In enzyme kinetic studies this is called the Lineweaver–Burk plot (41), and in spectroscopic studies of molecular complexes it is called the Benesi–Hildebrand plot (42), but we shall refer to it as the *double-reciprocal plot*.

Equation (2.86) also yields

$$\frac{x}{y} = \frac{e}{d} x + \frac{f}{d} \qquad (2.91)$$

We call this the *y-reciprocal plot*; it is shown in Figure 2.6.

Equation (2.92) is the third linear form, the *x-reciprocal plot*.

$$\frac{y}{x} = -\frac{e}{f} y + \frac{d}{f} \qquad (2.92)$$

In protein binding studies this is called a Scatchard plot (32a); it is the Eadie plot in enzyme kinetics (43). Equation (2.92) is plotted in Figure 2.7.

These three linear plotting forms are not equivalent in an experimental sense, and there is continuing controversy about their advantages and limitations. The salient features are these: When the values of the independent variable x are chosen to be equally spaced, as in Figure 2.2, the linearization transformation can change this relationship. The double-

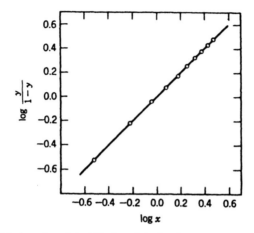

Figure 2.4. Log–log plot of binding data for the case $d/f = e/f$, Eq. (2.88).

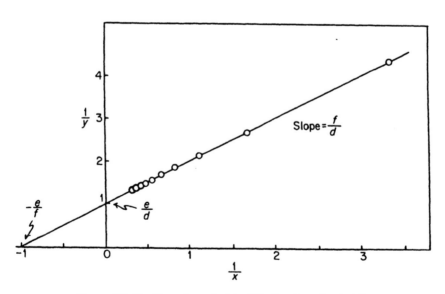

Figure 2.5. Double-reciprocal plot of binding data, Eq. (2.90).

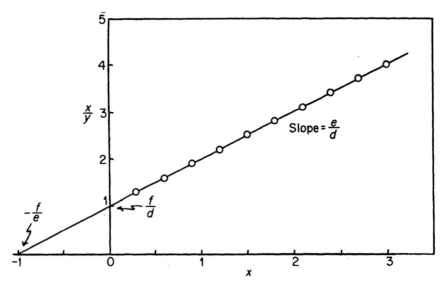

Figure 2.6. y-Reciprocal plot of binding data, Eq. (2.91).

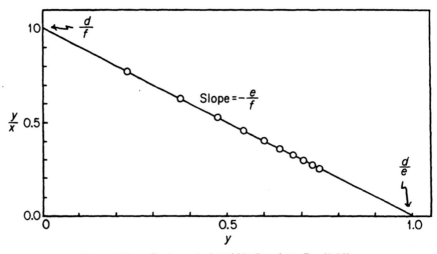

Figure 2.7. x-Reciprocal plot of binding data, Eq. (2.92).

reciprocal plot is notorious for this effect. Figure 2.5 shows that the points at low x are widely spaced, with those at high x being closely bunched. The result is that the placement of the line is extremely sensitive to the y value of the point having the smallest x. Some workers choose x such that points are evenly spaced on the $1/x$ axis. A good approach is to use a statistical weighting procedure, as developed in Chapter 3.

The double-reciprocal plot has the advantage that the variables x and y remain separated on the abscissa and ordinate; in the other plotting forms the variables become mixed. This is important because the uncertainty in the dependent variable y is usually much greater than that in the independent variable. Equations (2.90) and (2.91) have the advantage that the dependent variable appears only on the ordinate, so the uncertainty of the abscissa values is negligible, and this simplifies the statistical analysis. Note that in the y-reciprocal plot the equal spacing of x values is retained. The x-reciprocal plot, Figure 2.7, has the dependent variable on both axes, but it has an advantage in that it provides a closed scale representation of y on the abscissa, unlike the other two plots, which are open-ended.

Examining the Model

It is evident that at very low values of the independent variable, it is not possible to determine unambiguously whether or not the function is a rectangular hyperbola; it is necessary to explore a wide range of the binding isotherm. In principle it is impossible to measure the full extent of the curve, because of its asymptotic approach to the limit of $f_{11} = 1.00$. In practice it is often impossible to measure as much of the curve as may be desirable because of solubility limitations, nonideal behavior, or analytical difficulties. But we may at least explore the question of the extent to which the binding curve should be experimentally studied in order to confirm the simple chemical model.

Person (44) made the first critical analysis of this question. He pointed out that unless the study is extended into the strongly curved portion of the binding curve, it is not possible to assert that a chemical equilibrium exists. That is, the analytical response must be a nonlinear function of ligand concentration if Eq. (2.84) applies, and the concentration of the complex SL must be comparable to the concentration of uncomplexed substrate S (when L is in excess) in order to demonstrate this. Taking the definition of the point where definite curvature sets in as $f_{11} \approx 0.1$, and putting this into Eq. (2.84), shows that $K_{11}[L] \approx 0.1$ or $[L] \approx 0.1/K_{11}$. Person therefore concluded that the ligand concentration should be extended at least to the value $0.1/K_{11}$, and preferably higher, for a reported K_{11} value to be reliably interpreted as evidence for a chemical equilibrium.

Deranleau (45) broadened this analysis to include testing of the specific stoichiometric model and evaluation of the system parameters. (Both Person and Deranleau considered the spectrophotometric study of complexes, but their results are of general applicability.) Making use of Eq. (3.15) for the propagation of errors, solving Eq. (2.84) for K_{11}, and assuming the error in ligand concentration to be negligible, we find for the approximate relative error in K_{11}:

$$\frac{\sigma_K}{K_{11}} \approx \frac{\sigma_f}{f_{11}(1 - f_{11})} \qquad (2.93)$$

where σ_K, σ_f are standard deviations in K_{11} and f_{11}, respectively. Figure 2.8 is a plot of $1/f_{11}(1 - f_{11})$ as a function of f_{11}. This immediately shows that the relative error in K_{11} is realistically minimized by working in the range of $f_{11} \approx 0.2$ to 0.8. According to this result Person's criterion is not sufficiently demanding. The sharp increase in error at the extremes of the degree of saturation should be noted; Weber and Anderson (46) have pointed out that there is no information about K_{11} in data obtained at $f_{11} = 0$ or 1. Norheim has made similar calculations for complexes of higher stoichiometries (47).

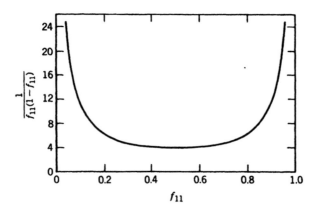

Figure 2.8. Plot of $1/f_{11}(1 - f_{11})$ (which is proportional to the relative error in K_{11}) as a function of f_{11}, the degree of saturation.

Deranleau (45a) also considered the question of the extent of the binding isotherm that must be experimentally investigated in order to establish with reasonable confidence that the simple 1:1 model leading to Eq. (2.84) is the appropriate stoichiometric model. He applied information theory to the binding curve, calculating the fraction of the total theoretical information that is accumulated as f_{11} increases from zero. This fraction I/I_{max} is given by

$$\frac{I}{I_{max}} = f_{11} - f_{11}^2 \ln f_{11} + (1 - f_{11})^2 \ln (1 - f_{11})$$

This fraction relative to f_{11} is then identified as the rate of accumulation of information. This quantity, $I/f_{11}I_{max}$, passes through a maximum at $f_{11} = 0.76$. Deranleau therefore suggested that measurements should be made up to about 75% of complete saturation to make an acceptable test of the model function.

Let us develop a geometrical criterion based on the properties of the rectangular hyperbola. As discussed in Appendix A, we know that, in the X, Y coordinate system, the vertex is at $X = a$ and the focus is at $X = c$, where $c^2 = 2a^2$. It is easy to show that the point $X = c$, $Y = a$ lies on the hyperbola. We therefore suggest as a criterion that if we follow $Y = f(X)$ as we go from $X = 0$ to $X = c$, and if at the point where $X = c$ we find that $Y = a$ (and that intermediate points are also satisfied), then we have confirmed that the curve is a rectangular hyperbola.

The coordinates $X = c$, $Y = a$ are converted into coordinates in the x, y frame by means of Eqs. (A.11) and (A.13). The result is

$$x = \frac{\sqrt{df}}{e} (1 + \sqrt{2}) - \frac{f}{e}$$

$$y = \frac{\sqrt{df}}{e} (1 - \sqrt{2}) + \frac{d}{e}$$

For the normalized function, $d = e = f = 1$, this gives $y = 1 - \sqrt{2} = 0.586$. Thus according to this criterion about 60% of the range in f_{11} should be experimentally examined. It is an interesting practical result that the ligand concentration must be doubled to pass from $f_{11} = 0.60$ to $f_{11} = 0.75$, as can be seen in Figure 2.2 and Table A.2.

These criteria specifying 60 to 75% coverage of the isotherm as a sound basis for testing the model are, though reasonable, still arbitrary, but the important point is that something over half of the theoretically accessible range in f_{11}, the degree of saturation, should be investigated. This may introduce other difficulties, particularly if the complex is quite weak so that such a high ligand concentration is required that a general medium effect is produced.

As a means of testing the simple model, Eq. (2.84) or Eq. (2.85), it might seem necessary only to make one of the plots according to Eqs. (2.90) to (2.92), examining it for linearity. This is certainly a necessary condition, but it is seldom sufficient. A fault of many reported studies has been the small range in degree of saturation that has been observed; if this range is very small, the plot will probably not reveal curvature even if Eq. (2.84) is inapplicable. The experimental uncertainty is a factor also, since this enters in determining whether or not a plot significantly deviates from linearity. Moreover, the several graphical forms of display are not equally useful, as was indicated earlier. The preference of plotting form depends on the purpose of the analysis, though workers still disagree in their choices.

Generally, there are two reasons to treat the data graphically: to test the model and to evaluate the parameters. In testing the model it is essential, as shown earlier, that a sufficiently extensive range in degree of saturation be covered, so the method of plotting should permit this to be discovered. At this stage the quantity f_{11} often cannot be calculated because the parameters are unknown, so one has available as the dependent variable only an experimental response. Three plotting forms are appropriate: the direct binding curve of y against x (which, however, has the disadvantage of being open-ended), the semilogarithmic plot, and the x-reciprocal plot (Scatchard plot). The double-reciprocal and y-reciprocal plots are not suitable for this task because their abscissa scales are arbitrarily adjustable. The x-reciprocal plot is useful in revealing the portion of the binding isotherm that has been experimentally examined, but it is necessary to construct the plot so the origin represents $f_{11} = 0$. Deranleau (45a) recommends the semilogarithmic and x-reciprocal plots because their abscissa scales are capable of containing all of the theoretically accessible data points and because the range of isotherm covered is visually apparent. Weber and Anderson (46) also prefer the semilogarithmic plot. Klotz (48) strongly advises that the semilog plot be used for this purpose and cautions against reliance on the x-reciprocal plot, showing examples of its misinterpretation. Probably the best course is to plot the data in several ways and to seek an interpretation consistent with all of the plots.

We return to a consideration of model testing in Section 2.6, on treating multiple equilibria, and in Chapter 3, on regression analysis of binding data.

Presuming that the model equation has been found to describe the data satisfactorily, we seek the model parameters, which include the binding constant and, possibly, quantities characteristic of the experimental technique, such as molar absorptivities. One method is to solve the isotherm equation, using measurements at as many ligand concentrations as there are parameters. The most popular method has been a graphical one, making use of the plots shown in Figures 2.5 to 2.7. In earlier work the lines were often drawn "by eye," the parameters being read from the slopes and intercepts of the graphs, but it is now more common to use linear least-squares regression analysis, including variable weighting of the data points, as described in detail in Chapter 3. Nonlinear regression of the binding data is also possible and is increasingly popular.

If the simple functional relationship of the rectangular hyperbola is not applicable because of higher stoichiometric ratios, the plotting functions will be different, and it may not be possible to find linear graphical forms. Nonlinear regression analysis can always be carried out.

2.6. VARIATIONS AND COMPLICATIONS

Multiple Equilibria

The problem of multiple equilibria—the formation of two or more complexes—has been dealt with in several fields. Classical studies in acid–base equilibria (49) and in metal ion–base coordination (50) have led to methods that are applicable to systems described by, typically, two to six stepwise equilibrium constants. Binding of small molecules to macromolecules, including enzymes and other proteins, nucleic acids, and synthetic polymers, is another area requiring consideration of multiple binding (32). In the field of small molecule–small molecule organic complexes it has often been assumed (perhaps too often) that only 1:1 stoichiometry need be considered, but careful investigation sometimes reveals that, even in these relatively simple systems, it is necessary to take account of additional complex species.

If the experimental technique yields $\bar{i}$, the average number of ligands bound per substrate molecule, then, as shown in Section 2.4, an appropriate isotherm describing multiple equilibria is

$$\bar{i} = \frac{\Sigma \, i\beta_{1i}[L]^i}{1 + \Sigma \, \beta_{1i}[L]^i} \tag{2.94}$$

where $\beta_{1i} = \Pi \, K_{1i}$, and the summations are from $i = 1$ to n. The immediate goal of the experimental study and the data analysis is the evaluation of all K_{1i}, the stepwise binding constants. In Section 2.5 the graphical behavior of Eq. (2.94) was treated for the case $n = 1$. Bardsley and Childs (51) have made an exhaustive analysis of the properties of the general function.

An important piece of information is n, the maximum value of i. This is equal to the number of binding sites on the substrate if each ligand occupies one site, so n may be obvious from structural considerations or may be inferred from the behavior of analogous systems. Thus, for example, in studying acid–base dissociation, the number of sites can usually be estimated by inspection of the structure together with knowledge of the acid–base properties of functional groups. The coordination numbers of metal ions are also known quantities. But n may be difficult to establish for some macromolecular systems, for which uncertain graphical extrapolations or rather weak statistical evidence may provide the only guidance.

Since least-squares methods are treated in Chapter 3, here we show some numerical and graphical approaches that have been devised for the evaluation of the K_{1i}. Additional techniques are reviewed elsewhere (50). A useful method is based on a suggestion of Scatchard, developed by Edsall et al. (52) and Poë (53). The quantity Q is defined by

$$Q = \frac{\bar{i}}{(n - \bar{i})[L]} \tag{2.95}$$

Comparison of Eq. (2.95) with Eq. (2.67) or Eq. (2.85) shows that $Q = k$, the site binding constant in the model of n independent identical sites. In general, however, Q is not a constant, but is a function of [L] (or $\bar{i}$). It is for this reason that Poë (53) calls Q the pseudo-constant. Combining Eqs. (2.94) and (2.95) yields

$$Q = \frac{\Sigma \, i\beta_{1i}[L]^{i-1}}{n + \Sigma(n - i)\beta_{1i}[L]^i} \tag{2.96}$$

We find, from Eq. (2.96),

$$\lim_{[L]\to 0} Q = \frac{K_{11}}{n} \tag{2.97}$$

$$\lim_{[L]\to\infty} Q = nK_{1n} \tag{2.98}$$

or, equivalently and often more conveniently, the limit of log Q as $\bar{i}$ goes to zero is log (K_{11}/n), and the limit of log Q as $\bar{i}$ goes to n is log (nK_{1n}). Thus the plot of log Q against $\bar{i}$ yields estimates of K_{11} and K_{1n} (with the extrapolation to $\bar{i} = n$ being the more uncertain one).

Initial estimates of the intervening stepwise constants can be found with this argument (50). The constant K_{1i} describes this equilibrium:

$$SL_{i-1} + L \rightleftharpoons SL_i$$

Now, if $(i-1) < \bar{i} < i$, we write that the concentration of SL_i is inversely proportional to $i - \bar{i}$ and that the concentration of SL_{i-1} is inversely proportional to $\bar{i} - (i-1)$, so $[SL_i]/[SL_{i-1}] = (\bar{i} - i + 1)/(i - \bar{i})$, or

$$K_{1i} = \left(\frac{\bar{i} - i + 1}{i - \bar{i}}\right) \frac{1}{[L]} \tag{2.99}$$

where [L] is the value corresponding to $\bar{i}$. Implicit in this argument is the assumption that SL_{i-1} and SL_i are the only substrate species present, which is acceptable only if K_{1i} differs from both $K_{1(i-1)}$ and $K_{1(i+1)}$ by at least two orders of magnitude. If additional species are present, Eq. (2.99) will probably give its best estimate when $\bar{i}$ is about equal to $i - \frac{1}{2}$, since this approximately minimizes the concentrations of SL_{i+1} and SL_{i-2}. Note that when $\bar{i} = i - \frac{1}{2}$ Eq. (2.99) becomes $K_{1i} = 1/[L]$; this is familiar in the context of equating a pK_a to the pH at half-neutralization.

Poë's method of successive approximation is as follows. Equation (2.94) can be put into the following form:

$$\sum_{i=0}^{n} (\bar{i} - i)\beta_{1i}[L]^i = 0 \tag{2.100}$$

where $\beta_{10} = 1$. This is solved for a particular β_{1i}, giving Eq. (2.101).

$$\beta_{1i} = \frac{\bar{i}}{(i - \bar{i})[L]^i}\left[1 + \sum_{j\neq i}\left(\frac{\bar{i} - j}{\bar{i}}\right)\beta_{1j}[L]^j\right] \tag{2.101}$$

Using the initial estimates of the β_{1j} obtained as described earlier, β_{1i} is calculated with (2.101) using an appropriate experimental value of $\bar{i}$. Evidently, as $\bar{i}$ approaches i, β_{1i} becomes indeterminate, so the best $\bar{i}$ will be about $i \pm \frac{1}{2}$ to $i \pm 1$. This procedure is carried out for all β_{1i}, giving the second round of estimates. Iteration is continued until the estimates converge. Experimental uncertainties in the parameter estimates can be obtained by a propagation of errors treatment.

We illustrate this method with the data (53) in Table 2.4 for the complexing of ammonia (the ligand) with Cu(II). Figure 2.9 is the semilog plot, which shows that $n = 4$. Figure 2.10 is a plot of log Q against $\bar{i}$. From the intercept at $\bar{i} = 0$ we estimate log $(K_{11}/n) = 3.52$, or log $K_{11} = 4.12$. The intercept at $\bar{i} = 4$ yields log $(nK_{14}) = 2.77$, or log $K_{14} = 2.17$. With Eq. (2.99) and Figure 2.9 the initial estimates log $K_{12} = 3.52$ and log $K_{13} = 2.85$ are obtained. Thus the first estimates of the β_{1i} are calculated to be

$$\beta_{11} = 1.32 \times 10^4 \ M^{-1}$$

$$\beta_{12} = 4.37 \times 10^7 \ M^{-2}$$

$$\beta_{13} = 3.09 \times 10^{10} \ M^{-3}$$

$$\beta_{14} = 4.57 \times 10^{12} \ M^{-4}$$

As an example, we calculate the second estimate of β_{11} using Eq. (2.101) and $\bar{i} = 0.486$; in this calculation $[L] = 4.624 \times 10^{-5} \ M$, $i = 1$, and the β_{1j} are listed earlier, where $j = 2, 3, 4$.

Table 2.4. Experimental Data and Derived Quantities for the Cu(II)–NH$_3$ System^a

$10^3[L]/M$	pL	$\bar{i}$	$10^{-3}Q$	log Q
0.02032	4.692	0.244	3.197	3.505
0.04624	4.335	0.486	2.991	3.476
0.1256	3.901	0.959	2.511	3.400
0.5346	3.272	1.877	1.654	3.219
2.286	2.641	2.784	1.002	3.001
8.630	2.064	3.437	0.7074	2.850
22.65	1.645	3.743	0.6430	2.808
247.7	0.606	4.002	–	–

^{a}In 2M NH$_4$NO$_3$ at 30°C.

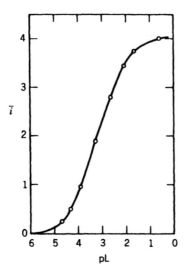

Figure 2.9. Semilogarithmic plot of data in Table 2.4, where pL = −log[L].

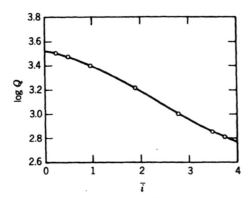

Figure 2.10. Plot of log Q against $\bar{i}$ for the data in Table 2.4.

$$\beta_{11} = \frac{0.486}{(1 - 0.486)[L]} \left(1 + \frac{0.486 - 2}{0.486}\beta_{12}[L]^2 + \frac{0.486 - 3}{0.486}\beta_{13}[L]^3\right.$$

$$\left. + \frac{0.486 - 4}{0.486}\beta_{14}[L]^4\right)M^{-1}$$

$$= 2.046 \times 10^4 (1 - 0.291 - 0.0158 - 0.0002)M^{-1}$$

$$= 1.418 \times 10^4\ M^{-1}$$

Hence, the improved estimate is $\log \beta_{11} = 4.152$. This process is carried through for all β_{1i}. Poë (53) gives these final estimates, which are in excellent agreement with results of least-squares analysis: $\log K_{11} = 4.139$, $\log K_{12} = 3.520$, $\log K_{13} = 2.884$; $\log K_{14} = 2.150$.

The following graphical method is described by Rossotti and Rossotti (54). Rearrangement of Eq. (2.100) gives Eq. (2.102).

$$\frac{\bar{i}}{(1-\bar{i})[L]} = \beta_{11} + \beta_{12}[L]\left[\frac{2-\bar{i}}{1-\bar{i}}\right] + \sum_{i=3}^{n}\left[\frac{i-\bar{i}}{1-\bar{i}}\right]\beta_{1i}[L]^{i-1} \qquad (2.102)$$

If $n = 2$, a plot of $\bar{i}/(1-\bar{i})[L]$ against $[L](2-\bar{i})/(1-\bar{i})$ will be a straight line with slope β_{12} and ordinate intercept β_{11}. If $n > 2$, the plot will be curved, but the intercept will still be β_{11}, and the limiting slope will be β_{12}. Note that, for the $n = 2$ case, points may fall in the first and third quadrants.

More generally, Eq. (2.100) is rearranged to Eq. (2.103).

$$\frac{\bar{i}}{(i-\bar{i})[L]^i} = \sum_{j=1}^{n}\left[\frac{j-\bar{i}}{i-\bar{i}}\right]\beta_{1j}[L]^{j-i} \qquad (2.103)$$

When $i = 1$, Eq. (2.102) is obtained. Setting $i = 2$ gives

$$\frac{\bar{i}}{(2-\bar{i})[L]^2} - \left[\frac{1-\bar{i}}{2-\bar{i}}\right]\frac{\beta_{11}}{[L]} = \beta_{12} + \beta_{13}[L]\left[\frac{3-\bar{i}}{2-\bar{i}}\right] + \sum_{j=4}^{n}\left[\frac{j-\bar{i}}{2-\bar{i}}\right]\beta_{1j}[L]^{j-2}$$

$$(2.104)$$

A plot of the left side of this equation against $[L](3-\bar{i})/(2-\bar{i})$ yields estimates of β_{12} and β_{13}. Evidently, this procedure can be repeated as required until all constants have been evaluated—that is, in principle. In practice this sort of "bootstrap" operation encounters the difficulty of error accumulation, and highly precise data are required for successful application beyond Eq. (2.102). Observe the similarity of functional form appearing in Eqs. (2.95), (2.99), (2.101), and (2.103).

It was pointed out in discussing the plotting forms based on Eqs. (2.90) to (2.92) that curvature in these plots rules out the simple 1:1 model, and experience has shown that the x-reciprocal or Scatchard plot ($\bar{i}/[L]$ versus $\bar{i}$) is most sensitive for this purpose (45b). Irving (55) has analyzed the properties of this plot, and the related plot of $\bar{i}/[L]$ versus $[L]$, when Eq. (2.94) describes the system. A MacLaurin's series expansion of Eq. (2.94) gives this polynomial:

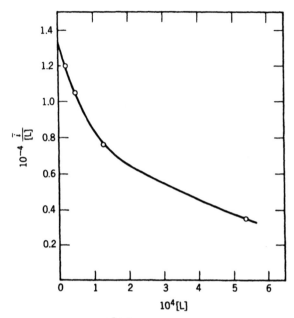

Figure 2.11. Plot of $\bar{i}/[L]$ against [L] for the data in Table 2.4.

$$\frac{\bar{i}}{[L]} = \beta_{11} + (2\beta_{12} - \beta_{11}^2)[L] + (3\beta_{13} - 3\beta_{11}\beta_{12} + \beta_{11}^3)[L]^2 + \cdots$$

$$(2.105)$$

It is obvious from Eq. (2.105) that the intercept on the ordinate axis of the $\bar{i}/[L]$ versus [L] plot is β_{11} and the initial slope is $2\beta_{12} - \beta_{11}^2$, which is equivalent to $K_{11}(2K_{12} - K_{11})$. Figure 2.11 is a plot of the data (at low ligand concentrations) from Table 2.4. It is evident that extrapolation of the curve and determination of the initial slope may be inaccurate. The estimates obtained graphically according to Eq. (2.105) are $\beta_{11} = 1.33 \times 10^4 \ M^{-1}$, $\beta_{12} = 5.3 \times 10^7 \ M^{-2}$.

A clear distinction must be made between extrapolation methods such as those based on Eqs. (2.96), (2.103), and (2.105), and a procedure in which K_{11} is estimated from data at low ligand concentrations by treating as negligible any contributions from higher-order complexes. This latter procedure is not defensible (45b). The extrapolations described earlier are (though subject to uncertainty, as are all extrapolations) based on sound principles and do not neglect the higher-order species, but instead take them into account in the extrapolation.

The greatest use of curved Scatchard plots has been in studies of binding to macromolecules, when such a plot with concave-upward curvature is often interpreted by means of Eq. (2.70) or its generalization, Eq. (2.106).

$$\bar{i} = \sum \frac{n_j k_j [L]}{1 + k_j [L]} \qquad (2.106)$$

According to this interpretation, the plot of $\bar{i}/[L]$ against $\bar{i}$ has an ordinate intercept equal to $\sum n_j k_j$ and an abscissa intercept of $\sum n_j$. The application of this equation requires the assumption that the substrate possesses two or more classes of binding sites, the n_j sites of class j being identical and independent and having site binding constant k_j. Figure 2.12 shows this plot for a hypothetical system of two classes of binding sites, with $n_1 = 6$, $n_2 = 10$, $k_1 = 10^4$, $k_2 = 10^3$. In the experimental situation it has been common practice to fit the data points by making use of Eq. (2.106), the extrapolated intercepts, and the minimum number of classes required to achieve a good fit (56). Cantor and Schimmel (57) describe this curve-fitting procedure.

This practice is uncertain for several reasons. First, the estimation of the intercepts may be subject to large error, especially if the curvature is considerable so that the approach of the curve to the axis is very gradual. Second, the choice of the number of classes of sites, if unconstrained by

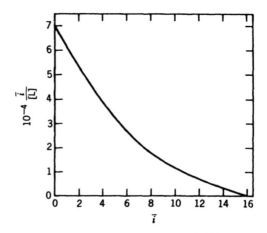

Figure 2.12. Scatchard plot (Eq. 2.92) of Eq. (2.105) with $n_1 = 6$, $n_2 = 10$, $k_1 = 10^4$, $k_2 = 10^3$.

structural considerations, is largely subjective; it is always possible to improve the curve fit by increasing the number of parameters. Third, and most important, is the possibility that some cause other than that implied by Eq. (2.106) may be responsible for the observed curvature; this is considered in the subsequent discussion. Klotz and co-worker (48, 58) have strongly argued against the decomposition of curved Scatchard plots in terms of Eq. (2.106). Nørby et al. (59) have described many literature examples of both inappropriate and incorrect analyses of this type.

There is a further level of interest in this problem. An understanding of binding at the molecular level requires an interpretation in terms of microscopic binding site constants, but this is difficult and may involve inappropriate assumptions (such as are made in the independent classes of sites analysis of curved Scatchard plots). An analysis of binding in terms of stepwise binding constants [the K_{1i} in Eq. (2.94)] is always thermodynamically sound, but by itself seldom can yield a satisfying molecular picture. Many authors have considered the possibility of relating the parameters of a Scatchard plot analysis to the stoichiometric stepwise binding constants and, ultimately, to microscopic binding constants (35, 45b, 46, 48, 58, 60). Although this may be valid in some cases, it does not appear to be generally applicable.

The goal of the experimentalist should be the evaluation of the K_{1i}, the stepwise binding constants. To achieve this, it is often helpful to make use of more than one method of data treatment. For example, a combination of intercepts from the plotting forms in Figures 2.10 and 2.11 allows n to be estimated from extrapolations to $[L] = 0$, and this quantity can be compared with the n estimated from Figure 2.9 by extrapolation to $[L] = \infty$. Consistency in comparisons of this type can be very reassuring.

Another valuable approach is to make measurements with more than one experimental technique to strengthen the test of the model and the evaluation of its parameters (4). This is especially useful when the different techniques respond to different functions of ligand concentration; thus one method may yield $\bar{i}$, whereas another may give $F_{\bar{s}}$ [see Eqs. (2.52) and (2.54)]. Another factor that comes into play in the study of multiple equilibria is that in some experimental methods the number of isotherm parameters increases as $2n$ rather than as n. If a technique yields $\bar{i}$, then evidently each additional species adds one parameter, its corresponding K_{1i}. Many methods, however, give an analytical response that is proportional to the extent of binding, and the proportionality constant is also incorporated into the isotherm. Thus with the spectrophotometric method each complex introduces both a binding constant and a molar absorptivity into the isotherm equation, with the result that the extraction of the parameter estimates rapidly becomes very difficult as n increases.

It may be asked whether these classical forms of data treatment retain much utility in the present, when computers have made nonlinear regression analysis a routine procedure. There are three reasons why the classical graphical and numerical methods are still valuable. First, for relatively simple systems the classical methods give results that are just as good as the nonlinear regression results. Second, nonlinear regression requires initial estimates of all parameters, and the classical techniques are capable of providing these. And third, the classical results provide criteria for decision making, in this sense: Nonlinear regression may, for some systems, yield illusory solutions that give excellent curve fits as a consequence of seeking local minima in the parameter space, and the classically derived results may help to rule these out.

Cooperativity

The concept of cooperative behavior is often encountered in the literature of macromolecular binding. This concept is not required for the determination of stability constants, so we merely outline the subject here. Let us start with the model of Section 2.5 in which the substrate possesses n identical independent binding sites, the ligand has one binding site, and there are no interactions between bound ligands. For this model it is possible to calculate the relative values of the stepwise binding constants by means of Eq. (2.62). These values are given in Table 2.5 for $n = 2$ to 6. According to Eq. (2.62), the stepwise constants decrease as i increases even though k, the microscopic site binding constant, is identical for each step; this is purely a statistical effect. We had earlier obtained the result

Table 2.5. Relative Values of K_{1i} with n Identical, Independent Binding Sites

n	K_{11}	K_{12}	K_{13}	K_{14}	K_{15}	K_{16}
2	2	$\frac{1}{2}$				
3	3	1	$\frac{1}{3}$			
4	4	$\frac{3}{2}$	$\frac{2}{3}$	$\frac{1}{4}$		
5	5	2	1	$\frac{1}{2}$	$\frac{1}{5}$	
6	6	$\frac{5}{2}$	$\frac{4}{3}$	$\frac{3}{4}$	$\frac{2}{5}$	$\frac{1}{6}$

for $n = 2$ by a different argument; see the discussion following Eq. (2.61). Note that Eq. (2.62) leads to $K_{11}/K_{1n} = n^2$.

There are many definitions of cooperativity, but they are all at least consistent with the following: A system is positively cooperative if the ratio $K_{1(i+1)}/K_{1i}$ is larger than the value calculated with Eq. (2.62), it is noncooperative if the ratio is equal to the statistical value, and it is negatively cooperative (anticooperative) if the ratio is smaller than the statistical value. These are statements about individual steps in the multistep process, and it is also possible to base a definition on the overall process. Some definitions are phenomenological, for example, the classification based on the value of the Hill coefficient (40c).

Figure 2.13 shows three manifestations of positive cooperativity. In 2.13*a* is shown a binding curve meeting the definition, but in which the ratio of stepwise constants, though exceeding the statistical value, is not so large that the qualitative shape of the curve is altered from that of a rectangular hyperbola. Figure 2.13*b* is a sigmoid binding curve, which according to some authors is the criterion of cooperativity. Figure 2.13*c* shows a hypothetical phase transition, the ultimate form of cooperative behavior; this is a true all-or-none phenomenon, not described by a mass-action relationship. Below a "transition ligand concentration" the substrate exists entirely in the unbound state, whereas above this concentration it is entirely bound. We wish to learn the conditions that define the three types of behavior shown in Figure 2.13.

Setting aside the phase transition phenomenon, so that mass-action relationships apply, Eq. (2.94) describes the system. [In the context of cooperative behavior, Eq. (2.94) is sometimes called Adair's equation

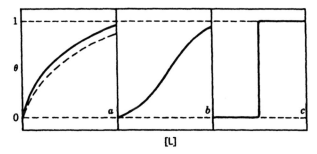

Figure 2.13. Forms of cooperative behavior. (*a*) Nonsigmoid, nonhyperbolic; the dashed line is a hyperbolic noncooperative binding curve. (*b*) Sigmoid binding curve. (*c*) A phase transition.

(61).] The question is, what produces a sigmoid binding curve? The necessary condition is the existence of an inflection point in the plot of $\bar{i}$ against [L]. It can be shown that an inflection point in the $\bar{i}$ versus [L] plot is equivalent to an extremum in the plot of $\bar{i}/[L]$ versus $\bar{i}$ or [L]. Irving (55) has discussed this by transforming Eq. (2.94) to Eq. (2.105). Differentiation then gives

$$\frac{d(\bar{i}/[L])}{d[L]} = (2\beta_{12} - \beta_{11}^2) + 2(3\beta_{13} - 3\beta_{11}\beta_{12} + \beta_{11}^3)[L] + \cdots$$

If $2\beta_{12} > \beta_{11}^2$ (i.e., $2K_{12} > K_{11}$), the initial slope is positive. Since the final slope in a plot of $\bar{i}/[L]$ versus [L] or $\bar{i}$ must be negative, the condition $2K_{12} > K_{11}$ suffices to produce a maximum irrespective of the value of n. This defines the passage from Figure 2.13a to Figure 2.13b. If n is greater than 2, there may be more than one extremum. Figure 2.14 shows Scatchard plots for the simplest system capable of exhibiting cooperative behavior, $n = 2$. (Notice that negative cooperativity is an alternative cause for concave-upward Scatchard plots.)

The essence of cooperativity is that the transition from the unbound state to the bound state occurs with a sharper gradient—over a smaller range of ligand concentration—than can occur in a noncooperative system. This effect is most marked when stepwise complex formation is replaced by a one-step process as described by the overall stability constant β_{1n} for this reaction.

$$S + nL \rightleftharpoons SL_n$$

$$\beta_{1n} = \frac{[SL_n]}{[S][L]^n}$$

Of course, for large n it is not likely that an $(n+1)$-body collision will take place leading to the formation of SL_n from S and L, so the stepwise mechanism still applies; but if K_{1i} is smaller than $K_{1(i+1)}$, the result will be that the SL_i $(i < n)$ will serve as reactive intermediates, and S and SL_n will be essentially the only substrate species present. Since β_{1n} is a product of n factors it may be very large even if K_{11} is small.

Figure 2.15 is a binding curve plot, for $n = 3$, of the noncooperative system $K_{11} = 90$, $K_{12} = 30$, $K_{13} = 10$ (see Table 2.5) and of the cooperative system $K_{11} = 1$, $K_{12} = 10$, $K_{13} = 2.7 \times 10^3$. Note that $\beta_{13} = 2.7 \times 10^4$ for both curves, but the sigmoid cooperative example fits the pattern of a

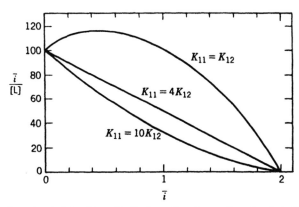

Figure 2.14. Scatchard plots of hypothetical $n = 2$ systems. From top to bottom, positively cooperative, noncooperative, and negatively cooperative systems. In each case $K_{11} = 100 \ M^{-1}$.

one-step overall binding process. Figure 2.16 shows Scatchard plots for these same systems. The fractions of the several substrate species S, SL, SL_2, SL_3 are graphed in Figures 2.17a (the noncooperative example) and 2.17b (the cooperative system). These plots show the behavior described in the preceding paragraph, with the sum of the concentrations of intermediate complexes SL and SL_2 barely reaching 2% of the total in the cooperative system.

The extent of cooperativity in cases controlled by the overall constant β_{1n} depends not only on the value of β_{1n} but also on n. This is shown in Figure 2.18 for hypothetical systems. This behavior is consistent with

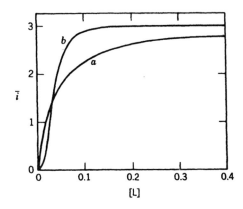

Figure 2.15. Binding curves for $n = 3$ (a) Noncooperative: $K_{11} = 90$, $K_{12} = 30$, $K_{13} = 10$. (b) Cooperative: $K_{11} = 1$, $K_{12} = 10$, $K_{13} = 2.7 \times 10^3$.

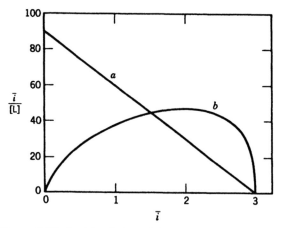

Figure 2.16. Scatchard plots of the systems in Figure 2.15.

T. L. Hill's statistical mechanical analysis of binding to proteins; Hill (62) found that at very large n it is possible for a phase transition to occur. The similarity of the curve for $n = 100$ in Figure 2.18 with the phase transition behavior in Figure 2.13c is very clear. Micelle formation is a cooperative process in which a large n is limited not by the number of binding sites but by an effect opposing association, namely, head-group repulsion. At one time a phase transition model of micelle formation was popular, but the mass-action model is now considered preferable. Cooperative behavior has been seen with very small n values; the classic example is the binding of oxygen to hemoglobin (61, 63), for which $n = 4$. Complex formation between 4-biphenylcarboxylate (the substrate) and α-cyclodextrin gives a 1:2 complex but no detectable 1:1 complex (64); thus β_{12} could be measured but β_{11} could not.

Many authors have discussed the experimental detection of cooperativity, theoretical models of cooperative phenomena, and molecular interpretations of the observed effects; Cantor and Schimmel (65) have reviewed this subject thoroughly. Among the measures of cooperativity or techniques for its detection are binding curves and Scatchard plots, as already discussed, Hill plots (40), the Bjerrum spreading factor (50), and various model-related parameters. One of these is the Schwartz q cooperativity parameter (66). This is based on a lattice model that allows two types of binding: There can be binding of an isolated ligand (nucleation), with constant K_n; and binding of a ligand at a

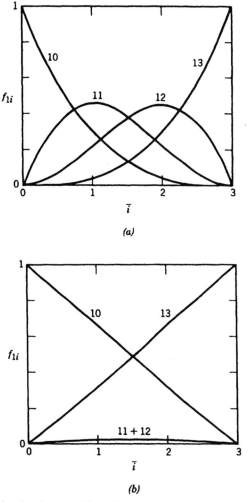

Figure 2.17. The fractional composition of $n = 3$ systems described in Figure 2.15; the numbers 10, 11, 12, 13, refer to species S, SL, SL_2, SL_3. (a) Noncooperative system. (b) Cooperative system.

site adjacent to a bound ligand (aggregation), with constant K_q. Then $q = K_q/K_n$; the model takes into account nearest-neighbor interactions, with positive and negative cooperativity giving q values greater and less than unity, respectively. Menter et al. (67) have applied the Schwartz model to the acridine orange–heparin system, finding that q is about 3, but that it is concentration dependent.

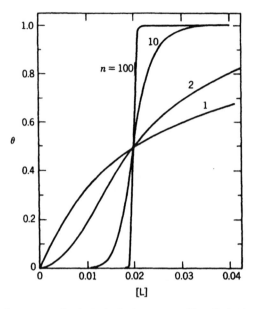

Figure 2.18. Binding curves for hypothetical one-step $(S \rightarrow SL_n)$ binding processes for systems having the indicated values of n. The β_{1n} were adjusted to place the curves on the same scale of ligand concentrations.

McGhee and von Hippel (68) have described a versatile lattice model with these characteristics. The lattice is one-dimensional and homogeneous (all binding sites identical). The ligand, on binding, occupies n consecutive lattice sites (in their symbolism). Then three binding situations are admitted: (1) an isolated bound ligand, in which the ligand binds with intrinsic constant K/M^{-1}, and the sites adjacent to the ligand are unoccupied; (2) singly contiguous bound ligand, with one nearest neighbor and constant $K\omega/M^{-1}$; (3) doubly contiguous bound ligand, with two nearest neighbors, and constant $K\omega^2/M^{-1}$. The cooperativity parameter ω is interpreted as a dimensionless equilibrium constant for moving a bound ligand from an isolated to a singly contiguous site or from a singly to a doubly contiguous site. Evidently, $\omega = 1$ for noncooperative binding, $\omega > 1$ for positive cooperativity, and $\omega < 1$ for negative cooperativity.

An interesting result of the theory is that if $\omega = 1$ but $n > 1$, a concave-upward Scatchard plot is produced. This is a steric or entropic effect, a result of the improbability of saturating the lattice; that is, saturation of the lattice would require a highly regular arrangement of bound ligands, which is entropically unfavorable. Of course, the combination $n = 1$, $\omega < 1$ also leads to a concave-upward Scatchard plot, denoting negative cooperativity. These results strengthen the argument against analysis of such plots in terms of classes of independent binding sites.

Klotz and Hunston (35, 58) rearrange Eq. (2.62) to

$$iK_{1i} = k(n + 1) - ki$$

and investigate cooperative behavior by plotting iK_{1i} against i. If the sites are identical and independent (noncooperative), this will give a straight line. Deviations can be analyzed in terms of stepwise cooperativity. This plot is called an affinity profile.

We conclude this discussion with a cautionary note illustrated with the data in Table 2.6 on complexes of α-cyclodextrin (the ligand) with sym-1,4-disubstituted benzenes (the substrate, XX). The two binding sites are identical, and with Eq. (2.61) the interaction parameter a_{XX} is calculated: $a_{XX} = 4K_{12}/K_{11}$. This quantity may be interpreted as an index of cooperativity. We have seen (Table 2.5) that for two independent sites $K_{11} = 4K_{12}$, hence in this case $a_{XX} = 1$. But it does not necessarily follow that if $a_{XX} = 1$ the sites are independent; such a diagnosis of noncooperative behavior may be erroneous. For this system the structural influences on the magnitude of a_{XX} are these:

1. *The electronic effect of L bound at site X' on the nature of site X.* If the sites in XX are electron deficient, upon interaction of one of them

Table 2.6. Stepwise Stability Constants for α-Cyclodextrin Complexes with Sym-1, 4-Disubstituted Benzenes, $X-C_6H_4-X^a$

X	K_{11}/M^{-1}	K_{12}/M^{-1}	a_{XX}
OCH_3	75.4	221	11.7
OC_2H_5	128	326	10.2
I	5060	6250	4.94
Br	913	397	1.74
Cl	232	90	1.55
CO_2H	1344	23.8	0.071
CO_2CH_3	454	106	0.93
CN	33.1	7.2	0.87
NO_2	35.8	4.6	0.51

^{a}At 25°C in 0.10 M NaCl (2).

with L to give X'X there will be a partial electron transfer from L to the binding site. This has the effect of increasing the charge density at site X in X'X relative to that at X in XX. Thus binding of the second ligand will be favored relative to that of the first one, and a_{xx} will be greater than unity through the operation of this effect. If the sites are electron rich the opposite drift of charge takes place, and a_{xx} will be less than unity. Thus a_{xx} may be expected to follow a Hammett plot with a negative slope.

2. *The repositioning effect.* In the 1:1 complex the relative position of ligand and binding site is optimal with regard to lowering the total free energy of the system. Formation of the 1:2 complex will result in adjustment of all three molecules to minimize the total free energy, since in the 1:2 complex X'X' the two bound sites are necessarily identical on average. This may require a repositioning of the substrate–ligand orientation that was reached in the 1:1 complex. Any such repositioning must therefore be destabilizing, since the orientation in the 1:1 complex is optimal, and will therefore lower a_{xx}.

3. *The ligand–ligand interaction effect.* In a 1:2 complex there is a possibility that the facing rims of the two cyclodextrin molecules may interact attractively. Such an effect could only be manifested as 1:2 complex stabilizing (increasing a_{xx}), because any destabilizing repulsive interactions would be accounted for in terms of the repositioning effect.

Thus a_{xx} is a resultant of these effects, and it is quite possible for them to combine fortuitously so as to yield the result $a_{xx} = 1$, even though the sites are far from independent; this appears to be the case for dimethyl terephthalate (Table 2.6). This cancellation of opposing effects was detected here because a series of related substrates was investigated. But in almost any study, the system of interest may be regarded as one member of a series produced by variation in structures, solvent, or conditions, so it is highly probable that opposing effects similar to these (or to those in the McGhee–von Hippel lattice model) will combine, and thus yield measures of cooperativity that are composite quantities.

Isomerism

If the substrate or the ligand possesses more than one binding site, and if these sites are not identical, then it is possible for more than one complex to form. These complexes are isomers; they have the same stoichiometric composition but different structures. For the case $m = 1$, the number of isomers having the formula SL_i is $n!/i!(n - i)!$ We shall consider only the simplest case, that of isomeric $1:1$ complexes. Let us obtain the binding isotherm for a system containing two such complexes, SL and LS, with stability constants K_{SL} and K_{LS}. The mass balance on substrate is

$$S_t = [S] + [SL] + [LS] = [S]\{1 + (K_{SL} + K_{LS})[L]\}$$

We have, in terms of fractions, $f_S + f_{SL} + f_{LS} = 1$, or, with the terminology of Section 2.4, $F_S + F_{\bar{S}} = 1$, where F_S is the fraction of substrate in the uncomplexed form and $F_{\bar{S}}$ is the fraction present as "not-substrate." We therefore get

$$F_{\bar{S}} = \frac{(K_{SL} + K_{LS})[L]}{1 + (K_{SL} + K_{LS})[L]} \tag{2.107}$$

Equation (2.107) is identical with Eq. (2.57) for a single $1:1$ complex if we set $K_{11} = K_{SL} + K_{LS}$. This can be generalized to any number of isomeric $1:1$ complexes; the observed K_{11} is equal to the sum of the stability constants for the individual complexes.

To extend this discussion we must introduce the specific features of the experimental technique. In Part II the validity of the preceding conclusion is demonstrated for the various methods; here we use the spectrophotometric method as an example. For a single $1:1$ complex the binding equation is

$$\frac{\Delta A}{b} = \frac{\Delta \epsilon_{11} K_{11} S_t [L]}{1 + K_{11}[L]} \tag{2.108}$$

where $\Delta A/b$ is the change in absorbance per centimeter when the free ligand concentration changes from zero to $[L]$, and $\Delta \epsilon_{11} = \epsilon_{11} - \epsilon_S - \epsilon_L$; Eq. (2.108) is derived in Chapter 4. The same development for a system containing two complexes SL and LS yields

$$\frac{\Delta A}{b} = \frac{(K_{SL} \Delta \epsilon_{SL} + K_{LS} \Delta \epsilon_{LS}) S_t [L]}{1 + (K_{SL} + K_{LS})[L]} \tag{2.109}$$

Linearization of Eq. (2.109) according to Eqs. (2.90) to (2.92) shows these correspondences:

$$K_{11} = K_{SL} + K_{LS} \tag{2.110}$$

$$\Delta\epsilon_{11} = \frac{K_{SL}\,\Delta\epsilon_{SL} + K_{LS}\,\Delta\epsilon_{LS}}{K_{SL} + K_{LS}} \qquad (2.111)$$

Thus the same result is reached, namely, that the observed K_{11} is the sum of binding constants for the isomeric constants; this has been pointed out by several authors (68, 69). Equation (2.111) shows that the experimental $\Delta\epsilon_{11}$ is now a weighted average. An important conclusion is that Eq. (2.110) applies even if one of the complexes shows no spectral change upon its formation. Thus the experimental technique may not respond to one of the complexes, yet the method still yields K_{11} as the sum of K_{SL} and K_{LS}. Of course, there is no way for the experimentalist to know, on the basis of this experimental result, whether the observed K_{11} describes a single complex or more than one complex. Orgel and Mulliken (69) suggest that the dependence of K_{11} on temperature may distinguish between these cases if the enthalpies of complex formation are sufficiently different for isomeric complexes.

The physical basis for the result given as Eq. (2.110) lies in the mass balance expression for the substrate, from which the binding isotherm is derived. This can be written

$$S_t = [S] + [\tilde{S}] \qquad (2.112)$$

where [S] is the concentration of free substrate and $[\tilde{S}]$ is the concentration of "not-substrate." This is independent of any particular observational technique; a molecule belongs to the class S if on reaction with L it becomes $\tilde{S}$. It is the reduction of [S] that is reflected in Eq. (2.110), irrespective of the properties of the complexes.

The literature contains many reports of studies of a complex system by more than one experimental technique. Often the equilibrium constants evaluated by two techniques are significantly different, and it has occasionally been suggested, in explanation, that one of the techniques is "more specific" than the other, and is responding selectively to one complex in a mixture of isomers, or that one method (often the spectral method) measures only "specific" interactions whereas another measures both specific and nonspecific effects. These are intuitive rather than analytical arguments. Upon careful comparison of such experiments it will usually be found that the different experimental techniques have not, after all, studied the same system; in the two kinds of experiment there will be significant and important differences in concentrations of interactants or in solvent composition that lead to the observed discrepancies. Another possible explanation is that the model is wrong; a difference between K_{11} values evaluated by different techniques is to be expected if complexes of higher stoichiometry are present (4).

Random Association

In a solution containing S and L solute molecules, some fraction of S and L will at any time be in mutual close proximity as a consequence of random collisions, even in the absence of specific complex formation. Several authors have developed this idea (70), and a treatment by Orgel and Mulliken (69) has stimulated much discussion. Most of this concern has been in the context of charge-transfer interactions, and the stimulus for the Orgel–Mulliken analysis was the observation that certain pairs of solutes can exhibit marked electronic spectral changes indicative of charge transfer even though they seem not to form stable molecular complexes. It was postulated that charge transfer can occur during statistically governed random encounters; this process was termed contact charge transfer, and other authors have spoken of contact complexes or contact pairs. (This terminology is not to be confused with the contact ion pairs that constitute the species at distances of closest approach in ion association.) We shall call the process random association.

First we review the basic argument of Orgel and Mulliken. These authors discussed the contact absorption in terms of $1:1$ association and the linearized double-reciprocal plot [Eq. (2.90), the Benesi–Hildebrand equation]. The isotherm is Eq. (2.108); in double-reciprocal form this is

$$\frac{S_t b}{\Delta A} = \frac{1}{\Delta \epsilon_{11} K_{11} [L]} + \frac{1}{\Delta \epsilon_{11}}$$

and the parameters are estimated from (are operationally defined by)

$$K_{11} = \frac{\text{intercept}}{\text{slope}}$$

$$\Delta \epsilon_{11} = \frac{1}{\text{intercept}}$$

The concentration of randomly associated S, L pairs will be directly proportional to the concentrations of S and L; Orgel and Mulliken therefore wrote $\Delta A = b \, \Delta \epsilon_{11} K_{11} S_t L_t$ and, in double-reciprocal form, $S_t b / \Delta A = 1 / \Delta \epsilon_{11} K_{11} L_t$. A plot according to this equation passes through the origin, with the result that $K_{11} = 0$ and $\Delta \epsilon_{11} = \infty$. Other authors have used either the double-reciprocal or other linearized forms to discuss this phenomenon.

It is not necessary to resort to a linearization, for the relation $\Delta A = b \, \Delta \epsilon_{11} K_{11} S_t L_t$ shows that the slope of the direct plot is not zero when a charge-transfer spectral change is observed, so evidently K_{11} is not zero; the preceding result was an artifact of the data treatment. More to the point is the matter of linearity in such a direct plot; this apparent linearity presumably is because it is difficult in such systems to cover enough of the isotherm to detect nonlinearity; as Eq. (2.108) shows, the initial slope is determined by the product $\Delta \epsilon_{11} K_{11}$, and $\Delta \epsilon_{11}$ may be very large, whereas the nonlinearity is controlled by the size of the product $K_{11}[L]$, which is very small. In any case, the result of the Orgel–Mulliken analysis is the conclusion that the randomly associated pairs make no contribution to the experimentally observed K_{11}.

We can generalize by considering the mass balance on substrate, Eq. (2.112), where $\bar{S}$ is read "not-substrate." When this equation forms the basis for the development of a binding isotherm, we first postulate the existence of species contributing to S_t and must then assign these species either to class S or to class $\bar{S}$. Having done this, the form and interpretation of the isotherm are fixed. If the concentration of a species is established by a mass-action relationship, and if we assign it to class $\bar{S}$, then it will contribute to K_{11}, just as we saw for isomeric complexes in the preceding topic. If, on the other hand, we assign a species to class S, it will make no contribution to K_{11}. This latter choice is the one that Orgel and Mulliken (69) made. In another example, Martire (71) concluded that NMR and UV measurements yield K_{11} values that do not include contact pairing; this was the result of assigning the contact pairs to class S. (Martire claimed that GLC measurements give K_{11} values that include the contact pairing, but this resulted because the contact pairs were assigned to $\bar{S}$.) Of course the experimental number K_{11} is in no way affected by these decisions; it is the interpretation of K_{11} that is affected.

A very simple lattice model is useful in discussing random association. The system includes substrate, ligand, and solvent molecules in a lattice whose coordination number (number of nearest neighbor sites) is α. Then the site fraction (which is equal to mole fraction) of random pairs, x_r, is given by $x_r = \alpha x_S x_L$. Using Eq. (2.20) to convert to molar concentrations gives this mass-action expression:

$$c_R = \left(\frac{\alpha M_1}{1000\rho_1}\right)c_S c_L$$

where M_1 and ρ_1 are the solvent molecular weight and density. The quantity $(\alpha M_1/1000\rho_1)$, which has units M^{-1}, has the character of a stability constant that is determined by statistical rather than energetic factors. The value of α would be 12 if the substrate, ligand, and solvent were close-packed spheres of equal size (69), but for nonspherical molecules of different sizes it will presumably be smaller. Using $\alpha = 6$, we calculate, for this random association, $0.11\ M^{-1}$ for water as solvent, $0.65\ M^{-1}$ for cyclohexane, and $0.58\ M^{-1}$ for carbon tetrachloride. With a different model of random association, Prue (72) estimated this quantity as $0.2\ M^{-1}$. The magnitudes of these estimates provide some basis for the statement in Section 1.1 that constants smaller than about $1\ M^{-1}$ may be of doubtful significance.

Since the interpretation of an experimental K_{11} value as either including or not including the random effect is a consequence of whether the random association product is classified as S or as $\tilde{S}$, it is reasonable to seek guidance in making this classification. Thus Orgel and Mulliken classified contact pairs as S, since this gave a description consistent with observations making use of the Benesi–Hildebrand equation. Foster (73) has pointed out that experimental K_{11} values have been reported that are smaller than Prue's estimate of $0.2\ M^{-1}$ for the random contribution; this would appear to require that the random effect is not included in K_{11}. However, this conclusion is uncertain because of the very approximate nature of the estimates of random association, and because such very small experimental K_{11} values are themselves uncertain. It might seem that molecular models of the process would provide insight, but such models require, in their formulation, the same decision, namely, whether the random contact pairs are S or $\tilde{S}$, that leads to the ambiguous interpretation of the experimental isotherm. For example, the lattice model of Scott (74) leads to the conclusion that the spectrophotometrically measured K_{11} does not include the random contribution; in effect, however, the random pairs are not counted as $\tilde{S}$, so they cannot contribute to K_{11}. The solvation model of Carter et al. (75) is formally equivalent to a lattice model, as Scott has pointed out (74), and it leads to the same result.

The energetics of association provides another viewpoint. A random contact pair devoid of interactions is a fiction, since as two molecules approach they experience each other's force fields; moreover, even strongly bound complexes are formed in consequence of collisions that are in part statistically determined. A complex species possesses vibrational energy levels that are not accessible to the separated species (76). It is a useful point of view to consider species that are "residents" of the same potential well as having the same structure, though a range of vibrational modes within this population may exist, and measured effects will reflect averages of contributions from the population. In extremely schematic form Figures 2.19 and 2.20 illustrate this idea; Figure 2.19 shows substrate–ligand pairs with some possible complex structures, the planar molecules being viewed in the molecular planes, and in Figure 2.20 these structures are reasonably located on the potential energy function connecting these structures. The high-energy "loose" complexes exemplified by c and d may be regarded as normal species in high-energy vibrational states.

From this point of view, random contact pairs may be identified as complexes in the c, d region of Figure 2.20. Because of the very gradual slope of the potential energy function at large intermolecular separations, there can be no clear-cut distinction between those species that are members of the complex population and those that are free, separated interactants—hence the difficulty of assigning species to the classes $\tilde{S}$ or S. But this description strongly suggests that if a physical or chemical property of the substrate changes upon introduction of the ligand, if the effect follows the mass-action law, and if one is committed to the complex hypothesis (in contrast to the nonideal behavior description), the

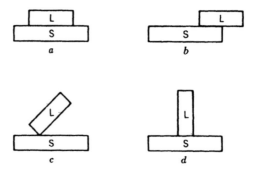

Figure 2.19. Hypothetical complex structures.

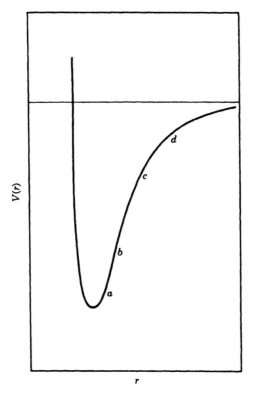

Figure 2.20. Hypothetical potential energy function for a complex species. The letters suggest possible energies of structures in Figure 2.19.

putative species should be classified as $\tilde{S}$ and interpreted as a member of the complex population.

If the substrate has two different binding sites, say X and Y, then there will be two definite isomeric complexes, each defined by its own potential function, and the preceding description applies to each of these species. Then the observed K_{11} will be the sum of the two isomeric site binding constants, as shown earlier, and these will each include the random association contribution if the preceding interpretation is used.

Medium Effects

The effect of the solvent on physical and chemical properties of solutes is a subject of great scope and difficulty, and here we can only comment on some features of particular pertinence in studies of binding constants.

These fall into four classes: (1) nonideality effects; (2) effects specific to an experimental technique; (3) specific solvation effects; (4) the nonspecific effect of solvent on binding.

Concern for nonideal behavior arises in studies of two kinds of system. One of these is the binding of ions, for the activity coefficients of ions are extremely sensitive to the medium, and in particular to its ionic strength. A common practice is to study such equilibria in a medium of high ionic strength, established with an indifferent electrolyte, so as to render negligible the small changes produced by the substrate and ligand as solutes. Another method is to measure the constant as a function of ionic strength and extrapolate to zero ionic strength. Thermodynamic constants are often estimated by calculation of activity coefficients by means of the Debye–Hückel equation or one of its empirical extensions (14, 77, 78). It is sound practice, when reporting such a calculated quantity, to specify precisely which equation was used, as well as the experimental quantity (e.g., K_{11}), so that a reader who wishes to rework the data may do so.

The other type of nonideal system is that in which a weak complex is formed. If the binding constant is small, a fairly high ligand concentration will have to be used in order to cover a sufficient portion of the binding isotherm, as discussed in Section 2.5. This means that the solvent composition varies significantly in the study, the ligand in effect behaving as a solvent component, and the activity coefficients may deviate from unity, as described in Section 2.1. To be specific we consider 1:1 complex formation; then we have

$$K_{th} = K_{11} \frac{\gamma_{SL}}{\gamma_S \gamma_L} \qquad (2.113)$$

where K_{11} is the usual 1:1 experimental concentration constant and K_{th} is the thermodynamic constant. In the usual procedure the ratio $\gamma_{SL}/\gamma_S \gamma_L$ is assumed to be unity or at least to remain constant. Several authors have examined this assumption and have proposed alternative treatments. These discussions have become entangled in the question of the preferred concentration scale because, as shown in Section 2.2, the activity coefficients are different on the several concentration scales. The experimental problem is that it is very difficult, and for practical purposes impossible, to measure all the individual activity coefficients γ_S, γ_L, γ_{SL}. Hanna and Rose (28f) studied the caffeine (S)–benzene (L) complex in carbon tetrachloride as solvent, using vapor pressure data on the binary system benzene–carbon tetrachloride to determine the benzene activity. The stability constant was determined (by linear plotting) with the use of ligand activity rather than ligand concentration. Slejko and Drago (28e) criticized this approach, pointing out that it assumes constancy of the ratio γ_{SL}/γ_S. Lim and Drago (79) urge caution in applying activity coefficient corrections in response to a nonconstant K_{11}, for this variation in K_{11} may be a consequence of use of an incorrect model, or it may reflect a solvent effect on a molar absorptivity, and so on. Christian and co-workers (28j, 80) have made use of more than one experimental technique, applying activity coefficient corrections and obtaining agreement of K_{11} values evaluated by different methods.

There is still no definitive approach to the problem. All of the cited studies made use of very weak complexes, typically with K_{11} values of $1 \, M^{-1}$ or smaller. We saw, in the preceding topic, that such quantities are of unknown significance, so attempts to account for nonideal behavior in such systems are frustrated by too many unknown factors. It may be more fruitful to study strong complexes in this context. Although stable complexes do not require high ligand concentrations on a molar basis, if the ligand has a high molecular weight its volume concentration may be quite high, and nonideality effects may be significant.

Turning to technique-specific solvent effects, we find that most attention has been directed to the spectrophotometric method. In the derivation of the isotherm [Eq. (2.108)] it is assumed that the absorptivities of all species are constant. Barb (81) and Emslie et al. (82) have considered the possibility that the molar absorptivity of the complex may vary with the solvent composition, and thus with ligand concentration. Barb calls attention to the distinction between ϵ_{max}, which may be substantially constant even though λ_{max} changes as ligand concentration varies, and ϵ at the fixed analytical wavelength, which is the needed experimental quantity. The analysis of Emslie et al. gives the result that if ϵ varies linearly with [L], the estimates of K_{11} and ϵ will be in error, although their product will be correct. Stamm et al. (83) discuss a similar effect, though in different terms, leading to variations in chemical shifts in nuclear magnetic resonance studies of complexing. Another type of effect would be an artifact such as an altered electrode response in a potentiometric study.

Specific solvation of solutes leading to erroneous estimates of system parameters is another type of medium effect that has been discussed in the literature, but it is a matter of interpretation whether the parameters (the stability constants) are in error or whether they correctly reflect molecular interactions in the system. Carter et al. (75) described $1:1$ complex formation in solvent M by this equilibrium

$$SM_a + LM_b \rightleftharpoons SLM_p + qM$$

where $q = a + b - p$. The equilibrium constant is defined

$$K_{11} = \frac{[SLM_p]x_M^q}{[SM_a][LM_b]}$$

where x_M is the solvent mole fraction. The question is whether K_{11} is the appropriate (correct) constant or whether it is in error by the factor x_M^q. Litt and Wellinghoff (84) observed nonhyperbolic isotherms in studies of the fumaronitrile (S)–styrene (L) system in several solvents and developed a molecular model very similar to the lattice model of Scott (74). Bertrand (85) criticized the Litt and Wellinghoff treatment in part on the basis that it invoked too many adjustable parameters and he described their data by means of a "nearly ideal binary solvent" treatment. Lamb and Purnell (86) have developed a "microscopic partition" theory of mixed solvents and applied it to complex equilibria; this theory postulates that binary solvents that are macroscopically miscible are microscopically heterogeneous, so that a solute molecule experiences an environment of either pure A or pure B in a binary solvent mixture of A and B.

In an alternative description of specific solvation effects, the solvent is seen as engaging in competitive mass-action reactions with one or more solute species (29c, 87). Considering $1:1$ stoichiometry for all equilibria, we write

$$S + L \overset{K_{11}^o}{\rightleftharpoons} SL$$

$$S + M \overset{K_{SM}}{\rightleftharpoons} SL$$

$$L + M \overset{K_{LM}}{\rightleftharpoons} LM$$

$$SL + M \overset{K_{SLM}}{\rightleftharpoons} SLM$$

where M is the solvent. Suppose we define the experimental stability constant by

$$K_{11} = \frac{\Sigma[\text{all SL species}]}{\Sigma[\text{all S species}] \cdot \Sigma[\text{all L species}]}$$

and proceed as follows:

$$K_{11} = \frac{[\text{SL}] + [\text{SLM}]}{([\text{S}] + [\text{SM}])([\text{L}] + [\text{LM}])}$$

Substitution with the equilibrium constant expressions gives (88, 89)

$$K_{11} = \frac{K_{11}^{\circ}(1 + K_{\text{SLM}}[\text{M}])}{(1 + K_{\text{SM}}[\text{M}])(1 + K_{\text{LM}}[\text{M}])} \tag{2.114}$$

where [M] is the molar concentration of solvent, and K_{11}° is the "true" binding constant. Equation (2.114) appears to relate K_{11} and K_{11}° in a straightforward manner, but (quite aside from the problem of evaluating the several solvation constants) its meaning is ambiguous for the reason already considered in the discussion of random association, namely, the form of the equation results from assignments of species to the classes S or $\tilde{\text{S}}$, and so on.

Despite this ambiguity, Eq. (2.114) represents an attempt to identify an intrinsic measure (K_{11}°) of the binding interaction. Experimental K_{11} values are commonly found to depend on the identity of the solvent, even in "inert" solvents. This is not surprising, whether the solvent is modeled as a continuous medium or as a molecular environment, because of differences in bulk and molecular properties (dielectric constant, density, dipole moment, polarizability, size, shape) that control molecular interactions. One goal of an interest in this effect is to extract a solvent-independent quantity that is essentially the binding constant in the gas phase. The approach of Gorodyskii and co-workers (90) is based on Eq. (2.115):

$$kT \ln K_g = kT \ln K_S + \delta_M \, \Delta f_{\text{solv}} + \delta_M \, \Delta f_{\text{cav}} + \delta_M \, \Delta f_c \tag{2.115}$$

where K_g is the binding constant in the gas phase, K_S is the constant in the solution phase, δ_M is the Leffler–Grunwald operator signifying a medium effect, Δf_{solv} is the work of solvation, Δf_{cav} is the work of creating a cavity in the solvent, and Δf_c is the work involved in the complex formation. Equation (2.115) is closely related to Eq. (1.115). A molecular model of the several components in Eq. (2.115) allows K_S to be related to solvent properties, and from this relationship K_g is estimated (90).

REFERENCES

1. J. T. Edsall and J. Wyman, *Biophysical Chemistry*, Vol. I, Academic, New York, 1958, Ch. 9.
2. K. A. Connors and D. D. Pendergast, *J. Am. Chem. Soc.*, **106**, 7607 (1984).
3. D. C. Teller and D. A. Deranleau, *Biochim. Biophys. Acta*, **421**, 416 (1976).
4. K. A. Connors and J. A. Mollica, *J. Pharm. Sci.*, **55**, 772 (1966).
5. A. Job, *Annales de Chimie* (10th series), **9**, 113 (1928).
6. K. C. Ingham, *Anal. Biochem.*, **68**, 660 (1975).
7. W. Likussar and D. F. Boltz, *Anal. Chem.*, **43**, 1265 (1971).
8. W. C. Vosburgh and G. R. Cooper, *J. Am. Chem. Soc.*, **63**, 437 (1941).
9. G. F. Atkinson, *J. Chem. Educ.*, **51**, 792 (1974).
10. A. S. Meyer and G. H. Ayres, *J. Am. Chem. Soc.*, **79**, 49 (1957).
11. C. D. Chriswell and A. A. Schilt, *Anal. Chem.*, **47**, 1623 (1975).
12. A. Beltran-Porter, D. Beltran-Porter, A. Cerville, and J. A. Ramirez, *Talanta*, **30**, 124 (1983).
13. J. E. Gordon, *The Organic Chemistry of Electrolyte Solutions*, Wiley-Interscience, New York, 1975, Ch. 1.
14. C. W. Davies, *Ion Association*, Butterworths, Washington, D.C., 1962, p. 41.
15. K. Denbigh, *The Principles of Chemical Equilibrium*, 2nd ed., Cambridge University Press, Cambridge, 1968, p. 274.
16. S. Glasstone, *Thermodynamics for Chemists*, Van Nostrand Reinhold, Princeton, 1947, p. 355.
17. K. Denbigh, *The Principles of Chemical Equilibrium*, 2nd ed., Cambridge University Press, Cambridge, 1968, pp. 275, 278.
18. A. Ben-Naim, *J. Phys. Chem.*, **82**, 792 (1978).
19. G. C. Pimental and A. L. McClellan, *The Hydrogen Bond*, Freeman, San Francisco, 1960, p. 348.
20. R. W. Gurney, *Ionic Processes in Solution*, McGraw-Hill, New York, 1953; reprinted by Dover Publications, 1962, Chs. 5, 6.
21. K. A. Connors and S. Sun, *J. Am. Chem. Soc.*, **93**, 7239 (1971).
22. W. P. Jencks, *Catalysis in Chemistry and Enzymology*, McGraw-Hill, New York, 1969, pp. 15, 373.
23. C. Tanford, *The Hydrophobic Effect*, 2nd ed., Wiley-Interscience, New York, 1980, pp. 5-6.
24. A. Ben-Naim, in *Solutions and Solubilities*, Part I, Ch. II, M. R. J. Dack, Ed., Wiley-Interscience, New York, 1975.

25. T. L. Hill, *An Introduction to Statistical Thermodynamics*, Addison-Wesley, Reading, MA, 1960, Ch. 4.

26. C. Tanford, *J. Phys. Chem.*, **82**, 1802 (1979).

27. T. L. Hill, *An Introduction to Statistical Thermodynamics*, Addison-Wesley, Reading, MA, 1960, Ch. 21.

28. (a) K. Denbigh, *The Principles of Chemical Equilibrium*, 2nd ed., Cambridge University Press, Cambridge, 1968, p. 298; (b) M. W. Hanna and A. L. Ashbaugh, *J. Phys. Chem.*, **68**, 811 (1964); (c) P. J. Trotter and M. W. Hanna, *J. Am. Chem. Soc.*, **88**, 3724 (1966); (d) I. D. Kuntz, F. P. Gasparro, M. D. Johnston, and R. P. Taylor, *J. Am. Chem. Soc.*, **90**, 4778 (1968); (e) F. Slejko and R. Drago, *J. Am. Chem. Soc.*, **94**, 6546 (1972); (f) M. W. Hanna and D. G. Rose, *J. Am. Chem. Soc.*, **94**, 2601 (1972); (f) C. Eon and B. L. Karger, *J. Chromatogr. Sci.*, **10**, 140 (1972); (g) J. H. Purnell and O. P. Srivastava, *Anal. Chem.*, **45**, 1111 (1973); (h) P. E. Rider and R. M. Hammacker, *Spectrochim. Acta*, **A29**, 501 (1973); (i) J. Homer and P. M. Whitney, *J. Chem. Soc. Faraday Trans.*, I, **69**, 1985 (1973); (j) E. H. Lane, S. D. Christian, and J. D. Childs, *J. Am. Chem. Soc.*, **96**, 38 (1974).

29. (a) R. M. Keefer and L. J. Andrews, *J. Am. Chem. Soc.*, **74**, 1891 (1952); (b) N. J. Rose and R. S. Drago, *J. Am. Chem. Soc.*, **81**, 6138 (1959); (c) M. Tamres, *J. Phys, Chem.*, **65**, 654 (1961); (d) R. P. Lang, *J. Am. Chem. Soc.*, **84**, 1185 (1962); (e) N. B. Jurinski and P. A. D. de Maine, *J. Am. Chem. Soc.*, **86**, 3217 (1964); (f) W. E. Wentworth, W. Hirsch, and E. Chen, *J. Phys. Chem.*, **71**, 218 (1967); (g) S. D. Christian, *J. Chem. Educ.*, **45**, 713 (1968); (h) J. Grundnes and S. D. Christian, *J. Am. Chem. Soc.*, **90**, 2239 (1968); (i) I. Moriguchi and N. Kaneniwa, *Chem. Pharm. Bull.*, **17**, 2173 (1969); (j) K. Kakemi, H. Sezaki, E. Suzuki, and M. Nakano, *Chem. Pharm. Bull.*, **17**, 242 (1969); (k) M. Nakano, N. I. Nakano, and T. Higuchi, *J. Phys. Chem.*, **71**, 3954 (1967).

30. K. A. Connors and D. D. Pendergast, *Anal. Chem.*, **56**, 1549 (1984).

31. T. L. Hill, *An Introduction to Statistical Thermodynamics*, Addison-Wesley, Reading, MA, 1960, pp. 144–145.

32. (a) G. Scatchard, *Ann. N.Y. Acad. Sci.*, **51**, 660 (1949); (b) C. Tanford, *Physical Chemistry of Macromolecules*, Wiley, New York, 1961, p. 532; (c) J. Steinhardt and J. A. Reynolds, *Multiple Equilibria in Proteins*, Academic, New York, 1969, p. 14.

33. T. L. Hill, *An Introduction to Statistical Thermodynamics*, Addison-Wesley, Reading, MA, 1960, Ch. 7.

34. J. T. Edsall and J. Wyman, *Biophysical Chemistry*, Vol. I, Academic, New York, 1958, p. 652.

35. I. M. Klotz and D. L. Hunston, *J. Biol. Chem.*, **250**, 3001 (1975).

36. P. Mukerjee and A. K. Ghosh, *J. Am. Chem. Soc.*, **92**, 6403 (1970).

37. C. Tanford, *The Hydrophobic Effect*, 2nd ed., Wiley-Interscience, New York, 1980, Ch. 7.

38. P. Mukerjee, *J. Pharm. Sci.*, **63**, 972 (1974).

39. A. V. Hill, *Biochem. J.*, **7**, 471 (1913).

40. (a) R. B. Loftfield and E. A. Eigner, *Science*, **164**, 305 (1969); (b) H. A. Saroff and A. P. Minton, *Science*, **175**, 1253 (1972); (c) L. D. Byers, *J. Chem. Ed.*, **54**, 352 (1977).

41. H. Lineweaver and D. Burk, *J. Am. Chem. Soc.*, **56**, 658 (1934).

42. H. Benesi and J. H. Hildebrand, *J. Am. Chem. Soc.*, **71**, 2703 (1949).

43. G. S. Eadie, *J. Biol. Chem.*, **146**, 85 (1942).

44. W. B. Person, *J. Am. Chem. Soc.*, **87**, 167 (1965).

45. (a) D. A. Deranleau, *J. Am. Chem. Soc.* **91**, 4044 (1969); (b) D. A. Deranleau, *J. Am. Chem. Soc.* **91**, 4050 (1969).

46. G. Weber and S. R. Anderson, *Biochemistry*, **4**, 1942 (1965).

47. G. Norheim, *Acta Chem. Scand.*, **23**, 2808 (1969).

48. (a) I. M. Klotz, *Acc. Chem. Res.*, **7**, 162 (1974); (b) I. M. Klotz, *Science*, **217**, 1247 (1982).

49. A. Albert and E. P. Serjeant, *The Determination of Ionization Constants*, 3rd ed., Chapman and Hall, London, 1984.

50. (a) F. J. C. Rossotti and H. S. Rossotti, *The Determination of Stability Constants*, McGraw-Hill, New York, 1961; (b) M. T. Beck, *Chemistry of Complex Equilibria*, Van Nostrand Reinhold, London, 1970; (c) F. R. Hartley, C. Burgess, and R. M. Alcock, Ellis Horwood, Chichester (Halsted Press), 1980.

51. W. G. Bardsley and R. E. Childs, *Biochem. J.*, **149**, 313 (1975).

52. J. T. Edsall, G. Felsenfeld, D. S. Goodman, and F. R. N. Gurd, *J. Am. Chem. Soc.*, **76**, 3054 (1954).

53. A. J. Poë, *J. Phys. Chem.*, **67**, 1070 (1963).

54. F. J. C. Rossotti and H. S. Rossotti, *Acta Chem. Scand.* **9**, 1166 (1956).

55. H. Irving, *J. Chem. Soc.*, 4056 (1962).

56. (a) G. Scatchard, I. H. Scheinberg, and S. H. Armstrong, *J. Am. Chem. Soc.*, **72**, 535 (1950); (b) F. Karush, *J. Am. Chem. Soc.*, **72**, 2705 (1950).

57. C. R. Cantor and P. R. Schimmel, *Biophysical Chemistry*, Part III, Freeman, San Francisco, 1980, p. 858.

58. (a) I. M. Klotz and D. L. Hunston, *Arch. Biochim. Biophys.*, **193**, 314 (1979); (b) D. L. Hunston, *Anal. Biochem.*, **63**, 99 (1975).

59. J. G. Nørby, P. Ottolenghi, and J. Jensen, *Anal. Biochem.*, **102**, 318 (1980).

60. (a) D. A. Deranleau, *J. Chem. Phys.*, **40**, 2134 (1964); (b) J. E. Fletcher, A. A. Spector, and J. D. Ashbrook, *Biochemistry*, **9**, 4580 (1970).

61. G. S. Adair, *J. Biol. Chem.*, **63**, 529 (1925).

62. T. L. Hill, *J. Phys. Chem.*, **57**, 324 (1953).

63. S. J. Edelstein, *Ann. Rev. Biochem.*, **41**, 209 (1975).

64. R. I. Gelb, L. M. Schwartz, C. T. Murray, and D. A. Laufer, *J. Am. Chem. Soc.*, **100**, 3553 (1978).

65. C. R. Cantor and P. R. Schimmel, *Biophysical Chemistry*, Part III, Freeman, San Francisco, 1980, Chs. 15, 17; see also B. Perlmutter-Hayman, *Acc. Chem. Res.*, **19**, 90 (1986).

66. G. Schwartz, *Eur. J. Biochem.*, **12**, 442 (1970).

67. J. M. Menter, R. E. Hurst, N. Nakamura, and S. S. West, *Biopolymers*, **18**, 493 (1979).

68. E. Grunwald and J. E. Leffler, cited by S. D. Ross, M. M. Labes, and M. Schwartz, *J. Am. Chem. Soc.*, **78**, 343 (1956).

69. L. E. Orgel and R. S. Mulliken, *J. Am. Chem. Soc.* **79**, 4839 (1957).

70. (a) R. W. Gurney, *Ionic Processes in Solution*, McGraw-Hill, New York, 1953; reprinted in Dover Publications, 1962, p. 61; (b) R. S. Mulliken, *Rec. Trav. Chim.*, **75**, 845 (1956); (c) L. Kennon and K.-S. Chen, *J. Pharm. Sci.*, **51**, 1149 (1962).

71. D. E. Martire, *Anal. Chem.*, **46**, 1712 (1974).

72. J. E. Prue, *J. Chem. Soc.*, 7534, 1965.

73. R. Foster, *Organic Charge-Transfer Complexes*, Academic, London, 1969, p. 165.

74. R. L. Scott, *J. Phys. Chem.*, **75**, 3843 (1971).

75. S. Carter, J. N. Murrell, and E. J. Rosch, *J. Chem. Soc.*, 2048, 1965.

76. J. E. Leffler and E. Grunwald, *Rates and Equilibria of Organic Reactions*, Wiley, New York, 1963, pp. 9–10.

77. G. H. Nancollas, *Interactions in Electrolyte Solutions*, Elsevier, Amsterdam, 1966, pp. 14, 78.

78. P. G. Daniele, C. Rigano, and S. Sammartano, *Ann. Chim.*, **73**, 741 (1983).

79. Y.-Y. Lim and R. S. Drago, *J. Am. Chem. Soc.*, **94**, 84 (1972).

80. S. D. Christian, J. D. Childs, and E. H. Lane, *J. Am. Chem. Soc.*, **94**, 6861 (1972).

81. W. G. Barb, *Trans. Faraday Soc.*, **49**, 143 (1953).

82. P. H. Emslie, R. Foster, C. A. Fyfe, and I. Horman, *Tetrahedron*, **21**, 2843 (1965).

83. H. Stamm, W. Lamberty, and J. Stafe, *J. Am. Chem. Soc.*, **102**, 1529 (1980).

84. M. H. Litt and J. Wellinghoff, *J. Phys. Chem.*, **81**, 2644 (1977).

85. G. L. Bertrand, *J. Phys. Chem.*, **83**, 769 (1979).

86. R. J. Laub and J. H. Purnell, *J. Am. Chem. Soc.*, **98**, 35 (1976).

87. R. E. Merrifield and W. D. Phillips, *J. Am. Chem. Soc.*, **80**, 2778 (1958).

88. R. J. Bishop and L. E. Sutton, *J. Chem. Soc.*, 6100 (1964).

89. E. N. Gur'yanova, I. P. Gol'dshtein, and I. P. Romm, *Donor–Acceptor Bond*, Keter Publishing House, Jerusalem (distributed by Halsted Press), 1975, p. 40.

90. (a) V. A. Gorodyskii and A. L. Fedorenkov, *Russ. J. Phys. Chem.* (Engl. transl.), **53**, 1317 (1979); (b) V. A. Gorodyskii, A. L. Fedorenkov, and V. V. Shchukareva, *Russ. J. Phys. Chem.* (Engl. transl.), **54**, 1729 (1980).

3

STATISTICAL
TREATMENT
OF DATA

3.1. LEAST-SQUARES REGRESSION

Many authors have reviewed least-squares analysis in the context of binding studies. Wilkinson (1) showed its application to enzyme kinetics, and Henderson (2) has also discussed this application. Briegleb (3) treated examples of charge-transfer complexing, and Wentworth (4) applied the method to complex formation as well as to other chemical problems. In Section 3.4 many additional contributions are cited.

The mathematical basis of least-squares analysis is presented in Appendix B, where it is shown that for a function of two variables (x, y) and three parameters (a, b, c) we obtain the set of *normal equations*, Eq. (3.1).

$$\sum w_i F_a^i F_a^i A + \sum w_i F_a^i F_b^i B + \sum w_i F_a^i F_c^i C = \sum w_i F_a^i F_0^i$$

$$\sum w_i F_b^i F_a^i A + \sum w_i F_b^i F_b^i B + \sum w_i F_b^i F_c^i C = \sum w_i F_b^i F_0^i \qquad (3.1)$$

$$\sum w_i F_c^i F_a^i A + \sum w_i F_c^i F_b^i B + \sum w_i F_c^i F_c^i C = \sum w_i F_c^i F_0^i$$

In Eq. (3.1) the symbols have these meanings: The running index i denotes the observation number (i goes from 1 to n, the total number of observations); $w_i = 1/L_i$, where

$$L_i = \frac{F_{xi}F_{xi}}{w_{xi}} + \frac{F_{yi}F_{yi}}{w_{yi}} \qquad (3.2)$$

and w_i is the weight of the ith observation; the parameters A, B, C are defined

$$A = a_0 - a$$
$$B = b_0 - b \qquad (3.3)$$
$$C = c_0 - c$$

where a_0, b_0, c_0 are the initial estimates of the parameters and a, b, c are the refined estimates, obtained from Eq. (3.3) and the values of A, B, C found in solving the normal equations. We define

$$F^i(x_i, y_i, a, b, c) = 0 \qquad (3.4)$$

Then

$$F_0^i = F^i(X_i, Y_i, a_0, b_0, c_0) \qquad (3.5)$$

where X_i, Y_i are the experimental values of the variables in the ith observation. The remaining quantities in Eq. (3.1) are partial derivatives, defined $F_{xi} = \partial F^i / \partial x_i$, $Fy_i = \partial F^i / \partial y_i$, $F_a^i = \partial F^i / \partial a$, $F_b^i = \partial F^i / \partial b$, $F_c^i = \partial F^i / \partial c$. Further details will be found in Appendix B.

To make this more concrete, let us apply Eq. (3.1) to this simple model function:

$$y = a + bx \qquad (3.6)$$

Dropping the running index for convenience, we have from Eq. (3.4),

$$F = y - a - bx = 0 \qquad (3.7)$$

and from Eq. (3.5),

$$F_0 = y - a_0 - b_0 x$$

where we understand that the experimental values are taken. The partial derivatives are, from Eq. (3.7), $F_x = -b$, $F_y = 1$, $F_a = -1$, $F_b = -x$. Substituting these into Eq. (3.1) gives

$$\sum wA + \sum wxB = -\sum w(y - a_0 - b_0 x)$$

$$\sum wxA + \sum wx^2 B = -\sum wx(y - a_0 - b_0 x)$$

Multiplying out the terms and using Eq. (3.3) gives

$$a \sum w + b \sum wx = \sum wy$$

$$a \sum wx + b \sum wx^2 = \sum wxy$$

(3.8)

Equations (3.8) are the normal equations for weighted linear least-squares regression.

Now suppose the weights are constant (independent of i), so that they cancel. We observe that $\sum a = na$, $\sum x/n = \bar{x}$, $\sum y/n = \bar{y}$, and we obtain

$$a + b\bar{x} = \bar{y}$$

$$a \sum x + b \sum x^2 = \sum xy$$

(3.9)

as the normal equations for unweighted linear regression; in Eqs. (3.9), $\bar{x}$ and $\bar{y}$ are the means of the x_i and y_i, respectively. Notice that the initial parameter estimates a_0 and b_0 do not appear in Eqs. (3.8) and (3.9). When, however, we apply Eq. (3.1) to a nonlinear model function (Section 3.3) we find that the initial parameter estimates appear in the normal equations.

The procedure is to develop the model function (which is the system binding isotherm) and to apply Eq. (3.1) to this function, obtaining the normal equations, which are solved for the parameters. If the function is nonlinear, preliminary parameter estimates $a_0, b_0, \ldots$ are needed, and the resulting parameter values $a, b, \ldots$ are used as the $a_0, b_0, \ldots$ in a second iteration, this calculation being repeated until the estimates converge.

3.2. CURVE FITTING

Weighting and the Propagation of Errors

Suppose we have a function $F(x, y)$, and we carry out a Taylor's series expansion about the point (x_0, y_0), thus

$$F(x, y) = F(x_0, y_0) + F_x(x) \cdot (x - x_0) + F_y(y) \cdot (y - y_0)$$

where we truncate the series at the linear term as shown. If we take x close to x_0, y close to y_0, the intervals $\Delta x = x - x_0$, $\Delta y = y - y_0$ will be very small, and the approximation is a reasonable one. In this equation F_x and F_y are partial first derivatives.

Defining $\Delta F = F(x, y) - F(x_0, y_0)$ gives

$$\Delta F = F_x \, \Delta x + F_y \, \Delta y \qquad (3.10)$$

We now identify the increments Δx and Δy as errors in x and y. Then Eq. (3.10) reveals, to a good approximation, how these errors are propagated into the error ΔF in the function F (5). Squaring Eq. (3.10) gives

$$(\Delta F)^2 = (F_x \, \Delta x)^2 + (F_y \, \Delta y)^2 + 2F_x F_y \, \Delta x \, \Delta y \qquad (3.11)$$

Now sum the errors over all possible values of their (small) ranges and take the averages. These quantities can be interpreted as in Eqs. (3.12) and (3.13):

$$\sigma_x^2 = \frac{\Sigma (x - \bar{x})^2}{n} \; ; \quad \sigma_y^2 = \frac{\Sigma (y - \bar{y})^2}{n} \qquad (3.12)$$

$$\sigma_{xy} = \frac{\Sigma (x - \bar{x})(y - \bar{y})}{n} \qquad (3.13)$$

where σ_x^2, σ_y^2 are the variances of x and y, and σ_{xy} is the covariance, often designated $\text{cov}(x, y)$, which measures the interaction between x and y. Then Eq. (3.11) becomes

$$\sigma_F^2 = F_x^2 \sigma_x^2 + F_y^2 \sigma_y^2 + 2F_x F_y \sigma_{xy} \qquad (3.14)$$

This argument obviously can be generalized to any number of variables.

Equation (3.14) describes the propagation of mean square error, or the propagation of variances and covariances.

If the errors in the variables are independent, then $\sigma_{xy} = 0$, and the propagation of error equation can be written

$$\sigma_F^2 = F_x^2 \sigma_x^2 + F_y^2 \sigma_y^2 \qquad (3.15)$$

In Appendix B we define weights as reciprocal functions of variances, or $w_x = k/\sigma_x^2$, $w_y = k/\sigma_y^2$, $w_F = k/\sigma_F^2$. Combining these definitions with Eq. (3.15) gives

$$\frac{1}{w_F} = \frac{F_x F_x}{w_x} + \frac{F_y F_y}{w_y} \qquad (3.16)$$

which may be compared with Eq. (3.2). Equation (3.16) describes the propagation of weights, subject to the condition that Eq. (3.15) is valid.

To carry out the least-squares analysis we need the w_i (Eq. 3.1), or, equivalently, the L_i, since $w_i = 1/L_i$. Equation (3.2) can also be written

$$L_i = \frac{1}{k} (F_{xi} F_{xi} \sigma_{xi}^2 + F_{yi} F_{yi} \sigma_{yi}^2) \qquad (3.17)$$

The partial derivatives F_{xi}, F_{yi} are evaluated at the points $(X_i, Y_i, a_0, b_0, c_0)$. Notice that the constant k cancels from the normal equations, Eq. (3.1). Evidently, estimates of the variances σ_{xi}^2, σ_{yi}^2 are needed. Several situations may arise:

1. $\sigma_{xi}^2 \ll \sigma_{yi}^2$; σ_{yi}^2 independent of i. This is the simplest possible case, and it is the one usually assumed (often without justification) in the application of least-squares regression. It signifies that the error in the X_i is negligible, and the X_i are therefore taken as exactly known quantities. Moreover, the error in the Y_i is considered constant, independent of X_i. Thus Eq. (3.17) becomes $L_i = F_{yi} F_{yi} \sigma_y^2/k$; since σ_y^2/k is a constant independent of i it cancels from the normal equations. Thus we have unweighted least-squares analysis.

2. $\sigma_{xi}^2 \ll \sigma_{yi}^2$; $\sigma_{yi}^2 = f(X_i)$. Again the X_i are considered to be exactly known, but in this case σ_{yi}^2 depends on i. Thus Eq. (3.17) becomes $L_i = F_{yi} F_{yi} \sigma_{yi}^2/k$; the variances do not cancel from the normal equations. This case is usually described as weighted least-squares analysis.

3. $\sigma_{xi}^2 = \sigma_{yi}^2$. Then the variances can be factored from Eq. (3.17); if they are constant (independent of i), they cancel from the normal equations.

4. $\sigma_{xi}^2 \neq \sigma_{yi}^2$, but they are comparable in magnitude. Either the σ_{xi}^2 or the σ_{yi}^2, or both, may be functions of i. This general case includes weighting of both variables, and the full Eq. (3.17) must be used.

In order to determine the case to which a system belongs, and then to apply the least-squares analysis, it is necessary to have estimates of σ_{xi}^2 and σ_{yi}^2. These can only arise from experimental observations of the system. For example, if the X_i are concentrations of standard solutions, σ_{xi}^2 can be estimated from experience with the method of preparation. If Y_i, the dependent variable, is, e.g., an absorbance measurement, σ_{yi}^2 can be measured by making several measurements of Y_i (on independently prepared solutions) at the same X_i. Considerable experimental effort may have to be expended in establishing these variances.

Weighting and the Transformation of Variables

It frequently happens that we plot or analyze data in terms of quantities that are transformed from the raw experimental variables. Our discussion of the propagation of error leads us to inquire into the distribution of errors in the transformed variables. We shall demonstrate with an important example. Many equations in chemistry are of the nonlinear form

$$y = ae^{-bx}$$

where a and b are parameters. The first-order rate law, the Arrhenius equation, and the van't Hoff equation are examples. It is common to linearize this equation by taking logarithms

$$\log y = \log a - \frac{b}{2.3}\, x \tag{3.18}$$

and then to obtain a and b from the straight-line plot of $\log y$ against x, or from a least-squares analysis in terms of these quantities. Since x is usually known with very good accuracy, only the error in the dependent variable is considered.

We define the transformed variable $y' = \log y$ and apply the propagation of error argument to find the variance of y' (6). From Eq. (3.15),

$\sigma^2_{y'} = (\partial y'/\partial y)^2 \sigma^2_y$, and since $d \ln u = du/u$ and $\ln u = 2.3 \log u$, we get

$$\sigma^2_{y'} = \sigma^2_y/(2.3y)^2$$

Thus in applying weighting factors w_i in the linear least-squares analysis according to Eq. (3.18), each $y'_i = \log y_i$ should be weighted inversely to $\sigma^2_y/(2.3y_i)^2$ rather than to σ^2_y.

We now apply this method to the 1:1 binding isotherm, written as in Eq. (3.19), where x represents free ligand concentration, y is a measure of extent of binding, and m, n are system parameters.

$$y = \frac{x}{m + nx} \qquad (3.19)$$

In Chapter 2 we saw that Eq. (3.19) can be put into several linear forms, which we called the double reciprocal (Eq. 2.90), y-reciprocal (Eq. 2.91), and x-reciprocal (Eq. 2.92) plots. We next obtain expressions for the variances of the transformed variables in each of these three plotting forms. Let x' and y' be the transformed abscissa and ordinate variables, respectively. Table 3.1 lists these variables in terms of x and y, the partial derivatives needed in Eq. (3.15), and the variances of the transformed variables in terms of the original variables.

We next need to obtain the $L_i = 1/w_i$ terms that appear in the normal equations. Since we have obtained the variances of the transformed variables (Table 3.1), we can work with Eq. (3.17). [Note carefully that the F_{xi}, F_{yi} in Eq. (3.17) are not the quantities listed in Table 3.1; in each use of this symbol it must be defined for the particular function being employed.] For example, the double-reciprocal plotting form is

$$\frac{1}{y} = \frac{m}{x} + n \qquad (3.20)$$

or

$$F = \frac{1}{y} - \frac{m}{x} - n = y' - mx' - n$$

Thus

$$F_{x'_i} = \frac{\partial F}{\partial x'_i} = -m$$

Table 3.1. Variances of Transformed Variables in Linear Plotting Forms of the Rectangular Hyperbola[a]

Plotting form		Ordinate			Abscissa	
	y'	Partial derivatives	$\sigma^2_{y'}$	x'	Partial derivatives	$\sigma^2_{x'}$
Double-reciprocal	$\dfrac{1}{y}$	$-\dfrac{1}{y^2}$ $\quad F_x = \dfrac{1}{y}$	$\dfrac{\sigma^2_y}{y^4}$	$\dfrac{1}{x}$	$-\dfrac{1}{x^2}$	$\dfrac{\sigma^2_x}{x^4}$
y-reciprocal	$\dfrac{x}{y}$	$F_y = \dfrac{-x}{y^2}$	$\dfrac{\sigma^2_x}{y^2} + \dfrac{x^2\sigma^2_y}{y^4}$	x	1	σ^2_x
x-reciprocal	$\dfrac{y}{x}$	$F_x = -\dfrac{y}{x^2}$ $\quad F_y = \dfrac{1}{x}$	$\dfrac{y^2\sigma^2_x}{x^4} + \dfrac{\sigma^2_y}{x^2}$	y	1	σ^2_y

[a] In general each quantity in this table should be subscripted with an (i), and all quantities may vary with i.

$$F_{y_i} = \frac{\partial F}{\partial y'_i} = 1$$

Combining these results with those in Table 3.1 we get

$$kL_i = \frac{m^2\sigma^2_{xi}}{x_i^4} + \frac{\sigma^2_{yi}}{y_i^4} \tag{3.21}$$

The y-reciprocal form is

$$\frac{x}{y} = nx + m \tag{3.22}$$

$$F = \frac{x}{y} - nx - m = y' - nx' - m$$

$$F_{x_i} = \frac{\partial F}{\partial x'_i} = -n$$

$$F_{y_i} = \frac{\partial F}{\partial y'_i} = 1$$

$$kL_i = n^2\sigma_{xi}^2 + \frac{\sigma_{xi}^2}{y_i^2} + \frac{x_i^2\sigma_{yi}^2}{y_i^4} \qquad (3.23)$$

In the same way, for the x-reciprocal form,

$$\frac{y}{x} = -\frac{n}{m}y + \frac{1}{m} \qquad (3.24)$$

$$F = \frac{y}{x} + \frac{n}{m}y - \frac{1}{m} = y' + \frac{n}{m}x' - \frac{1}{m}$$

$$F_{x_i} = \frac{\partial F}{\partial x_i'} = \frac{n}{m}$$

$$F_{y_i} = \frac{\partial F}{\partial y_i'} = 1$$

$$kL_i = \frac{n^2}{m^2}\sigma_{yi}^2 + \frac{\sigma_{yi}^2}{x^2} + \frac{y^2\sigma_x^2}{x^4} \qquad (3.25)$$

If the $\sigma_{x_i}^2$ are negligible these L_i become somewhat simpler, but they remain fairly complicated weighting functions.

Let us now find the weighting factors for the untransformed rectangular hyperbola:

$$y = \frac{x}{m + nx} \qquad (3.26)$$

$$F = y - \frac{x}{m + nx}$$

$$F_{x_i} = \frac{\partial F}{\partial x_i} = -\frac{m}{(m + nx_i)^2}$$

$$F_{y_i} = \frac{\partial F}{\partial y_i} = 1$$

From Eq. (3.17),

$$kL_i = \frac{m^2\sigma_{x_i}^2}{(m + nx_i)^4} + \sigma_{yi}^2 \qquad (3.27)$$

When the error in the independent variable is negligible, we simply get

$L_i = \sigma_{yi}^2/k$. We reach the interesting result that the weighting of the nonlinear form is simpler than the weighting of the transformed linear forms.

Testing Goodness of Fit

One of the goals of the curve-fitting process is to test the adequacy of the stoichiometric model that was (tentatively) adopted. We define the residual in the dependent variable by Eq. (3.28),

$$e_{yi} = Y_i - y_i \tag{3.28}$$

where Y_i is the observed value and y_i is the value calculated from the regression line, both at the same value of the independent variable. The e_{yi} therefore represent the deviations of the experimental points from the regression line. If the model is an adequate representation of the physical system, the residuals should not exhibit any alarming behavior, in the sense that we expect the residuals to be randomly distributed about the line. Several graphical means of examining the residuals have been devised. Thus the e_{y_i} may be plotted against the X_i; the e_{y_i} should be scattered about their mean value in a horizontal band of more or less uniform width. Obvious trends (such as greater residuals at large X_i) indicate a poor fit. Another plot is of e_{y_i} against y_i (not against Y_i, for the e_{y_i} and Y_i are correlated) (7).

Another technique is to calculate a measure of the dispersion of the experimental points about the regression line and to compare this with an independent experimental measure of the dispersion in the variable. If the fit is good, evidently the scatter about the line should not be markedly greater than the scatter that would be observed in replicate observations. Let us consider this calculation for systems in which the error in the independent variable is negligible. We calculate the sum of squares of the weighted residuals:

$$S = \sum w_{y_i} e_{y_i}^2 \tag{3.29}$$

[If there is error in both variables, Eq. (B.11) is used.] Suppose we have fitted an equation with p parameters to n observations; then there remain $n - p$ degrees of freedom. We take as a measure of the dispersion of points about the regression line the quantity

$$\sigma^2_{ext} = \frac{S}{n - p} \tag{3.30}$$

where σ^2_{ext} is the "external" variance in Deming's nomenclature (8).

A quantity σ^2_{int}, the variance measuring "internal consistency" must be obtained by experimental measurements of the dependent variable. If σ^2_{int} is a constant of the population, and the errors are normally distributed, these variances can be compared by means of the F test.

$$F = \frac{\sigma^2_{ext}}{\sigma^2_{int}}$$

This test for goodness of fit depends on the normal distribution applying to the errors. We cannot usually expect to make a rigorous F test of goodness of fit in nonlinear modeling, in the sense that we cannot attach a specific level of significance to an F ratio. The test can still be of semiquantitative or comparative value, however; for example, it could be used to decide which of two models gives a better fit of the data.

Variances of the Parameters

In Appendix B it is shown that the least-squares normal equations can be written and solved as a matrix equation. Appendix C discusses the matrix solution of simultaneous equations. An advantage of the matrix solution is that it yields, besides the parameters, estimates of the variances and covariances of the parameters. Equation (B.30) shows that the solution of the matrix equation requires evaluation of the inverse matrix M^{-1}, where M is defined by Eq. (B.26). It can be shown (9) that the inverse matrix M^{-1} arising from the solution of the normal equations has the form

$$M^{-1} = \begin{bmatrix} C_{11} & C_{12} & C_{13} \\ C_{21} & C_{22} & C_{23} \\ C_{31} & C_{32} & C_{33} \end{bmatrix}$$

where the C_{ij} are given by

$$C_{11} = \frac{\sigma^2_a}{\sigma^2} \tag{3.31a}$$

$$C_{22} = \frac{\sigma_b^2}{\sigma^2} \qquad (3.31b)$$

$$C_{33} = \frac{\sigma_c^2}{\sigma^2} \qquad (3.31c)$$

$$C_{12} = C_{21} = \frac{\sigma_{ab}}{\sigma^2} \qquad (3.32a)$$

$$C_{13} = C_{31} = \frac{\sigma_{ac}}{\sigma^2} \qquad (3.32b)$$

$$C_{23} = C_{32} = \frac{\sigma_{bc}}{\sigma^2} \qquad (3.32c)$$

and σ^2 is the variance of the dependent variable. If we have an estimate of σ^2, therefore (usually σ_{ext}^2), we can estimate the variances and covariances of the parameters. The inverse matrix M^{-1} is therefore called the variance–covariance matrix or simply the covariance matrix.

Although we are in this way able to estimate the uncertainties in the parameters, these estimates should be regarded as approximate because (aside from possible effects of nonnormal error distributions) the non-linear least-squares analysis has been developed on the basis of a linear approximation.

This is an appropriate point at which to discuss the selection of the proportionality constant k that appears in Eqs. (B.23) and (3.17). It is evident that this constant cancels from the normal equations; thus its value is arbitrary in calculating the parameter estimates, $a, b, c, \ldots$. However, k contributes to the matrix elements as a constant multiplier; hence the variance ratios in Eq. (3.31) depend on the value of k. Nevertheless, the estimates of the variances of the parameters, σ_a^2, σ_b^2, σ_c^2, will be unaffected by the value of k provided the *same value* of k is used in solving the normal equations and in calculating σ_{ext}^2 (as an estimate of σ^2).

Let us use a two-parameter linear model to demonstrate these statements. The normal equations are (Eq. 3.8)

$$a \sum w_i + b \sum w_i x_i = \sum w_i y_i$$
$$a \sum w_i x_i + b \sum w_i x_i^2 = \sum w_i x_i y_i$$

As is shown in Section 3.3, the variance ratios of Eq. (3.31) are as follows:

Quantity	*Unweighted*		*Weighted*
$\dfrac{\sigma_a^2}{\sigma^2} =$	$\dfrac{\Sigma\, x^2}{n\,\Sigma\, x^2 - (\Sigma\, x)^2}$	or	$\dfrac{\Sigma\, wx^2}{\Sigma\, w\,\Sigma\, wx^2 - (\Sigma\, wx)^2}$
$\dfrac{\sigma_b^2}{\sigma^2} =$	$\dfrac{n}{n\,\Sigma\, x^2 - (\Sigma\, x)^2}$	or	$\dfrac{\Sigma\, w}{\Sigma\, w\,\Sigma\, wx^2 - (\Sigma\, wx)^2}$

Suppose $\sigma_{x_i}^2$ is negligible and we need consider only the error in the dependent variable. Then $w_i = k/\sigma_{y_i}^2$, and the formulas for the weighted case show immediately that k does not cancel; hence its value is not arbitrary in estimating the variance ratios. Next consider the evaluation of σ_{ext}^2, which is our estimate of σ^2. From Eq. (3.30) we have

$$\sigma_{\text{ext}}^2 = \frac{\Sigma\, we^2}{n - p} = \frac{k\,\Sigma\,(1/\sigma_{y_i}^2)e^2}{n - p}$$

Again we find that the value of k is not arbitrary. However, if we compute σ_a^2 by taking the product $\sigma_a^2 = (\sigma_a^2/\sigma^2)\sigma_{\text{ext}}^2$, the k's cancel provided the same value was used in both calculations.

Some workers set $k = 1$, giving, from Eq. (B.23), $w_i = 1/\sigma_i^2$. Another convention is to normalize the weights by setting $\Sigma\, w_i = 1$, with the result that $k = 1/\Sigma\,(1/\sigma_i^2)$, so that $w_i = (1/\sigma_i^2)/\Sigma\,(1/\sigma_i^2)$. We may consider whether there is a "best" value for k, since as we have seen its selection is not universally arbitrary. We can specify a criterion that identifies the correct k value (at least for the linear model) as follows: We make the reasonable requirement that the numerical estimates of the parameter ratios σ_a^2/σ^2 and σ_b^2/σ^2 for a weighted regression be identical to those for unweighted regression when the w_i are all equal.

It is obvious that this criterion means that $w_i = 1$ and $\Sigma\, w_i = n$; therefore $\Sigma\, w_i = k\,\Sigma\,(1/\sigma_i^2) = n$, or

$$k = \frac{n}{\Sigma\,(1/\sigma_i^2)}$$

Thus with this choice,

$$w_i = \frac{n/\sigma_i^2}{\Sigma\,(1/\sigma_i^2)} \tag{3.33}$$

Convergence

The normal equations for nonlinear models require preliminary estimates a_0, b_0, c_0 of the parameters. Solution of the equations for A, B, C then yields the further estimates a, b, c through Eqs. (3.3). If we now allow these latter values to play the role of a_0, b_0, c_0, we can repeat the calculation and obtain refined parameter estimates. This iterative process can be continued until no further change in the parameter estimates occurs.

Several features of this process are of interest. First, we need a criterion for terminating the iteration process. We might decide that if the parameters in the $(j + 1)$st iteration do not differ by more than some small amount or fraction from those in the jth iteration, the calculation may be terminated. Or we could use the sum of squares of the residuals rather than the parameters as the quantity to monitor.

Second, we consider the initial estimates a_0, b_0, c_0. Evidently, we should wish these to be as close to the final values as possible so as to reduce the required number of iterations. In some instances (see later) it happens that if the initial estimates are seriously in error, the iterations will not converge on the desired values. The selection of a_0, b_0, c_0 therefore deserves careful attention. Any knowledge the experimenter has about the system should be made use of. Perhaps prior work in the laboratory, or literature reports, may provide rough estimates. Structural analogs, linear free energy relationships, and theoretical considerations are tools that may help to assign values to a_0, b_0, c_0. But the best source of initial estimates is the data themselves. If the model function is intrinsically linear (as is the rectangular hyperbola), a straight-line plot will yield reasonable estimates; it may even be worthwhile to carry out a weighted linear least-squares regression for this purpose. If the model is intrinsically nonlinear, with p parameters, then solution of p simultaneous equations written for p sets of data will yield estimates; it is necessary to choose representative data points for this purpose, and it may be a good plan to plot the data, draw a smooth curve through the points, and select the p points from this curve. Another approach is to obtain parameter estimates from limiting regions of the observations. For example, at very low X_i it often happens that the function approaches linearity, and the limiting slope may yield a parameter estimate; at very high X_i a different parameter may be estimated. (A plot of Y_i or $\log Y_i$ against $1/X_i$ may provide a useful way to extrapolate Y to infinite X by extrapolating to

$1/X = 0$.) The graphical and numerical techniques described in Section 2.6 can be very helpful in generating sound initial parameter estimates in multiple equilibria systems.

Finally, we consider very briefly a graphical interpretation of least-squares regression and its application to the nonlinear problem. Box et al. (10) and Draper and Smith (11) give good discussions of this subject. Suppose, first, that we have a one-parameter model function, $y = f(x, a)$. According to the least-squares principle we minimize the sum of squares of the residuals,

$$S = \sum [Y_i - f(X_i, a)]^2 \qquad (3.34)$$

We can do this by setting $\partial S/\partial a = 0$ and solving for a, but alternatively we can assign a value to a and calculate the corresponding S by inserting the experimental X_i, Y_i in Eq. (3.34). Repeating this operation for numerous values of a (chosen in the vicinity of the "correct" value) provides a set of (S, a) points that can be presented as a plot of S against a. Equation (3.34) is the equation of a parabola, and the plot will reveal a minimum in S that corresponds to the value of a that minimizes the sum of squares of the residuals. In this way it is possible to find the least-squares solution numerically for the one-parameter case.

For a two-parameter linear model we can carry out the numerical evaluation by assigning pairs of values to the parameters a, b, and calculating sums of squares. Now the resulting sets of values (S, a, b) define a sum of squares surface; if S is plotted along the z coordinate, with a and b on the x and y coordinates, the surface will possess a minimum corresponding to the least-squares values of the parameters. Any cross section of the surface at constant S will be an ellipse in a two-dimensional *parameter space*, and this is a more convenient form of presentation. The results of the numerical calculation can then be displayed as a series of sum of squares elliptical contours in the plot in a, b coordinate space. (For three parameters it is necessary to hold one parameter constant while varying the other two, and the visualization obviously requires much more calculational effort.)

When this same approach is taken with a two-parameter nonlinear model, we find that the sum of squares contours in parameter space are nonelliptical, and may be "banana-shaped." Moreover, some functions define surfaces having more than one minimum. For such surfaces the minimum corresponding to the absolute minimum value of S is called the global minimum, and this corresponds to the desired least-squares solution; other minima are called local minima. The Taylor's series linearization that we carried out to produce linear normal equations in effect approximates the sum of squares surface in the vicinity of the point (a_0, b_0, c_0) by the surface of a linear model. That is, the distorted ellipses of the sum of squares contours are approximated by true ellipses. This approximation may be one reason why the iteration may fail to converge, or why it may converge very slowly.

This geometric view of the least-squares search for the minimum sum of squares also shows why our initial estimates should be as close as possible to the least-squares estimates, for if a_0, b_0, c_0 are not in the region of the global minimum, the sum of squares minimization may result in convergence to a local minimum, which may have no physical significance. Some writers suggest that different initial estimates should result in the same final least-squares parameters, but if the surface contains multiple minima it is quite possible for different initial estimates to converge to different parameter values.

The shapes of the linearized elliptical contours can determine the pattern and success of the iteration process. Draper and Smith (11) treat this problem.

3.3. APPLICATION TO BINDING

Linear Regression

Equation (3.6) is the model function; we used the general approach with this function as an example of the method, obtaining Eq. (3.8) for weighted linear regression and Eq. (3.9) for unweighted linear regression. Weighting was discussed in detail in Section 3.2.

Let us find the variance–covariance matrix for this model. We first consider the unweighted case. According to Appendix B, the normal equations can be written in matrix form as $\mathbf{MP} = \mathbf{Q}$, where

$$\mathbf{M} = \begin{bmatrix} n & \Sigma\, x \\ \Sigma\, x & \Sigma\, x^2 \end{bmatrix}$$

$$\mathbf{P} = \begin{bmatrix} a \\ b \end{bmatrix}$$

$$\mathbf{Q} = \begin{bmatrix} \Sigma\, y \\ \Sigma\, xy \end{bmatrix}$$

We need the inverse matrix $\mathbf{M}^{-1}$. By the technique described in Appendix C we find

$$\mathbf{M}^{-1} = \frac{1}{\det \mathbf{M}} \begin{bmatrix} \Sigma\,x^2 & -\Sigma\,x \\ -\Sigma\,x & n \end{bmatrix}$$

where $\det \mathbf{M} = n\,\Sigma\,x^2 - (\Sigma\,x)^2$. But this can also be written $\det \mathbf{M} = n\,\Sigma\,(x - \bar{x})^2$, so

$$\mathbf{M}^{-1} = \begin{bmatrix} \dfrac{\Sigma\,x^2}{n\,\Sigma\,(x-\bar{x})^2} & \dfrac{-\bar{x}}{\Sigma\,(x-\bar{x})^2} \\[2mm] \dfrac{-\bar{x}}{\Sigma\,(x-\bar{x})^2} & \dfrac{1}{\Sigma\,(x-\bar{x})^2} \end{bmatrix} \tag{3.35}$$

Thus, from Eqs. (3.31) and (3.32),

$$\sigma_a^2 = \frac{\sigma^2\,\Sigma\,x^2}{n\,\Sigma\,(x-\bar{x})^2} \tag{3.36}$$

$$\sigma_b^2 = \frac{\sigma^2}{\Sigma\,(x-\bar{x})^2} \tag{3.37}$$

$$\sigma_{ab} = \frac{-\sigma^2\bar{x}}{\Sigma\,(x-\bar{x})^2} \tag{3.38}$$

and σ^2 is estimated by Eq. (3.30), which for this system is

$$\sigma_{\text{ext}}^2 = \frac{\Sigma\,[y_{\text{obs}} - (a + bx)]^2}{n - 2} \tag{3.39}$$

If the weighting factors do not cancel they must be incorporated into the inverse matrix, which becomes

$$\mathbf{M}^{-1} = \frac{1}{\det \mathbf{M}} \begin{bmatrix} \Sigma\,wx^2 & -\Sigma\,wx \\ -\Sigma\,wx & \Sigma\,w \end{bmatrix} \tag{3.40}$$

where $\det \mathbf{M} = \Sigma\,w\,\Sigma\,wx^2 - (\Sigma\,wx)^2$, and the variances and covariances are obtained from Eqs. (3.31), (3.32), and (3.40) as in the preceding example.

We are now able to apply the least-squares technique to the linear transformations of the rectangular hyperbola. The transformed variables are listed in Table 3.1, and the weighting factors L ($= 1/w$) are given in

Eqs. (3.21), (3.23), and (3.25) for the three linear forms. In many applications it will be permissible to assume $\sigma_x^2 = 0$, where x here refers to the original independent variable. Then, as Table 3.1 shows, the weighting involves only the transformed dependent variable for the double-reciprocal and y-reciprocal plots. However, for the x-reciprocal plot, weighting is required on both axes, and the procedure is cumbersome. For the purpose of estimating parameters, therefore, the x-reciprocal plot is less desirable. (On the other hand, as we saw in Sections 2.5 and 2.6, the x-reciprocal plot is the preferred one in establishing the range of the binding curve that has been experimentally studied.)

Let us write out the normal equations for the double-reciprocal plot, assuming $\sigma_x^2 = 0$. If we use primed quantities to represent the transformed variables, $y' = a + bx'$, where $y' = 1/y$, $x' = 1/x$, $a = n$, $b = m$ (where n and m are the parameters in Eq. 3.19). The normal equations are

$$a \sum w + b \sum wx' = \sum wy'$$
$$a \sum wx' + b \sum wx'^2 = \sum wx'y'$$

(3.41)

and w is calculated with Eq. (3.33).

We illustrate the analysis with the data of Kramer and Connors (12) on complex formation between sodium cinnamate (the substrate) and theophylline (the ligand). Spectrophotometric observations were made at constant total substrate concentration $S_t = 1.00 \times 10^{-3} M$ and varying total ligand concentration L_t. The wavelength was 314 nm and the path length was 10.0 cm. Five replicate runs were made. These were not merely repeat observations on the same solutions; rather, fresh solutions were prepared for each replicate run, and in some instances the stock solutions were also freshly prepared. The absorbance measurements are given in Table 3.2. These are written A_{ui}, where $u = 1, 2, \ldots, 5$ and $i = 1, 2, \ldots, 7$. Equation (3.42) is the isotherm, which is not being tested here because the range of L_t values was too limited to provide a good test of the model.

$$\frac{\Delta A}{b} = \frac{K_{11}S_t \, \Delta\epsilon \, L_t}{1 + K_{11}L_t}$$

(3.42)

In Eq. (3.42), $\Delta A = A_{L_t} - A_{L_t=0}$, K_{11} is the stability constant for complex

Table 3.2. Absorbance Measurements on the Sodium Cinnamate–Theophylline System at 25°C in a 10-cm Cell at 315 nm

			A_{ui}				
L_t/M	i	$u =$ 1	2	3	4	5	
0.0250	1	1.063	1.064	1.034	1.063	1.041	
0.0200	2	0.958	0.973	0.970	0.972	0.950	
0.0167	3	0.903	0.917	0.904	0.913	0.892	
0.0143	4	0.854	0.863	0.855	0.871	0.848	
0.0125	5	0.827	0.828	0.818	0.838	0.817	
0.0111	6	0.794	0.792	0.784	0.797	0.789	
0.0000	7	0.531	0.530	0.530	0.529	0.530	

formation (this is the parameter sought), $\Delta\epsilon$ is the difference in molar absorptivities between the complexed and uncomplexed forms of the substrate, and we make the approximation $L_t = [L]$, which is reasonable in this case. We take the L_t as exactly known.

From Table 3.2 the quantity $\Delta A/b$ is calculated; these data are given in Table 3.3. Since we will use the double-reciprocal plot, we make these definitions:

$$x = L_t$$
$$y = \frac{\Delta A}{b}$$
$$x' = \frac{1}{x}$$
$$y' = \frac{1}{y}$$
$$a = \frac{1}{S_t \Delta\epsilon}$$
$$b = \frac{1}{K_{11} S_t \Delta\epsilon}$$
$$y' = a + bx'$$
$$y = \frac{x}{b + ax}$$

The mean $\bar{y}$ was used for y in the calculations. In Table 3.4 are listed x', y, y', and σ_y^2. It is immediately obvious that σ_y^2 is not a constant; hence

Table 3.3. The Observations x and y (see Table 3.2)

			$y = \Delta A/b$		
$x = L_t$	1	2	3	4	5
0.0250	0.0532	0.0534	0.0504	0.0534	0.0511
0.0200	0.0427	0.0443	0.0440	0.0443	0.0420
0.0167	0.0372	0.0387	0.0374	0.0384	0.0362
0.0143	0.0323	0.0333	0.0325	0.0342	0.0318
0.0125	0.0296	0.0298	0.0288	0.0309	0.0287
0.0111	0.0263	0.0262	0.0254	0.0268	0.0259

weighted least-squares analysis is required. From Table 3.1, $\sigma_{y'}^2 = \sigma_y^2/y^4$, and w is then calculated with Eq. (3.33); these quantities are also given in Table 3.4.

The summations needed to solve the normal equations (3.41) are tabulated in Table 3.5. (It is usually necessary to carry many figures in such calculations to minimize rounding errors, and different workers may reach slightly different results.) Now we write the matrix equation $MP = Q$, where

$$M = \begin{bmatrix} 6 & 350.55 \\ 350.55 & 22265.7 \end{bmatrix}$$

$$P = \begin{bmatrix} a \\ b \end{bmatrix}$$

$$Q = \begin{bmatrix} 156.61 \\ 9826.82 \end{bmatrix}$$

We find det $M = 10708.9$, and the inverse matrix, obtained as shown in Appendix C, is

Table 3.4. Quantities Required for the Weighted Least-Squares Analysis

x'	y	y'	$10^6\sigma_y^2$	$\sigma_{y'}^2$	$\sigma_{y'}^{1/2}$	w
40.0	0.0523	19.12	2.070	0.277	3.610	1.684
50.0	0.0435	22.99	1.103	0.308	3.247	1.514
60.0	0.0376	26.60	1.002	0.501	1.996	0.931
70.0	0.0328	30.49	0.887	0.767	1.304	0.608
80.0	0.0296	33.78	0.793	1.033	0.968	0.451
90.0	0.0261	38.31	0.267	0.575	1.739	0.811

Table 3.5. Summations for the Normal Equations, Eqs. (3.41)

x'	wx'	wy'	wx'^2	$wx'y'$
40	67.36	32.20	2694.4	1287.92
50	75.70	34.81	3785.0	1740.34
60	55.86	24.76	3351.6	1485.88
70	42.56	18.54	2979.2	1297.65
80	36.08	15.23	2886.4	1218.78
90	72.99	31.07	6569.1	2796.25
Sums	350.55	156.61	22265.7	9826.82

$$\mathbf{M}^{-1} = \frac{1}{\det \mathbf{M}} \begin{bmatrix} 22265.7 & -350.55 \\ -350.55 & 6 \end{bmatrix}$$

or

$$\mathbf{M}^{-1} = \begin{bmatrix} 2.079177 & -0.032734 \\ -0.032734 & 0.000560282 \end{bmatrix}$$

The solution is therefore $\mathbf{P} = \mathbf{M}^{-1}\mathbf{Q}$, or

$$\mathbf{P} = \begin{bmatrix} a \\ b \end{bmatrix} = \begin{bmatrix} 3.94878 \\ 0.379318 \end{bmatrix}$$

giving $a = 3.949$ and $b = 0.379$. From the definitions of these parameters in terms of the model quantities we see

$$K_{11} = \frac{a}{b} = 10.41 \ M^{-1}$$

We can now evaluate the residuals and the sum of squares of the residuals, since $e = y'_{obs} - y'_{calc}$, and $y'_{calc} = 3.949 + 0.379x'$. Table 3.6 lists

Table 3.6. Sum of Squares of the Residuals of the Linear Double-Reciprocal Regression

x'	y'_{calc}		we^2
40	19.12		0
50	22.91		0.00875
60	26.71		0.01066
70	30.50		0.00006
80	34.29		0.11869
90	38.09		0.04069
		Sum	0.17885

the results. An estimate of the variance of the curve fit is given by Eq. (3.30):

$$\sigma^2_{ext} = \frac{0.17885}{4} = 0.04471$$

We will use this estimate to find the variances of the parameters. Since M^{-1} is the variance–covariance matrix, we find, from Eqs. (3.31) and (3.32),

$$\sigma^2_a = (2.079177)(0.04471) = 0.09296$$
$$\sigma^2_b = (0.00056028)(0.04471) = 0.00002505$$
$$\sigma_{ab} = -(0.032734)(0.04471) = -0.001464$$

Since $K_{11} = a/b$, by Eq. (3.14) we have

$$\sigma^2_{K_{11}} = \left(\frac{\partial K_{11}}{\partial a}\right)^2 \sigma^2_a + \left(\frac{\partial K_{11}}{\partial b}\right)^2 \sigma^2_b + 2\left(\frac{\partial K_{11}}{\partial a}\right)\left(\frac{\partial K_{11}}{\partial b}\right)\sigma_{ab}$$

or

$$\sigma^2_{K_{11}} = \frac{\sigma^2_a}{b^2} + \frac{\sigma^2_b}{a^2} + \frac{2\sigma_{ab}}{ab} \tag{3.43}$$

which gives

$$\sigma^2_{K_{11}} = 0.6461 + 0.000002 - 0.00196 = 0.6441$$

Thus $\sigma_{K_{11}} = 0.803$. Notice that the principal contributor to the uncertainty in K_{11} comes from parameter a (the intercept). It is interesting that the covariance term slightly reduces the variance of K_{11}.

Nonlinear Regression

We write the rectangular hyperbola as

$$y = \frac{x}{b + ax}$$

to be consistent with the preceding section. Then

$$F = y - \frac{x}{b + ax}$$

and the partial derivatives are

$$F_x = \frac{-b}{(b + ax)^2}$$

$$F_y = 1$$

$$F_a = \frac{x^2}{(b + ax)^2}$$

$$F_b = \frac{x}{(b + ax)^2}$$

and we define

$$F_0 = y - \frac{x}{b_0 + a_0 x}$$

Note that the partial derivatives contain the parameters, but they are to be evaluated at the point a_0, b_0, where a_0 and b_0 are the initial estimates of the parameters. We consider that the error in x is negligible, so we do not need F_x. Then substitution of the partial derivatives into Eqs. (3.1) gives the normal equations for this system:

$$\sum \frac{wx^4}{z^4} A + \sum \frac{wx^3}{z^4} B = \sum \left(\frac{wx^2 y}{z^2} - \frac{wx^3}{z^3} \right)$$

$$\sum \frac{wx^3}{z^4} A + \sum \frac{wx^2}{z^4} B = \sum \left(\frac{wxy}{z^2} - \frac{wx^2}{z^3} \right)$$

(3.44)

where $z = b_0 + a_0 x$.

The parameter estimates that we obtained from the linear regression would be excellent initial estimates for the nonlinear regression. Let us, however, deliberately take rounded values that are probably not so close to the final estimates. We choose $a = 4.0$ and $b = 0.40$. Then the sums in Eqs. (3.44) are evaluated with the quantities $x = L_t$, $y = \Delta A / b$, and w is calculated with Eq. (3.33); see Table 3.7. We set up the matrix solution in the usual way, finding the inverse matrix, and solving to obtain

$$A = 0.061731$$

$$B = 0.021590$$

Table 3.7. Variables, Variances, and Weights for the Nonlinear Regression

x	y	$10^6\sigma_y^2$	$10^{-6}/\sigma_y^2$	w
0.0250	0.0523	2.070	0.4831	0.3402
0.0200	0.0435	1.103	0.9066	0.6383
0.0167	0.0376	1.002	0.9980	0.7027
0.0143	0.0328	0.887	1.1274	0.7938
0.0125	0.0296	0.793	1.2610	0.8879
0.0111	0.0261	0.267	3.7453	2.6371
		Sums	8.5214	6.0000

From the definitions of A and B in Eq. (3.3), and the initial estimates of a_0 and b_0, we calculate $a = 3.9383$ and $b = 0.3784$.

We now repeat the process, letting these new values of a and b serve as the estimates a_0 and b_0. We find

$$A = -0.00365$$

$$B = -0.000988$$

Thus we get $a = 3.9420$ and $b = 0.3794$. The progress of the iteration is shown in this table:

Iteration	a	b
0	4	0.4
1	3.9383	0.3784
2	3.9420	0.3794

Table 3.8 gives the residuals and weighted sum of squares after this last iteration.

We now make use of the inverse matrix to find the parameter variances. From Table 3.8 we estimate $\sigma_{ext}^2 = \Sigma\, we^2/4 = 4.546 \times 10^{-8}$. As shown earlier, we find the variances and covariance from the inverse matrix elements, obtaining

$$\sigma_A^2 = 0.62521$$

$$\sigma_B^2 = 1.685 \times 10^{-4}$$

$$\sigma_{AB} = -0.9848 \times 10^{-2}$$

Table 3.8. Residuals and Sum of Squares of the Nonlinear Regression

x	y_{obs}	y_{calc}	$10^4 e$	$10^8 e^2$	$10^8 w e^2$
0.0250	0.0523	0.0523	0	0	0
0.0200	0.0435	0.0436	−1	0	0.6383
0.0167	0.0376	0.0375	+1	1	0.7027
0.0143	0.0328	0.0328	0	0	0
0.0125	0.0296	0.0292	+4	16	14.2664
0.0111	0.0261	0.0262	−1	1	2.6371
				Sum	18.1844

These variances are for A, B in the normal equations, and, from Eqs. (3.3), since $a_0 \approx a$ we can set $\sigma_{a_0}^2 = \sigma_a^2$ and write $\sigma_A^2 = 2\sigma_a^2$. Then with Eq. (3.43) we find the variance of K_{11}.

$$\sigma_{K_{11}}^2 = 2.165$$

$$\sigma_{K_{11}} = 1.471$$

Since $K_{11} = a/b$, we calculate

$$K_{11} = 10.39 \ M^{-1}$$

The estimate of K_{11} is the same as that obtained by the linear regression of the double-reciprocal form, and the standard deviation is somewhat larger. It is interesting to compare the weights in Tables 3.4 and 3.7.

The very detailed presentation of theory and numerical examples in this chapter and in Appendix B is intended to provide a firm basis for understanding and applying regression analysis. Of course, computer programs are widely available that will carry out linear and nonlinear regression, and even hand calculators can perform linear regression. But it is unsound to make use of equations or techniques whose physical or mathematical bases are not understood. To apply least-squares analysis effectively, and to recognize its opportunities, limitations, and options, requires such understanding.

3.4. TRENDS IN DATA ANALYSIS

Because binding phenomena are important in many fields, and because the underlying mathematical relationships are the same for all these phenomena irrespective of the field, it might have been expected that workers in the several fields would be in effective communication concerning their data analysis. This has not happened in any extensive way. Instead, the data analysis within each field (enzyme kinetics, organic molecular complex formation, metal ion complexing, protein binding) has tended to be developed independently. In this section we survey the methods that have been developed, incorporating work from several fields. The methods themselves have already been described in Chapter 2 (Sections 2.3–2.6), Appendix B, and Sections 3.1–3.3.

Testing the Model

In Section 2.5 we saw that it is necessary to explore enough of the binding isotherm to see the nonlinearity in order to conclude that an equilibrium process is operative, and in order to test the stoichiometric model. Dowd and Riggs (13) surveyed enzyme kinetic data to find the typical ranges of saturation that are covered in such studies. Few studies extended beyond 50% saturation. In a study of aniline–tetracyanoethylene complexes studied by spectrophotometry for the purpose of comparing calculational methods, Farrel and Ngo (14) extended one system up to 42% saturation. For determining the range of saturation studied (i.e., before data analysis allows calculation of this quantity), graphical display is useful. Derenleau (15) recommends the direct binding isotherm plot, the logarithmic plot, and the x-reciprocal (Scatchard) plot. Klotz (16), however, argues that the x-reciprocal plot can be misleading, and prefers the logarithmic plot, as does Weber (17).

Rather than focusing on the concentration scale in complexing studies, Carta et al. (18) define a quantity

$$G = K^{-1}[(a + b + K^{-1})^2 - 4ab]^{-1/2}$$

where a and b are substrate and reagent concentrations, and G arises in the description of the 1:1 complexing system; they suggest that a well-designed experiment should have a wide dispersion in G values.

As a quantitative criterion of the goodness of fit Deming (8) described the ratio $\sigma_{ext}^2/\sigma_{int}^2$, where σ_{ext}^2 is the variance of the experimental points about the fitted curve, and σ_{int}^2 is the directly measured variance of the experimental points. This ratio can be subjected to the F test, if we are optimistic enough to assume that the errors satisfy the required conditions. Sprague et al. (19) call this ratio F (absolute). They also define F (relative) $= \sigma_{ext}^2(1)/\sigma_{ext}^2(2)$ for two models being fitted to the same data and use this as a criterion for the rejection of one of the models.

Another test of the model that has been widely used is the adherence to linearity of the data when plotted according to one of the three linear transformations. There is now sufficient experience with these plotting forms to assert that linearity is a necessary but not a sufficient condition for acceptance of the rectangular hyperbola as a model function. There are three factors at work here: (1) the dispersion (error) in the data points; (2) the range of concentrations studied (i.e., the extent of binding); (3) the effect of the transformation of variables on the graphical presentation. The double-reciprocal plot is particularly weak, not to say misleading, as a criterion for model testing, as many workers have shown. This plot may fail to show curvature when it exists (20), and it tends to distort the effects of outliers, either obscuring them or overemphasizing them depending on the concentration range (13, 14, 21, 22).

Finally, we should note that the selection of models for testing, and their tentative acceptance after testing, is controlled largely by chemical intuition and knowledge. A model based on a physicochemical concept of wide applicability (such as the rectangular hyperbola in a 1:1 binding system) is more convincing than an empirical curve-fitting result. Modifications of the initial model should themselves be based on physical principles. It is wise to keep in mind, however, that alternative models may exist, and that statistics alone may not distinguish between them. For example, we study a presumed 1:1 binding system by varying ligand concentration and measuring a solution property. Suppose the statistical result is a poor curve fit to the model function. One possibility is that there is another complex present with different stoichiometry; the corresponding model may be tested. An alternative possibility is that nonideality has set in, that is, that activity coefficients change as the ligand concentration changes. The choice between these explanations is usually based on chemical, not statistical, grounds. The statistical results of Lewis and Knott (23) suggest caution in adopting a model without a chemical basis; these authors simulated data on a nonideal isodesmic system, an isodesmic association being an unlimited self-association in which the addition of each successive monomer to the polymer involves the same free energy change. They found an excellent curve fit to the data with a nonideal 1–2–4–8 model, that is, an "imposter" model in which only monomer, dimers, tetramers, and octamers are permitted.

Graphical and Linear Methods

We have seen that the rectangular hyperbola, Eq. (3.45), can be transformed to three linear plotting forms:

$$y = \frac{x}{m + nx} \tag{3.45}$$

Double-reciprocal

$$\frac{1}{y} = \frac{m}{x} + n$$

y-Reciprocal

$$\frac{x}{y} = nx + m$$

x-Reciprocal

$$\frac{x}{y} = -\frac{n}{m} y + \frac{1}{m}$$

Extensive, and nearly mutually exclusive, literatures have developed about these three equations in the fields of enzyme kinetics, molecular complex formation, and protein binding. This will not be a complete review of the subject, but will provide sufficient citation for our present purpose.

These three linear forms were first described by Barnett Woolf in unpublished work that was described in 1932 by Haldane (24). They have since been rediscovered, many times, and other workers' names are attached to them. In enzyme kinetics the double-reciprocal plot is called the Lineweaver–Burk plot (25); in the field of spectrophotometric studies of complex formation it is the Benesi–Hildebrand equation (26). The x-reciprocal plot is known as the Scatchard plot in protein binding (27) or the Eadie plot in enzyme kinetics (28). The y-reciprocal plot is the Scott plot in complex studies (29). Numerous variations have been described to suit particular experimental situations, but these do not affect our general discussion. The double-reciprocal plot has been, and continues to be, the most widely used method for parameter estimation in enzyme kinetics and complex studies, whereas the x-reciprocal plot is preferred in protein binding. Several authors have suggested that the reason for the preference for the double-reciprocal plot is because it is so forgiving, in the sense that deviations from linearity tend to be obscured, and this tendency is aggravated by the experimentalist's ability to compress the $1/x$ scale arbitrarily. Perhaps another reason this plot has been favored is that it preserves the separation of the independent and dependent variables; the cause-and-effect relationship is not confounded as it may seem to be with the other plots.

Despite its widespread use, the double-reciprocal plot has been severely criticized. Some of this criticism is unmerited: the Benesi–Hildebrand plot has been faulted by many workers for the approximation $L_t = [L]$ that has often been made, but this is not a characteristic of the plotting form, and it can be corrected if necessary, as shown in Section 2.3. The real objection to the double-reciprocal plot is the transformation effect on the variables. It is essential to distinguish between methods in which the parameters are estimated *graphically*, by physically drawing a line on the paper and measuring the slope and intercept, and those in which the parameters are extracted *statistically* by carrying out a linear least-squares regression on the transformed variables. Deming (30) has commented that the eye can be trained, to some extent, to take account of variable weighting in drawing a line, but this is subjective and probably the "eyeball" fitting of straight lines is responsible for much variability of reported parameters. Perhaps the strongest objection to the graphical treatment is that it yields no estimates of parameter variances. In the fields of enzyme kinetics and protein binding there is continuing controversy (31) over the utility of plotting forms, particularly the Scatchard plots.

The linear least-squares regression on the transformed variables should provide reliable results, *provided the proper weighting is used*. In Table 3.1 and the associated discussion we have considered this question in detail. Many authors have made comparisons of the different linear transformations, but in some of these treatments the least-squares analyses were unweighted (14, 22). When properly weighted, linear least-squares analysis appears to give good results (13, 32, 33). At the present time many workers would prefer to use a least-squares analysis on a linear transformation, because of their familiarity with linear regression; as we have seen earlier, however, incorporation of the appropriate weights complicates the regression analysis considerably, and a case could be made that nonlinear regression is actually simpler. Probably the x-reciprocal plot should be avoided for parameter estimation by linear regression because of the complication of the dependent variable error on both axes (though this plot retains its advantage for examining the binding range and for detecting nonlinearity). It is clear that weighting is much more important to the success of the transformed linear equations than to the untransformed nonlinear analysis, and the difficulty of obtaining the correct weights may be a good reason not to use linear regression.

[There is, incidentally, a nice historical irony in the general neglect of a paper (34) by Lineweaver, Burk, and the statistician Deming, that used statistical analysis together with the double-reciprocal plot *before* the appearance of the paper universally cited as introducing the Lineweaver–Burk plot.]

A more recent plotting method has been described by Cornish-Bowden and Eisenthal (21, 35) for the analysis of enzyme kinetic data. A graph is constructed in parameter space by laying out axes in units of K_m and $V_{\max}$. Each experimental point (s_i, v_i) generates a point on the corresponding axis and thereby defines a line in parameter space. The point of intersection common to all of these lines gives K_m and $V_{\max}$ for the system. Because of experimental error, of course, the lines do not all intersect at a single point, but instead generate a collection of intersections. The best estimates of K_m and $V_{\max}$ are taken to be the medians of the intersections projected on the appropriate axes. This method requires no calculations, either to make the plot or to obtain the estimates of the parameters. Since it takes the median rather than the mean as the best estimate, it is sometimes called a nonparametric method. It appears to be robust in that the parameter estimates are relatively insensitive to outliers. Porter and Trager (36) have developed the theory of the method to enable confidence intervals to be estimated. Several workers have compared the Cornish-Bowden direct linear plot (as it is called) with other forms of parameter estimation and find it to be competitive in performance (22, 33, 37, 38). This method is very similar to the plotting method of Rose and Drago, introduced for the analysis of spectrophotometric data (39).

Nonlinear Least-Squares Regression

The general least-squares analysis presented in Appendix B and applied in this chapter was developed in its most systematic form by Deming (5, 40). Most authors expand the nonlinear function about the parameters alone, whereas Deming expands (as we have done) about the parameters *and* the variables, and this leads directly to weighting factors. Deming was the first to present a method for weighting variables on both axes.

Many workers have used nonlinear regression to analyze data in terms of the rectangular hyperbola. A brief survey, approximately chronological, will be adequate to show the level of activity. Deming's work of the 1930s (40) was expanded and presented in book form in 1943 (5) and reprinted in 1964 (46). Although this was not the only source of guidance for the experimentalist, his book was very influential, and most later treatments are similar to Deming's.

In 1958 Grunwald and Coburn (41) used what they called a trial-and-error form of nonlinear least-squares analysis to obtain a complex stability constant from infrared data. This method consisted of the assumption of numerous reasonable values for the constant and calculation of the corresponding sums of squares of the residuals; a plot of the sums against the K values gave the best estimate as that K corresponding to the minimum sum. (We described this technique earlier in this chapter.) Grundnes and Christian used a similar method (42). In 1961 Briegleb used nonlinear regression to analyze spectrophotometric data on molecular complexes (3), and in the same year Wilkinson (1) applied Deming's nonlinear regression method to enzyme kinetics. Wilkinson suggested that a weighted linear least-squares analysis be used to obtain the initial parameter estimates, and he described the appropriate weighting factor for the double-reciprocal plot (though he preferred the y-reciprocal plot for this purpose). Cleland (43) described computer programs for several model functions important in enzyme kinetics.

Wentworth in 1965 (4) applied Deming's method directly to several chemical problems, and in 1967 demonstrated its use for studying organic complex equilibria. Landsbury et al. (44) used a similar method to analyze silver (I)–amine complex systems.

Cohen (45) argued that a better least-squares criterion is to minimize the sum of the squares of the relative error in velocity (in enzyme kinetics) rather than the absolute error; Atkins and Nimmo (22) found that this modification was not an improvement.

Rosseinsky and Kellawi (46) treated literature spectral data on complexing by Deming's method, but they did not weight the data; rather they rejected outliers, which may be a weakness in their procedure. They concluded that the number of different chemical solutions studied (not simply the number of measurements at different wavelengths) is as important as the total range in concentration in determining the reliability of parameters.

Fletcher et al. (47) fitted data on the binding of palmitic acid to human serum albumin to a model of stepwise binding constants. Their final result yielded the most reliable parameters when the model possessed eight stepwise binding constants. This conclusion appears to be at least in part a matter of taste, and this approach, which in effect postulates an open-ended model, would be strengthened by some extrastatistical constraints.

Farrel and Ngo (14) compared nonlinear regression with linear analysis for stability constant estimation and preferred the nonlinear regression; as noted earlier, however, they used unweighted regression, and Christian et al. (32) showed that the weighted linear regression result was equivalent to the nonlinear result.

Madsen and Robertson (48) and Priore and Rosenthal (49) have considered another aspect of the problem. Madsen and Robertson used simulated protein binding data to compare nonlinear regression with linear regression. They concluded that nonlinear regression is always superior to linear regression on the transformed forms, and that nonlinear regression can be improved by *defining* as the dependent variable that one having the greater uncertainty; they then weight only this variable. Priore and Rosenthal tested experimental binding data graphically to find which variable yields the most symmetrical error distribution. Letting B and U represent bound and unbound concentrations, they plotted B/U, $B/(B + U)$, and $\log(B/U)$ against $(B + U)$, finding that the plot of fraction bound gave the best-behaved error distribution. They then carried out unweighted nonlinear regression analysis. The parameters were significantly different when evaluated by the three forms.

Atkins and Nimmo (22) compared the performance of the three linear forms, Cohen's nonlinear method (45), the direct linear plot (35), Merino's plot (50), and nonlinear least-squares regression (1) using simulated data. Unfortunately, the regression methods were unweighted. The performance depended on the type of error (small or large, constant relative or constant absolute). The direct linear plot and the nonlinear least-squares regression seemed to perform most reliably. Nimmo and Atkins (51) believe that the parameter variances estimated from the nonlinear analysis are not very useful because they cannot be interpreted rigorously. They consider it more economical to derive one's own normal equations for specific problems than to use general computer programs. Carta and co-workers (18, 52) applied nonlinear regression to the reevaluation of complex stability constants from literature spectral data. The results were in some instances grossly different from reported values, which they attributed to earlier neglect of weighting factors in the transformed variables. Carta and Crisponi (52) note that the use of nonlinear least-squares analysis makes weighting less important (than when linear regression of transformed variables is undertaken), and that when weighting is needed it can be more easily invoked. These authors have discussed the error contours of the parameter estimates. Northrop (53) has pointed out, in the context of enzyme kinetics, that the physically significant parameters may not coincide with the parameters as defined by the model function, and that alternative forms of expression may lead to different error estimates.

Many nonlinear regression computer programs have been described for the analysis of binding data, including macromolecular binding (54), molecular complexes (55, 56), acid–base equilibria (57, 58), metal ion equilibria (59), and biochemical applications (60).

Three further points are pertinent. One of these is a change in attitude that took place as computer treatment of data became possible. The classical approach to describing physical phenomena quantitatively is to seek a functional relationship between an experimental quantity (or a quantity that can be derived from experiment) and an independent variable, such as ligand concentration. This relationship provided physical insight and guidance for evaluation of system parameters. It is no longer necessary to proceed in this way. All that need be done is to set up all the appropriate independent equations (definitions of binding constants, mass balance expressions, etc.), together with the data, and to solve the simultaneous equations for the parameters. The numerical results are identical with the analysis based on the derived functional relationship, but the physical insight is missing. Evidently one has a choice of style.

The second matter concerns the widespread abandonment of graphical presentation of data as a consequence of modern computational capabilities. It is no longer necessary to graph data. Again, however, something has been lost, because there is a great deal of information available in graphs, and the human eye–brain combination is a very powerful tool in extracting qualitative knowledge from graphical displays. It is particularly useful for detecting trends and deviations. It seems likely that the continuing development of computer graphics capabilities will result in a return to pictorial display as an aid to data analysis.

The third point is the question of the number of adjustable parameters in the model function. In testing models and seeking "understanding" of the system we should attempt to minimize the number of parameters, whereas in merely describing the data points we achieve the best curve fit by using a large number of parameters. To illustrate the two points of view we take as an example the first-order integrated rate equation for a reaction followed spectrophotometrically:

$$\log \left[\frac{A_t - A_\infty}{A_0 - A_\infty} \right] = - \frac{kt}{2.303}$$

In this equation A_t and t are variables and k is a parameter of the model. But are A_0 and A_∞ adjustable parameters or are they observable quantities? Both viewpoints are currently held. If one is seeking to establish the order of the reaction, it will be advisable to measure A_∞, not to treat it as a parameter in the regression analysis.

REFERENCES

1. G. N. Wilkinson, *Biochem. J.*, **80**, 324 (1961).

2. P. J. F. Henderson, *Techniques in Protein and Enzyme Biochemistry*, Part II, Elsevier/North-Holland Biomedical Press, Amsterdam, 1978, p. B113.

3. G. Briegleb, *Elektronen-Donator-Acceptor-Komplexe*, Springer-Verlag, Berlin, 1961, Ch. XII.

4. (a) W. E. Wentworth, *J. Chem. Educ.*, **42**, 96 (1965); (b) W. E. Wentworth, *J. Chem. Educ.*, **42**, 162 (1965); (c) W. E. Wentworth, W. Hirsch, and E. Chen, *J. Phys. Chem.*, **71**, 218 (1967).

5. W. E. Deming, *Statistical Adjustment of Data*, Wiley, New York, 1943; reprinted by Dover Publications, New York, 1964, Ch. III.

6. W. E. Deming, *Statistical Adjustment of Data*, Wiley, New York, 1943; reprinted by Dover Publications, New York, 1964, p. 45.

7. N. R. Draper and H. Smith, *Applied Regression Analysis*, 2nd ed., Wiley, New York, 1981, Ch. 3.

8. W. E. Deming, *Statistical Adjustment of Data*, Wiley, New York, 1943; reprinted by Dover Publications, New York, 1964, pp. 27–29.

9. N. R. Draper and H. Smith, *Applied Regression Analysis*, 2nd ed., Wiley, New York, 1981, pp. 82–83.

10. G. E. P. Box, W. G. Hunter, and J. S. Hunter, *Statistics for Experimenters*, Wiley, New York, 1978, pp. 483–487.

11. N. R. Draper and H. Smith, *Applied Regression Analysis*, 2nd ed., Wiley, New York, 1981, Ch. 10.

12. P. A. Kramer and K. A. Connors, *Am. J. Pharm. Educ.*, **33**, 193 (1969), and unpublished data.

13. J. E. Dowd and D. S. Riggs, *J. Biol. Chem.*, **240**, 863 (1965).

14. P. G. Farrel and P.-N. Ngo, *J. Phys. Chem.*, **77**, 2545 (1973).

15. D. A. Deranleau, *J. Am. Chem. Soc.*, **91**, 4044 (1969).

16. (a) I. M. Klotz, *Acc. Chem. Res.*, **7**, 162 (1974); (b) I. M. Klotz, *Science*, **217**, 1247 (1982).

17. (a) G. Weber, *Molecular Biophysics*, B. Pullman and M. Weissbluth, eds., Academic, New York, 1965, p. 369; (b) G. Weber and S. R. Anderson, *Biochemistry*, **4**, 1942 (1965).

18. G. Carta, G. Crisponi, and V. Nurchi, *Tetrahedron*, **37**, 2115 (1981).

19. E. D. Sprague, C. E. Larrabee, and H. B. Halsall, *Anal. Biochem.*, **101**, 175 (1979).

20. G. D. Johnson and R. E. Bowen, *J. Am. Chem. Soc.*, **87**, 1655 (1965).

21. A. Cornish-Bowden and R. Eisenthal, *Biochem. J.*, **139**, 721 (1974).

22. G. L. Atkins and I. A. Nimmo, *Biochem. J.*, **149**, 775 (1975).

23. M. S. Lewis and G. D. Knott, *Biophys. Chem.*, **5**, 171 (1976).

24. (a) J. B. S. Haldane, *Nature (London)*, **179**, 832 (1957); (b) J. B. S. Haldane and K. Stern, *Allgemeine Chemie der Enzyme*, Steinkopf, Leipzig, 1932, pp. 119–120.

25. H. Lineweaver and D. Burk, *J. Am. Chem. Soc.*, **56**, 658 (1934).

26. H. Benesi and J. H. Hildebrand, *J. Am. Chem. Soc.*, **71**, 2703 (1949).

27. G. Scatchard, *Ann. N.Y. Acad. Sci.*, **51**, 660 (1949).

28. G. S. Eadie, *J. Biol. Chem.*, **146**, 85 (1942).

29. R. L. Scott, *Rec. Trav. Chim.*, **75**, 787 (1956).

30. W. E. Deming, *Statistical Adjustment of Data*, Wiley, New York, 1943; reprinted by Dover Publications, New York, 1964, p. 141.

31. (a) I. H. Scheinberg, *Science*, **215**, 312 (1982); (b) R. A. Weisiger, J. L. Gollan, and R. K. Ockner, *Science*, **215**, 313 (1982); (c) P. J. Munson and D. Rodbard, *Science*, **220**, 979 (1983); (d) I. M. Klotz, *Science*, **220**, 981 (1983); (e) K. E. Light, *Science*, **223**, 76 (1984); (f) B. H. Hofstee, *Nature (London)*, **184**, 1296 (1959); (g) M. Dixon and E. C. Webb, *Nature (London)*, **184**, 1298 (1959).

32. S. D. Christian, E. H. Lane, and F. Gasland, *J. Phys. Chem.*, **78**, 557 (1974).

33. G. L. Atkins and I. A. Nimmo, *Anal. Biochem.*, **104**, 1 (1980).

34. H. Lineweaver, D. Burk, and W. E. Deming, *J. Am. Chem. Soc.*, **56**, 225 (1934).

35. R. Eisenthal and A. Cornish-Bowden, *Biochem. J.*, **139**, 715 (1974).

36. W. R. Porter and W. F. Trager, *Biochem. J.*, **161**, 293 (1977).

37. R. C. Kohberger, *Anal. Biochem.*, **101**, 1 (1979).

38. I. A. Nimmo, *Biochem. J.*, **157**, 493 (1976).

39. N. J. Rose and R. S. Drago, *J. Am. Chem. Soc.*, **81**, 6138 (1959).

40. W. E. Deming, *Phil. Mag.*, **11**, 146 (1931); **17**, 804 (1934); **19**, 389 (1935).

41. E. Grunwald and W. C. Coburn, *J. Am. Chem. Soc.*, **80**, 1322 (1958).

42. J. Grundnes and S. D. Christian, *J. Am. Chem. Soc.*, **90**, 2239 (1968).

43. W. W. Cleland, *Nature (London)*, **198**, 463 (1963).

44. R. C. Landsbury, V. E. Price, and A. G. Smeeth, *J. Chem. Soc.*, **1965**, 1896.

45. S. R. Cohen, *Anal. Biochem.*, **22**, 549 (1968).

46. D. R. Rosseinsky and H. Kellawi, *J. Chem. Soc.*, **1969**, 1207.

47. J. E. Fletcher, A. A. Spector, and J. D. Ashbrook, *Biochemistry*, **9**, 4580 (1970).

48. B. W. Madsen and J. S. Robertson, *J. Pharm. Pharmacol.*, **26**, 807 (1974).

49. R. L. Priore and H. E. Rosenthal, *Anal. Biochem.*, **70**, 231 (1976).

50. F. de M. Merino, *Biochem. J.*, **143**, 93 (1974).

51. (a) I. A. Nimmo and G. L. Atkins, *Biochem. J.*, **157**, 489 (1976); (b) G. L. Atkins and I. A. Nimmo, *Anal. Biochem.*, **104**, 1 (1980).

52. G. Carta and G. Crisponi, *J. Chem. Soc., Perkin Trans. 2*, 53 (1982).

53. D. B. Northrop, *Anal. Biochem.*, **132**, 457 (1983).

54. P. J. Munson and D. Rodbard, *Anal. Biochem.*, **107**, 220 (1980).

55. S. D. Christian and E. E. Tucker, *Am. Lab.*, September, 1984, p. 112.

56. J. P. Davis and I. I. Schuster, *J. Solution Chem.*, **13**, 167 (1984).

57. R. J. Motekaitis and A. E. Martell, *Can. J. Chem.*, **60**, 168 (1982).

58. C. Mongay, G. Ramis, and M. C. Garcia, *Spectrochim Acta*, **38A**, 247 (1982).

59. (a) F. R. Hartley, C. Burgess, and R. M. Alcock, *Solution Equilibria*, Ellis Horwood, Chicester, 1980, Ch. 5; (b) M. Cromer-Morin, J. P. Scharff, and R. P. Martin, *Analysis*, **10**, 92 (1982); (c) R. J. Motekaitis and A. E. Martell, *Can. J. Chem.*, **60**, 2403 (1982); (d) D. J. Leggett, S. L. Kelly, L. R. Shiue, Y. T. Wu, D. Chang, and K. M. Kadish, *Talanta*, **30**, 579 (1983); (e) M. Meloun and J. Cermak, *Talanta*, **31**, 947 (1984).

60. R. G. Duggleby, *Anal. Biochem.*, **110**, 9 (1981).

PART II

THE METHODS

Experimental methods for the determination of binding constants can be classified as involving either one-phase (homogeneous) systems or two-phase (heterogeneous) systems. One-phase methods include spectrophotometry, reaction kinetics, and potentiometry; in making these assignments neither the experimental probe (e.g., an electrode) nor the containing vessel is counted as a phase of the system. Two-phase methods are exemplified by solubility, liquid–liquid extraction, and chromatography. The two-phase techniques introduce possibilities for complications (e.g., polymorphism in the solubility method, mutual dissolution of solvents in the liquid–liquid extraction method) that are absent in the one-phase techniques.

Another way to consider these methods is as equilibrium or nonequilibrium methods. Even presuming that the association–dissociation rates are much faster than any rates (such as mixing) of the experimental procedure, there is a possibility for the binding equilibrium to be perturbed by the act of observing it. Thus the spectrophotometric method is, in this sense, a nonequilibrium method, for the system is observed by means of the interaction of photons with the components of the system, and it is conceivable that a significant perturbation of the equilibrium may result. As an example of an equilibrium method we have the solubility technique, in which the system is brought to equilibrium with respect to solubility and binding, the solution phase is then physically separated from the excess solute phase, and further analytical manipulations may be carried out without concern for the equilibrium. Equilibrium methods are noninvasive; nonequilibrium methods are invasive.

Yet another division of methods is into the direct and competitive classes. In direct methods we typically allow substrate and ligand to interact and measure a physical or chemical property that changes as a consequence of binding. In competitive methods a binding equilibrium is established between the ligand and an "indicator" substrate. This equilibrium is then perturbed by adding the substrate of interest, and the extent of the perturbation is monitored by following a property of the indicator substrate.

Although the main goal of investigations into binding is the estimation of binding constants, usually some further information is concomitantly obtained, and this additional knowledge may also be of great interest, either because it has independent standing as a physicochemical constant or because it may bear on the interpretation of the binding constant. Among these kinds of additional information are molar absorptivities, chemical shifts, and solubilities of complex species. Of course, from measurements of the binding constants as a function of temperature the enthalpy and entropy of binding also can be obtained.

4

OPTICAL ABSORPTION SPECTROSCOPY

Judging from the literature of the field, optical spectroscopy is the most widely applied experimental method for the study of binding constants. The essential requirement for its use is that a significant spectral change occur as a consequence of complex formation. In this chapter we consider the methods, both direct and competitive, that make use of observations in the ultraviolet, visible, and infrared regions of the spectrum.

4.1. DIRECT ULTRAVIOLET–VISIBLE SPECTROPHOTOMETRY

The Number of Stoichiometric States

Several spectrophotometric procedures have been described to determine the number of species in solution, as the problem is usually expressed. For example, if only a 1:1 complex is formed, it is said that the solution contains two species, namely, S and SL. It is more appropriate to speak of the number of stoichiometric states of the system, for we certainly have additional species (e.g., L), which may themselves absorb, and isomeric complexes are possible. Thus the system containing S, SL, and LS has three substrate species, but two stoichiometric states. Spectrophotometry may give information about the number of states. In some instances the stoichiometric ratio may be accessible by application of the continuous variations or mole ratio methods described in Section 2.1, "Determination of Stoichiometry."

Consider a two-state system whose two states have different absorption spectra. If the spectra, expressed on a molar basis, are superimposed, and if they intersect, then evidently the spectrum of any mixture of the two states, also expressed on a molar basis, will pass through the point of intersection. This point of common absorption intensity is called an isosbestic point. (Spectra can be placed on a molar basis by either plotting molar absorptivity ϵ against wavelength or measuring all spectra at the same total substrate concentation S_t.) Figure 4.1 shows an isosbes-

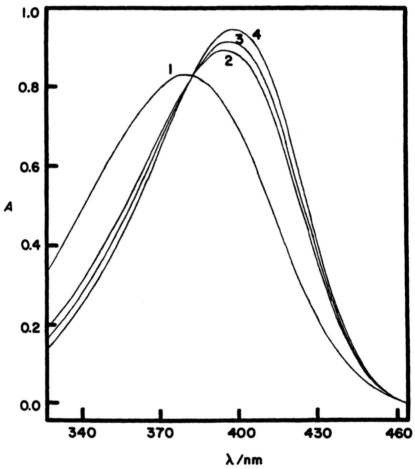

Figure 4.1. Absorption spectra of 4-nitroaniline at various concentrations of α-cyclodextrin; $S_t = 7.11 \times 10^{-5}$ M, pH = 9.18, ionic strength = 0.10 M. Cyclodextrin concentrations: 1, zero; 2, 3.39×10^{-3} M; 3, 5.68×10^{-3} M; 4, 1.67×10^{-2} M.

tic point at 382 nm in the system 4-nitroaniline (substrate): α-cyclodextrin (ligand) (1). Of course, it is possible for a two-state system to exhibit more than one isosbestic point, or even no such point if the spectra nowhere intersect. But if there is an intersection of any two members of such a family of spectra, and if the system possesses just two states, then all possible members must pass through the isosbestic point.

In making use of this spectral property the argument is reversed. If an isosbestic point is observed over a wide range of composition (i.e., corresponding to a substantial portion of the binding isotherm), it is often concluded that the system has just two states; in other words, one complex is formed. The corollary is that if two spectra of a family intersect, but not all of the spectra pass through a common point, the system must possess more than two states, that is, more than one complex stoichiometry. Figure 4.2 shows a set of spectra for the *trans*-cinnamic acid: α-cyclodextrin system (2); evidently there exist regions near 230 and 295 nm where intersections occur, but over the range of ligand concentrations studied isosbestic points are not maintained, suggesting the formation of more than one complex.

Implicit in the preceding arguments is the assumption that the observed spectral shifts are ascribable entirely to the changes in concentrations of the absorbing species as the ligand concentration is changed; that is, it is assumed that the spectral characteristics of the absorbing species are constant, unaffected by the inevitable changes in solution composition. However, it is evident that a system might possess only two states yet fail to exhibit a sharp isosbestic point if a spectral solvent dependence were operative. It is also possible that effects may combine to generate an isosbestic point even though the system possesses more than two states. Such possibilities have been analyzed in detail, and some authors have concluded that neither the presence nor the absence of isosbestic points has any practical value in diagnosing the number of states in the system (3). Nevertheless, a result like that shown in Figure 4.1 would be interpreted by most workers in the simplest manner, namely, that this is a two-state system showing no spectral medium effect. Isosbestic point observations constitute only one piece of evidence in the study of a complexing system, and they can be useful if taken together with other data to generate a consistent description.

Aside from the factors previously mentioned that may control the sharpness of isosbestic point intersections, there is the variability introduced by sample preparation and spectrophotometer performance. Skul-

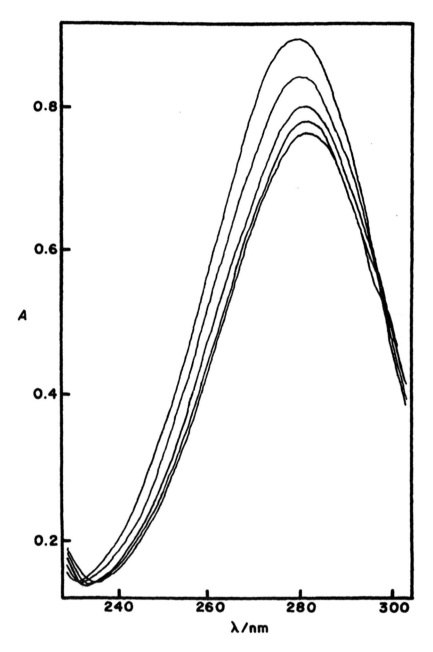

Figure 4.2. Absorption spectra of *trans*-cinnamic acid: α-cyclodextrin solutions; $S_t =$ 4.61×10^{-5} *M*, pH = 2, ionic strength = 0.01 *M*. Cyclodextrin concentrations: 1, zero; 2, 2.0×10^{-4} *M*; 3, 5.0×10^{-4} *M*; 4, 2.0×10^{-2} *M*; 5, 3.0×10^{-2} *M*.

ski and Plucinski (4) have described experimental designs for the sensitive detection of isosbestic points.

The study of isosbestic points obviously generates a great deal of spectral data, which can be represented as a matrix of elements A_{ij}, where each A_{ij} is the absorbance of solution j (characterized by its solution composition) at wavelength i. This matrix contains information about the number of states of the system; indeed, the matrix is simply a representation of the set of absorption spectra. This idea has been quantitatively developed by Coleman et al. (5), who describe graphical methods for testing the matrix to determine the number of states. We outline here a simplified treatment to show the approach.

The most useful conclusion we can reach is to establish whether the system includes one state (i.e., no complexes), two states (a single complex), or more than two states (multiple equilibria). First suppose we have a one-state system that exhibits a spectral change as some alteration is made to the system (change in L_t, pH, temperature). Assume Beer's law is followed. Then for solution j at wavelengths 1 and 2 we write $A_{1j} = \epsilon_{1j}[S]$ and $A_{2j} = \epsilon_{2j}[S]$, where [S] is the molar concentration of absorbing solute and the ϵ_{ij} are molar absorptivities; for convenience we take unit path length. Then we get

$$A_{1j} = \frac{\epsilon_{1j}}{\epsilon_{2j}} A_{2j} \tag{4.1}$$

Now suppose the solution composition (j) is altered, and assume that the ratio $\epsilon_{1j}/\epsilon_{2j}$ is a constant, independent of j. A plot of A_{1j} against A_{2j}, for different j, will be a straight line passing through the origin. This plot can be made for other pairs of wavelengths to give a family of such lines. This behavior indicates a one-state system.

Next consider a two-state system, the only absorbing species being unbound substrate S and complex C. Assuming additivity of absorbances we have

$$A_{1j} = \epsilon_{1j}^{S}[S]_j + \epsilon_{1j}^{C}[C]_j$$
$$A_{2j} = \epsilon_{2j}^{S}[S]_j + \epsilon_{2j}^{C}[C]_j$$

We write analogous equations for solution k, use the conservation equation $[S]_j + [C]_j = [S]_k + [C]_k = S_t$ (where S_t is constant), and make

Table 4.1. Spectral Data for the Nitrazene Yellow : α-Cyclodextrin System[a,b]

i(i')	\multicolumn{6}{c}{$A_{j(k)}$}					
	0	1	2	3	4	5
1	0.903	0.775	0.760	0.728	0.690	0.665
2	1.158	1.000	0.976	0.940	0.895	0.867
3	1.476	1.298	1.270	1.230	1.180	1.145
4	1.777	1.640	1.620	1.590	1.550	1.522
5	1.722	2.016	2.060	2.111	2.193	2.260
6	1.136	1.650	1.728	1.828	1.976	2.082
7	0.623	1.098	1.170	1.260	1.395	1.495
8	0.321	0.625	0.670	0.730	0.818	0.882

[a] From reference 6.
[b] $S_t = 4.85 \times 10^{-5}$ M, pH = 9.2.

the assumption that the spectra are not medium dependent, so $\epsilon_{1j}^{S} = \epsilon_{1k}^{S}$, and so on. Thus the result is

$$(A_{1j} - A_{1k}) = (\epsilon_{1j}^{C} - \epsilon_{1j}^{S})([C]_j - [C]_k)$$
$$(A_{2j} - A_{2k}) = (\epsilon_{2j}^{C} - \epsilon_{2j}^{S})([C]_j - [C]_k)$$

Combining these gives

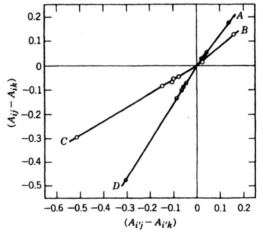

Figure 4.3. Plots of data in Table 4.1 according to Eq. (4.2). Line A, $i = 3$, $i' = 4$; line B, $i = 1$, $i' = 2$; line C, $i = 5$, $i' = 6$; line D, $i = 7$, $i' = 8$.

$$(A_{1j} - A_{1k}) = \frac{\epsilon_{1j}^{C} - \epsilon_{1j}^{S}}{\epsilon_{2j}^{C} - \epsilon_{2j}^{S}} (A_{2j} - A_{2k}) \qquad (4.2)$$

which is analogous to Eq. (4.1). If the ratio of differences of absorptivities is a constant as the medium is changed, plots of $(A_{ij} - A_{ik})$ versus $(A_{i'j} - A_{i'k})$, where $i \neq i'$ and $j \neq k$, will give straight lines passing through the origin if the system has two states.

Table 4.1 lists a matrix of spectral data for the system Nitrazene Yellow (substrate): α-cyclodextrin (ligand), and plots according to Eq. (4.2) are shown in Figure 4.3, which suggests that this is a two-state system. The treatment of Coleman et al. (5) is more general than this and describes additional plotting functions, but the underlying assumptions are the same.

The 1:1 Binding Isotherm

Consider the system in which a single 1:1 complex SL is formed, with the complex and the free substrate S having significantly different absorption spectra. We assume that Beer's law is followed by all species. A wavelength is selected at which the molar absorptivities ϵ_S and ϵ_{11} are different. Then at total concentration S_t of substrate, in the absence of ligand, the solution absorbance is

$$A_0 = \epsilon_S b S_t$$

In the presence of ligand at total concentration L_t, the absorbance of a solution containing the same total substrate concentration is

$$A_L = \epsilon_S b[S] + \epsilon_L b[L] + \epsilon_{11} b[SL]$$

which, combined with the mass balances on S and L, gives

$$A_L = \epsilon_S b S_t + \epsilon_L b L_t + \Delta\epsilon_{11} b[SL]$$

where $\Delta\epsilon_{11} = \epsilon_{11} - \epsilon_S - \epsilon_L$. By measuring the solution absorbance against a reference containing ligand at the same total concentration L_t, the measured absorbance becomes

$$A = \epsilon_S b S_t + \Delta\epsilon_{11} b[SL] \qquad (4.3)$$

Combining Eq. (4.3) with the stability constant definition, $K_{11} = $ [SL]/[S][L], gives

$$\Delta A = K_{11} \, \Delta \epsilon_{11} \, b[S][L] \qquad (4.4)$$

where $\Delta A = A - A_0$. From the mass balance expression $S_t = [S] + [SL]$ we get $[S] = S_t/(1 + K_{11}[L])$, which is used in Eq. (4.4), giving Eq. (4.5) as the relationship between the observed absorbance change per centimeter and the system variables and parameters.

$$\frac{\Delta A}{b} = \frac{S_t K_{11} \, \Delta \epsilon_{11} \, [L]}{1 + K_{11}[L]} \qquad (4.5)$$

Equation (4.5) is the binding isotherm, which shows the hyperbolic dependence on free ligand concentration that is treated in detail in Appendix A and Sections 2.5 and 3.3. Equation (4.5) has been used in illustration of points discussed in Sections 2.3 and 2.6, "Isomerism."

Note that the reference solution contains the same total ligand concentration L_t as the sample solution. Even if the ligand does not absorb at the analytical wavelength, it is good practice to include it in the reference solution so that the reference and sample have the same refractive index.

Now suppose that two isomeric 1:1 complexes, SL and LS, are formed, with stability constants K_{SL} and K_{LS}. The concentration of free substrate is then given by

$$[S] = \frac{S_t}{1 + (K_{SL} + K_{LS})[L]}$$

Development from the Beer's law expressions as for the simple case leads to Eq. (4.6),

$$\frac{\Delta A}{b} = \frac{S_t (K_{SL} \, \Delta \epsilon_{SL} + K_{LS} \, \Delta \epsilon_{LS})[L]}{1 + (K_{SL} + K_{LS})[L]} \qquad (4.6)$$

where $\Delta \epsilon_{SL} = \epsilon_{SL} - \epsilon_S - \epsilon_L$ and $\Delta \epsilon_{LS} = \epsilon_{LS} - \epsilon_S - \epsilon_L$. Equation (4.6) should be compared with Eq. (4.5). Because of the way the parameters appear in these equations, we may write identities as follows:

$$K_{11} = K_{SL} + K_{LS} \qquad (4.7)$$

$$\Delta\epsilon_{11} = \frac{K_{SL}\,\Delta\epsilon_{SL} + K_{LS}\,\Delta\epsilon_{LS}}{K_{SL} + K_{LS}} \qquad (4.8)$$

These results have been discussed in Section 2.6 [Eqs. (2.110), (2.111)]. They can be generalized to any number of 1:1 complexes. Thus the experimentally determined 1:1 binding constant K_{11} is the sum of all 1:1 binding constants, and the molar absorptivity difference $\Delta\epsilon_{11}$ is a weighted average.

It is particularly important to note that Eq. (4.7) applies as long as ΔA in Eq. (4.6) is significantly different from zero. That is, even if one of the complexes has an absorption spectrum identical with that of the free substrate, its binding constant will be included in the spectrally measured K_{11}. Thus it is incorrect to conclude that the spectral method only "sees" or "counts" those complexes that exhibit a spectral difference; the method includes all of them in K_{11} as long as at least one of them produces a significant ΔA. The physical basis of this result is discussed in Section 2.6, "Random Association."

The isotherm, Eq. (4.5), is expressed as a function of free ligand concentration, which is not known a priori. We can relate [L] to the known total ligand concentration L_t by means of the mass balance on ligand, $L_t = [L] + [SL]$ (written here for a single 1:1 complex), or

$$L_t = [L] + \frac{S_t K_{11}[L]}{1 + K_{11}[L]} \qquad (4.9)$$

Equations (4.5) and (4.9) together describe the system; they are specific examples of Eqs. (2.35) and (2.36) in Section 2.3.

Evaluation of Parameters

Many authors have reviewed the evaluation of parameters (7). In this section we restrict attention to the 1:1 case, so Eq. (4.5) is the binding isotherm. The experimental data consist of ΔA values as a function of total ligand concentration L_t, usually at fixed total substrate concentration S_t. [If S_t is not held fixed in the series of measurements, Eq. (4.5) can be rewritten with $\Delta A/b\,S_t$ as the dependent variable.] The fixed wavelength is usually chosen so as to give the largest possible values of ΔA, and the range of L_t should be large enough to cover much of the binding isotherm (say 60–75% of saturation) or as large as is permitted by the experimental situation.

Figure 4.4 shows absorption spectra obtained in a study of a 1:1 binding system (8); the ligand concentrations were chosen to generate comparable increments in absorbance throughout the binding curve. In this example complex formation results in a marked decrease in absorption in the visible region. The isosbestic point at 414 nm is suggestive of a 1:1 system. Table 4.2 gives absorbance values at 508 nm from this set of spectra.

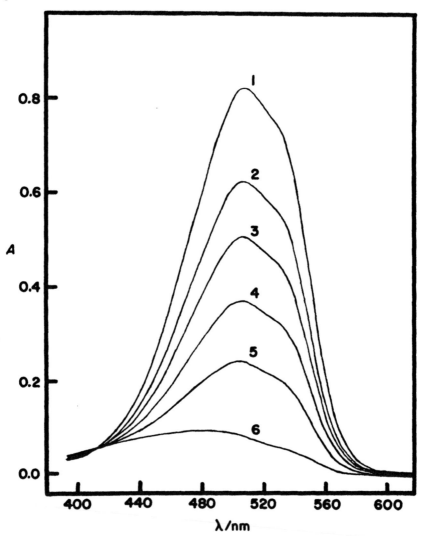

Figure 4.4. Absorption spectra of the methyl orange (substrate): α-cyclodextrin (ligand) system. Conditions are given in Table 4.2.

Table 4.2. Spectral Data for the Methyl Orange: α-Cyclodextrin System (Figure 4.4)[a]

$10^3\ L_t/M$	A_{508}	$-\Delta A_{508}$
(0.000)	0.807	–
0.478	0.617	0.190
0.637	0.571	0.236
0.972	0.495	0.312
1.944	0.360	0.447
3.999	0.234	0.573
20.00	0.077	0.730

[a] $S_t = 1.67 \times 10^{-5}\ M$ in 0.08 M HCl; 25°C; $b = 1$ cm.

A plot of ΔA against [L] is the direct binding isotherm plot. The raw data, however, consist of A as a function of L_t. Figure 4.5 shows a plot of A against L_t, at two values of S_t, for the system methyl *trans*-cinnamate (S):8-chlorotheophylline (L) (9).

We have seen that Eqs. (4.5) and (4.9) describe the spectral data for a 1:1 complexing system. In Sections 2.3, 2.5, 3.2, 3.3, and 3.4 the solution of such equations has been discussed in detail, and it is not necessary to add much more. It may be helpful, however, to continue the presentation of actual data, and a short survey of the extensive literature in this area will also be given.

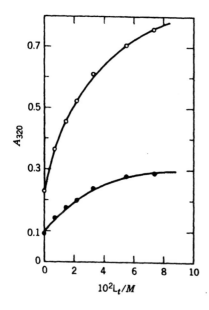

Figure 4.5. Spectral data for the methyl *trans*-cinnamate:8-chlorotheophylline system; pH 8.47, 25°C. Upper curve, $S_t = 6.16 \times 10^{-4}\ M$; lower curve, $S_t = 2.38 \times 10^{-4}\ M$.

The double-reciprocal form of plotting the rectangular hyperbola [Eq. (2.90)] is based on the linearization of Eq. (4.5) according to Eq. (4.10).

$$\frac{b}{\Delta A} = \frac{1}{S_t K_{11} \, \Delta\epsilon \, [L]} + \frac{1}{S_t \, \Delta\epsilon_{11}} \qquad (4.10)$$

Equation (4.10) is called the Benesi–Hildebrand equation (10). A rigorous application of the equation requires evaluation of [L] by one of the methods of Section 2.3. We shall treat the data of Table 4.2; in this system L_t was much larger than S_t, and it is appropriate to set $[L] = L_t$. Figure 4.6 is the Benesi–Hildebrand plot of these data. The parameters are evaluated from this plot according to Eqs. (4.11) and (4.12).

$$K_{11} = \frac{(y\text{-intercept})_{4.10}}{(\text{slope})_{4.10}} \qquad (4.11)$$

$$\Delta\epsilon_{11} = \frac{1}{S_t (y\text{-intercept})_{4.10}} \qquad (4.12)$$

where the subscripts identify the plotting form. Alternatively, K_{11} is equal to the negative of the intercept on the x axis.

The y-reciprocal form (Eq. 2.91) is given in Eq. (4.13).

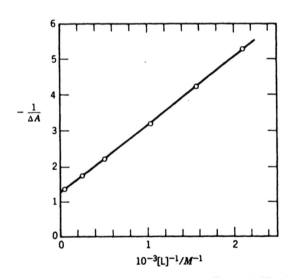

Figure 4.6. Double-reciprocal plot of the data in Table 4.2 [Eq. (4.10)].

$$\frac{b[L]}{\Delta A} = \frac{[L]}{S_t \Delta \epsilon_{11}} + \frac{1}{S_t K_{11} \Delta \epsilon_{11}} \qquad (4.13)$$

This is sometimes called the Scott equation (11); a plot of the data in Table 4.2 according to Eq. (4.13) is shown in Figure 4.7. The parameters are given by Eqs. (4.14) and (4.15).

$$K_{11} = \frac{(\text{slope})_{4.13}}{(y\text{-intercept})_{4.13}} \qquad (4.14)$$

$$\Delta \epsilon_{11} = \frac{1}{S_t (\text{slope})_{4.13}} \qquad (4.15)$$

Also, K_{11} is given by the negative of the reciprocal of the intercept on the x axis. Note the spacing of the points in Figures 4.6 and 4.7; this experiment was designed to yield fairly regular spacing in the double-reciprocal plot, so the spacing is not optimal in the y-reciprocal plot.

The x-reciprocal form (Eq. 2.92) is written

$$\frac{\Delta A}{b[L]} = -\frac{K_{11} \Delta A}{b} + S_t K_{11} \Delta \epsilon_{11} \qquad (4.16)$$

and the plot is shown in Figure 4.8. The parameters are evaluated with Eqs. (4.17) and (4.18).

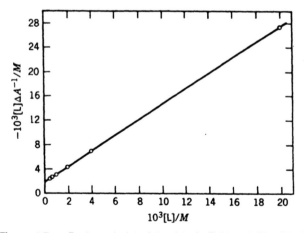

Figure 4.7. y-Reciprocal plot of the data in Table 4.2 [Eq. (4.13)].

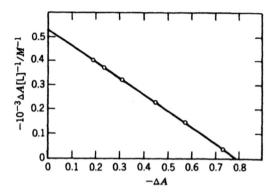

Figure 4.8. x-Reciprocal plot of the data in Table 4.2 [Eq. (4.16)].

$$K_{11} = -(\text{slope})_{4.16} \tag{4.17}$$

$$\Delta\epsilon_{11} = \frac{(x\text{-intercept})_{4.16}}{S_t} \tag{4.18}$$

This plot was apparently first described by Foster et al. (12) in the context of spectroscopy; it is also called a Scatchard plot (13).

As can be seen from the distribution of points in these plots, an experiment designed to be analyzed by one plotting form may not be optimized for one of the other plots. The double-reciprocal plot gives (8), for this complexing system, $K_{11} = 673\ M^{-1}$ and $\Delta\epsilon_{11} = -4.72 \times 10^4\ M^{-1}\,cm^{-1}$.

Figure 4.9 is the semilogarithmic plot of the Table 4.2 data. With the symbolism of Section 2.4 we write f_{10} and f_{11} as the fractions of substrate present as free substrate S and complex SL, respectively. From Eq. (2.57) we have

$$f_{11} = \frac{K_{11}[L]}{1 + K_{11}[L]} \tag{4.19}$$

Combining this with Eq. (4.5) yields $\Delta A/b = S_t\,\Delta\epsilon_{11}\,f_{11}$, showing that ΔA is directly proportional to f_{11}. Thus Figure 4.9 is essentially a plot of f_{11} against $pL = -\log[L]$. This is analogous to a fractional distribution diagram in acid–base chemistry. By taking the second derivative $d^2 f_{11}/d\,pL^2$ and setting this equal to zero (14) we find

$$pK_{11} = -pL_{\text{infl}} \tag{4.20}$$

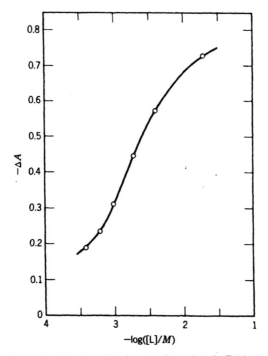

Figure 4.9. Semilogarithmic plot of the data in Table 4.2.

Thus K_{11} can be estimated by locating the inflection point in the semilog plot. (The methods devised in titrimetric analysis for locating end points may be useful; the first derivative is a maximum and the second derivative is zero at this point.)

The fractional distribution diagram is also useful in displaying the range of the binding isotherm that has been investigated experimentally. K_{11} having been evaluated, f_{11} can be calculated as a function of pL by means of Eq. (4.19). Figure 4.10 shows the sigmoid curves of f_{10} and f_{11} for the methyl orange: α-cyclodextrin system that has been used throughout this discussion as an example. Superimposed on the f_{11} curve are the points corresponding to the experimental pL values, and from this presentation the range of the degree of saturation that has been studied is immediately obvious. Note that at the intersection of the two curves $f_{10} = f_{11}$, so [S] = [SL] and $K_{11} = 1/[L]$.

There is a considerable literature on the spectroscopic method of studying stability constants, and some amplification of the preceding

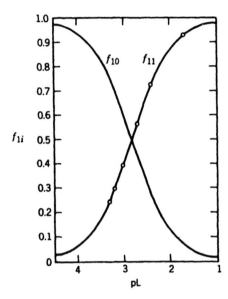

Figure 4.10. Fractional distribution curves for the methyl organge: α-cyclodextrin sysem.

treatment is necessary to appreciate the basis for much of this literature. We have seen that Eqs. (4.5) and (4.9) together provide a complete description of the spectral study of a $1:1$ complexing system (aside from the question of activity coefficient effects). Historically, however, the problem has been approached by solving special cases of the general question. For example, the Benesi–Hildebrand equation was first applied to a system in which the simplifications $[L] = L_t$ and $\Delta\epsilon = \epsilon_{11}$ could be written, the resulting equation [see Eq. (4.10)] then being

$$\frac{b}{A} = \frac{1}{S_t K_{11}\epsilon_{11}L_t} + \frac{1}{S_t\epsilon_{11}}$$

Subsequent workers referred to this form as the Benesi–Hildebrand equation and developed more general expressions. Much of the literature consists of such developments. The view adopted here is that it is more efficient to begin with the general expression for the binding isotherm, Eq. (4.5), and to introduce simplifications as appropriate. Moreover, the essence of the Benesi–Hildebrand treatment is the double-reciprocal plotting form, not the particular assumptions first made by Benesi and Hildebrand (10). Earlier reviews (7) of this field provide access to this portion of the literature. In Section 2.3 we treated methods for finding the free ligand concentration [L] when the assumption $[L] = L_t$ is not justified; reference (29) of Chapter 3 leads to versions that have been developed for spectroscopic studies, and Eq. (2.41) gives an expression for [L] based on a Taylor's series expansion.

There are some methods of data treatment that are not obvious variations of Eqs. (4.10), (4.13), or (4.16). The method of Rose and Drago (15) is a graphical technique that does not require the assumption $[L] = [L_t]$. Combining Eqs. (4.5) and (4.9) gives $[L] = L_t - \Delta A / \Delta \epsilon_{11}$, which is substituted into Eq. (4.5) to give

$$\frac{1}{K_{11}} = \frac{\Delta A}{\Delta \epsilon_{11}} - S_t - L_t + \frac{S_t L_t \Delta \epsilon_{11}}{\Delta A} \qquad (4.21)$$

Equation (4.21) is an example of procedure 3 in Section 2.3. Now for a given experimental $\Delta A, L_t$ pair, a range of plausible $\Delta \epsilon_{11}$ values is substituted into Eq. (4.21), which is solved for the corresponding K_{11}^{-1}. A plot is made of K_{11}^{-1} against $\Delta \epsilon_{11}$. This calculation is repeated for additional data points. In a well-behaved system the family of lines thus generated will intersect at the point corresponding to the actual K_{11}^{-1} and $\Delta \epsilon_{11}$ for the system. In practice, of course, the several lines produce a dispersion of intersections (15, 16), and the method is not always successful. Nevertheless, this scatter of points of intersection is an indication of the uncertainty of the parameter estimates and may also be helpful in detecting unreliable data points (15).

Liptay (17) described a method for assessing data reliability and adherence to the 1:1 stoichiometric model in which absorbance measurements are made at several wavelengths and a matrix of these absorbances, normalized with respect to a reference wavelength, is examined for consistency and variability.

In Sections 2.5 and 3.4 the advantages and drawbacks of the several plotting forms of the 1:1 binding isotherm were described in general terms, and we now can comment on features that are particular to spectroscopic studies. Many workers have noted that linear plotting of spectral data yields estimates of the product $K_{11} \Delta \epsilon_{11}$ that are more reliable than are the separate parameters K_{11} and $\Delta \epsilon_{11}$, especially for weak complexes. The use of the double-reciprocal plot, Eq. (4.10), is typical. The product $K_{11} \Delta \epsilon_{11}$ is obtained from the slope, whereas the intercept gives $\Delta \epsilon_{11}$, and K_{11} is obtained by a combination of these results, as shown in Eq. (4.11). In the other linear forms, however, the product $K_{11} \Delta\epsilon_{11}$ is obtained from an intercept. In linear plotting, a slope is usually estimated more reliably than is an intercept (unless the slope is close to zero or infinity), because an intercept requires extrapolation, often over long distances; moreover, an intercept close to the origin may have a high relative error. It is therefore helpful to make use of more than one of the linear plots [according to Eqs. (4.10), (4.13), (4.16)], seeking estimates of K_{11} and $\Delta\epsilon_{11}$ that are consistent with all of the plots and placing emphasis on the graphical features that are most reliably estimated. This approach may require separate data sets (obtained at different ligand concentrations), each optimized for its plotting form. The lines should be drawn by linear least-squares regression, with weighting carried out as described in Sections 3.2 and 3.3.

There are several other ways to obtain K_{11} and $\Delta\epsilon_{11}$ individually without proceeding by means of their product. The sigmoid semilog plot yields K_{11} through Eq. (4.20) if the isotherm is studied through enough of its range to detect the inflection point. Nagakura (18) has used the following method to obtain K_{11} independently of $\Delta\epsilon_{11}$. We measure ΔA at fixed S_t but at two levels of L_t, say L_t and L'_t. Equation (4.5) is then written $\Delta A = S_t f_{11} \Delta\epsilon_{11}$ and $\Delta A' = S_t f'_{11} \Delta\epsilon_{11}$, or $\Delta A f'_{11} = \Delta A' f_{11}$. Using Eq. (4.19) and solving for K_{11} yields

$$K_{11} = \frac{\Delta A' [L] - \Delta A [L]'}{(\Delta A - \Delta A')[L][L]'} \qquad (4.22)$$

Nagakura set $[L] = L_t$ and $[L]' = L'_t$ to find K_{11} by this method. Seal et al. (19) have generalized this method, obtaining a cubic equation in K_{11} that is solved numerically.

The method of corresponding solutions is based on this idea: Let the complexing system be studied spectrophotometrically at two sets of conditions of total substrate and ligand concentrations, adjusted such that the absorption spectra, on a molar basis, of the two solutions are identical. Now fix attention on a single wavelength, and write Eq. (4.5) for the two solutions as $\Delta A = S_t f_{11} \Delta\epsilon_{11}$ and $\Delta A' = S'_t f'_{11} \Delta\epsilon_{11}$. We have stipulated that $\Delta A/S_t = \Delta A'/S'_t$, hence $f_{11} = f'_{11}$, and $[L] = [L]'$. The solutions are said to be corresponding solutions. Now $f_{11} = [SL]/S_t = (L_t - [L])/S_t = (L'_t - [L])/S'_t$, or

$$f_{11} = \frac{L_t - L'_t}{S_t - S'_t} \qquad (4.23)$$

and

$$[L] = L_t - f_{11}S_t \qquad (4.24)$$

With Eqs. (4.23) and (4.24), f_{11} and $[L]$ are calculated for this pair of solutions. This procedure is repeated for other pairs of corresponding solutions. With f_{11} known as a function of $[L]$, K_{11} is readily found [see Eq. (4.19)]. The condition of corresponding solutions is located by plotting $\Delta A / S_t$ against L_t for two values of S_t, and reading off L_t values at equal $\Delta A / S_t$. Although the development here was for a 1:1 complex, for which $f_{11} = \bar{i}$, the mean number of ligands bound per substrate, the technique has more potential for the study of multiple equilibria, when $\bar{i}$ is determined by all of the complexes present (20).

It is possible to find $\Delta\epsilon_{11}$ independently of K_{11} (19, 21). Solving Eq. (4.5) for K_{11} gives (for $b = 1$ cm)

$$K_{11} = \frac{\Delta A}{[L](S_t\, \Delta\epsilon_{11} - \Delta A)}$$

For another set of initial conditions S_t' and L_t' we measure $\Delta A'$ and write a similar equation. These are set equal to eliminate K_{11}, and the identities $[L] = L_t - \Delta A / \Delta\epsilon_{11}$, $[L]' = L_t' - \Delta A' / \Delta\epsilon_{11}$ (derived in the development of the Rose–Drago method) are used, to yield the following quadratic equation in $\Delta\epsilon_{11}$:

$$\begin{aligned}
(\Delta A\, L_t'S_t' - \Delta A'\, L_tS_t)(\Delta\epsilon_{11})^2 \\
+ \Delta A\, \Delta A'\, (S_t - S_t' + L_t - L_t')\, \Delta\epsilon_{11} \\
+ \Delta A\, \Delta A'\, (\Delta A' - \Delta A) = 0 \qquad (4.25)
\end{aligned}$$

Minor rearrangements of the y-reciprocal [Eq. (4.13)] and x-reciprocal [Eq. (4.16)] equations allow the product $K_{11}\, \Delta\epsilon_{11}$ to be separated into its factors as follows (22). These equations essentially consist of interchanging the independent and dependent variables in Eqs. (4.13) and (4.16).

$$[L] = \frac{S_t\, \Delta\epsilon_{11}\, b[L]}{\Delta A} - \frac{1}{K_{11}} \qquad (4.26)$$

$$\frac{\Delta A}{b} = -\frac{\Delta A}{bK_{11}[L]} + S_t\, \Delta\epsilon_{11} \qquad (4.27)$$

Nonlinear least-squares regression analysis using Eqs. (4.5) and (4.9) is an alternative to the preceding methods. The principles of nonlinear regression are dealt with in Appendix B and Chapter 3, and an example is given in Section 3.3 of nonlinear regression applied to spectroscopic data. In simple 1:1 binding systems, linear and nonlinear treatments should be equally successful. The nonlinear regression requires preliminary parameter estimates, and the linear methods are valuable means for obtaining these.

We have stated that Eqs. (4.5) and (4.9) give an exact description (aside from activity coefficient effects on K_{11}) of a 1:1 complexing system. Implicit in this statement, however, is the assumption that $\Delta \epsilon_{11}$ is a constant, independent of the medium changes that inevitably occur as L_t is altered in the spectral study. Medium effects acting on stability constants and absorptivities were discussed in Section 2.6, "Medium Effects." There appears to be evidence for significant changes in absorptivities in weakly complexing systems, which require a large change in the composition of the solution (because L_t must be large to form a significant fraction of complex). Such effects may be manifested as deviations from sharply defined isosbestic points, nonlinearity in the 1:1 linear plotting forms, or inconstancy of K_{11} estimated at different wavelengths. Unfortunately, all these phenomena would also be produced by a system containing complexes of higher stoichiometry, so there is ambiguity in interpreting such observations. In practice the magnitudes of the deviations are important, and interpretation is on an individual case basis. If K_{11} is large so that the range of L_t studied is small, it is more reasonable to interpret deviations from simple 1:1 behavior in terms of higher complexes than when K_{11} is small and L_t must extend to substantial fractions of the solvent composition.

The choice of concentration scale most appropriate to spectroscopic studies has been discussed at length in the literature; see Sections 2.2 and 2.6, "Medium Effects," and reference 28 in Chapter 2. Note that according to Beer's law, $A = abc$, the absorptivity a has the units $(\text{concentration})^{-1} \cdot (\text{length})^{-1}$, so the magnitude of the absorptivity depends on the concentration scale used in its evaluation. When the molar scale is used, the absorptivity is called the molar absorptivity, and is designated ϵ. The units of K_{11} and $\Delta \epsilon_{11}$ must be consistent with Eq. (4.5), which shows that K_{11} will have the reciprocal unit in which [L] is expressed, and $\Delta \epsilon_{11}$ will have the reciprocal unit in which S_t is expressed.

Multiple Equilibria

Let us first consider the kinds of evidence that might lead us to examine a system for the presence of complexes having stoichiometric ratios other than 1:1. It is common practice to attempt, first, to treat a system as a simple 1:1 case, and to evaluate K_{11}. Evidently nonlinear plots made according to Eqs. (4.10), (4.13), or (4.16) indicate that the system is not simply 1:1. This is not a very sensitive criterion, however (especially with the double-reciprocal form), as Johnson and Bowen (23) have shown. Multiple spectral intersections instead of a precise isosbestic point, as seen in Figure 4.2, are another indication of complications; yet the spectral characteristics of the system may generate a good isosbestic point even in the presence of multiple complexes, as shown by Halpern and Weiss (24) for the iodine:tetramethyl-1,6-hexanediamine system. If the system is purely 1:1, then the value of K_{11} should be independent of wavelength, so a dependence on wavelength is good evidence for higher complexes (23, 25). A concentration dependence of K_{11} is also suggestive, as is a significant difference in K_{11} values measured by different experimental methods (25). There may be causes other than higher-order complexes for these kinds of observations—nonideal behavior and self-association are possibilities—but multiple equilibria are often responsible (26).

Spectroscopy is not in general a very powerful means for studying multiple equilibria. This is because the analytical response (absorbance) is not a direct measure of extent of binding, but rather is proportional to it. As a consequence, each stoichiometric state in the system adds two unknown parameters—a binding constant and a molar absorptivity—to the isotherm, so the number of parameters is twice the number of complex states. For this reason we consider a system containing only two complex states, namely, 1:1 (SL) and 1:2 (SL_2). The development follows the lines of the treatment given earlier for the 1:1 system, the result being

$$\frac{\Delta A}{b} = \frac{S_t(\beta_{11}\,\Delta\epsilon_{11}\,[L] + \beta_{12}\,\Delta\epsilon_{12}\,[L]^2)}{1 + \beta_{11}[L] + \beta_{12}[L]^2} \qquad (4.28)$$

$$L_t = [L] + \frac{S_t(\beta_{11}[L] + 2\beta_{12}[L]^2)}{1 + \beta_{11}[L] + \beta_{12}[L]^2} \qquad (4.29)$$

where $\beta_{11} = K_{11}$ and $\beta_{12} = K_{11}K_{12}$. These equations may be compared with Eqs. (4.5) and (4.9). The best way to evaluate the parameters is by nonlinear regression analysis of Eq. (4.28) as outlined in Chapter 3, the free ligand concentration being obtained with Eq. (4.29) by methods described in Section 2.3. However, nonlinear regression requires preliminary estimates of the parameters, so we consider this problem. We work in the spirit of the methods developed for multiple equilibria in Section 2.6, in which extrapolations are based on the principles embodied in Eq. (4.28), the theoretical description of the system.

We define $\Delta A / b\, S_t = Z$ (an apparent molar absorptivity). Then from Eq. (4.28) we find that the slope of the plot of Z against [L] at [L] = 0 is $\beta_{11}\, \Delta\epsilon_{11}$. Equation (4.28) can also be written

$$Z = \Delta\epsilon_{12} + \frac{(\Delta\epsilon_{11} - \Delta\epsilon_{12})\beta_{11}[L] - \Delta\epsilon_{12}}{1 + \beta_{11}[L] + \beta_{12}[L]^2}$$

As [L] $\rightarrow \infty$, this becomes

$$Z \rightarrow \Delta\epsilon_{12} + \frac{(\Delta\epsilon_{11} - \Delta\epsilon_{12})\beta_{11}}{\beta_{12}[L]} \qquad (4.30)$$

Thus a plot of Z against 1/[L] gives $\Delta\epsilon_{12}$ as the intercept when 1/[L] = 0. Equation (4.28) can be rearranged to

$$\frac{1}{[L]} - \frac{\beta_{11}\, \Delta\epsilon_{11}}{Z} = \beta_{12}[L]\left(\frac{\Delta\epsilon_{12}}{Z} - 1\right) - \beta_{11} \qquad (4.31)$$

With the prior estimates of $\beta_{11}\, \Delta\epsilon_{11}$ and $\Delta\epsilon_{12}$, a plot is made according to Eq. (4.31), the slope and intercept yielding estimates of β_{12} and β_{11}. All these plots require [L] values, which can often be approximated by L_t. If this approximation is clearly invalid, [L] must be estimated with the use of Eq. (4.29) and β_{11}, β_{12} estimates obtained from experience on related systems.

The parameters estimated can now be used as initial values in a nonlinear regression, or they can be further refined by rearranging Eq. (4.28) to

$$Z(1 + \beta_{11}[L] + \beta_{12}[L]^2)[L]^{-1} = \beta_{11}\, \Delta\epsilon_{11} + \beta_{12}\, \Delta\epsilon_{12}[L] \qquad (4.32)$$

From a plot of the left side against [L] new estimates of $\beta_{11}\, \Delta\epsilon_{11}$ and $\Delta\epsilon_{12}$

are obtained. These can now be used iteratively in the plot according to
Eq. (4.31) to refine the β_{11} and β_{12} estimates.

Let us apply these methods to the experimental results in Table 4.3 for
the system *trans*-cinnamic acid (S): α-cyclodextrin (L). Figure 4.11 is the
semilog plot of these data, and they clearly show that the system cannot
be described solely as a 1:1 system; compare Figure 4.11 with Figure 4.9.
Figure 4.12 is the double-reciprocal plot. The marked curvature in this
plot, and the maximum in the semilog plot, result because $\Delta\epsilon_{12}$ is smaller
than $\Delta\epsilon_{11}$ and cannot be generally anticipated in a system with multiple
equilibria.

In analyzing this system the most reasonable model to adopt is the
1:1 + 1:2 model, because $L_t > S_t$ and because it is chemically reasonable,
the substrate having two potential binding sites (the side chain and the
phenyl group). Figure 4.13 is the plot of $\Delta A/S_t$ against L_t. Obviously,
some guesswork is needed to sketch the slope at $L_t = 0$, but this is not a
serious drawback because this is only the first step in an iterative process.
From this slope we estimate $\beta_{11}\,\Delta\epsilon_{11} = 1.2 \times 10^7\ M^{-2}\ \text{cm}^{-1}$. Figure 4.14 is
the plot according to Eq. (4.30); it yields from the intercept the estimate

Table 4.3. Spectral Data for the *trans*-Cinnamic Acid: α-Cyclodextrin System[a,b]

$10^2\ L_t/M$	ΔA[c]
0.0245	0.155
0.0324	0.168
0.0597	0.233
0.0978	0.263
0.158	0.300
0.309	0.317
0.479	0.322
1.00	0.307
2.00	0.275[d]
10.0	0.203[d]

[a] From reference 2.
[b] $S_t = 8.48 \times 10^{-5}\ M$; 25.0°C; pH 2.2.
[c] 279 nm; $b = 1$ cm.
[d] These ΔA values were corrected for a
general medium effect as determined using
an equivalent weight/volume concentration
of α-methylglucoside.

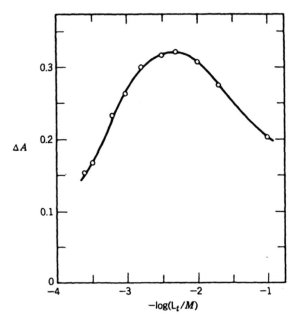

Figure 4.11. Semilogarithmic plot of data in Table 4.3.

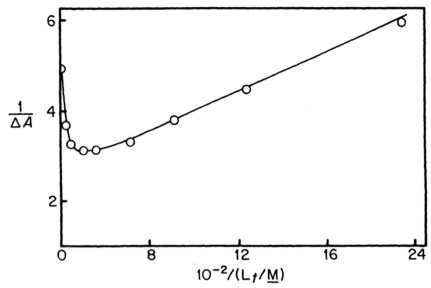

Figure 4.12. Double-reciprocal plot of data in Table 4.3.

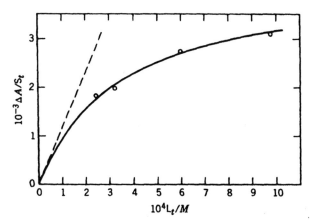

Figure 4.13. Graphical estimation of β_{11}, $\Delta\epsilon_{11}$ for the data in Table 4.3.

$\Delta\epsilon_{12} = 2 \times 10^3\ M^{-1}\ cm^{-1}$. With these estimates the plot of Eq. (4.31) is made, as shown in Figure 4.15. From this plot we estimate $\beta_{11} = 2580\ M^{-1}$ and $\beta_{12} = 1.25 \times 10^5\ M^{-2}$. These quantities are then used to construct a plot of Eq. (4.32), seen in Figure 4.16, from which we find $\beta_{11}\Delta\epsilon_{11} = 1.15 \times 10^7\ M^{-2}\ cm^{-1}$ and $\Delta\epsilon_{12} = 1.98 \times 10^3\ M^{-2}\ cm^{-1}$. In this way reasonable initial parameter estimates are obtained for use in a nonlinear regression analysis.

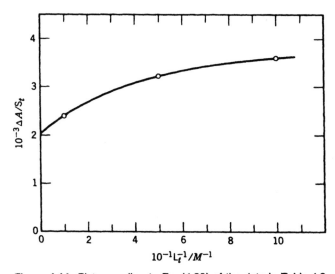

Figure 4.14. Plot according to Eq. (4.30) of the data in Table 4.3.

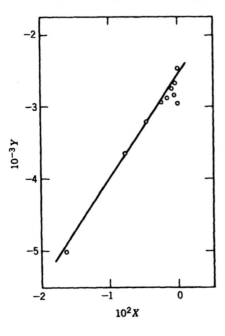

Figure 4.15. Plot according to Eq. (4.31) of the data in Table 4.3. X and Y are plotting variables defined by Eq. (4.31).

Other extrapolation schemes and methods for extracting parameters have been described, and the preceding treatment can be generalized (27). Yet another way to estimate the parameters is to combine data from more than one experimental technique. This is the way the data of Table 4.3 were originally analyzed (28); they were subjected to curve-fitting with the constraint that β_{11} and β_{12} had also to satisfy independently generated solubility data.

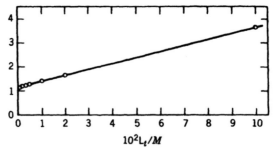

Figure 4.16. Plot according to Eq. (4.32); the vertical axis gives values of the left side of Eq. (4.32) in units of 10^7.

The method of corresponding solutions offers a means for evaluating a binding function that does not include the molar absorptivities. This method was described earlier for a $1:1$ complexing system. If two solutions of the same system $(1:1 + 1:2$ in the present case) have identical absorption spectra on a molar basis, at all wavelengths, they are called corresponding solutions. The fractional distribution of species in this system is completely determined by the free ligand concentration [see Eq. (2.55)], so if the solutions have identical apparent molar absorptivities, their free ligand concentrations are identical. Let one solution have total substrate and ligand concentrations S_t and L_t, and the other, S'_t and L'_t. With Eq. (2.46) we define

$$\bar{i} = \frac{L_t - [L]}{S_t} = \frac{L'_t - [L]}{S'_t}$$

Solving for both $\bar{i}$ and $[L]$ gives

$$\bar{i} = \frac{L_t - L'_t}{S_t - S'_t} \tag{4.33}$$

$$[L] = \frac{L_t S'_t - L'_t S_t}{S'_t - S_t} \tag{4.34}$$

Thus $\bar{i}$ can be evaluated as a function of $[L]$, and the stability constants evaluated by methods already treated in Section 2.6. Evidently, $[L]$ and L_t must be significantly different in this method.

If the absorption spectra of the several species do not overlap at all wavelengths, it may be possible to measure a species concentration directly, and thus to simplify the binding function. An example is provided by Halpern and Weiss (24), who studied the complexing of iodine (S) with several diamines (L). In this system $1:1$ and $2:1$ (S_2L) complexes are formed, and the concentration of free iodine, $[S]$, can be measured spectrophotometrically. With the symbolism of Section 2.4 we define the average number of S molecules bound per molecule of L:

$$\bar{h} = \frac{S_t - [S]}{L_t} \tag{4.35}$$

Combination with the stability constant expressions yields

$$\bar{h} = \frac{K_{11}[S] + 2K_{11}K_{21}[S]^2}{1 + K_{11}[S] + K_{11}K_{21}[S]^2} \tag{4.36}$$

Since [S] is directly measured, $\bar{h}$ can be found as a function of [S]. Rearrangement of Eq. (4.36) gives this linear plotting form:

$$\frac{\bar{h}}{(1-\bar{h})[S]} = K_{11} + K_{11}K_{21}\frac{[S](2-\bar{h})}{(1-\bar{h})} \tag{4.37}$$

Equation (4.37) is a special case of Eq. (2.102), modified to express $\bar{h}$ as a function of [S] rather than $\bar{i}$ as a function of [L].

Equation (4.28) is an isotherm for a $1:1 + 1:2$ system expressed in [L], and Eq. (4.36) is an isotherm for a $1:1 + 2:1$ system expressed in [S]; hence these equations are functionally the same. The $1:1 + 2:1$ system can also be described as a function of [L] by means of a development like that leading to Eq. (4.5). The result is

$$\frac{\Delta A}{b} = \frac{S_t(K_{11}\Delta\epsilon_{11}[L] + K_{11}K_{21}\Delta\epsilon_{21}[S][L])}{1 + K_{11}[L] + 2K_{11}K_{21}[S][L]} \tag{4.38}$$

where $\Delta\epsilon_{21} = \epsilon_{21} - 2\epsilon_s - \epsilon_L$. Comparing Eq. (4.38) with Eqs. (4.5) and (4.28), we see that the apparent molar absorptivity $\Delta A/b\,S_t$ is a function of substrate concentration in Eq. (4.38), but not in the other expressions. This dependence provides an experimental test for higher-order complexes in S. This test must be made carefully, because it is valid only if the apparent molar absorptivities are compared (for two values of S_t) at the same free ligand concentration [L].

4.2. VARIATIONS

The Solubility-Spectral Method

In the spectral method as ordinarily carried out, and as described in Section 4.1, the *total* substrate concentration S_t is held constant while ligand concentration L_t is varied. By carrying out the spectroscopic measurements on solutions saturated with respect to the substrate, the *free* substrate concentration [S] can be held constant, and this procedure allows the complex molar absorptivity to be separately evaluated. The method is developed here for a $1:1$ complex (29, 30). The Beer's law expression for the absorbance of the solution is

$$A_L = \epsilon_s b[S] + \epsilon_L b[L] + \epsilon_{11}b[SL]$$

But the system contains solid substrate at solubility equilibrium, thus maintaining free substrate concentration (actually activity) constant at $[S] = s_0$. Using this substitution with the mass balance on L, gives

$$A_L = \epsilon_s b s_0 + \epsilon_L b L_t + (\epsilon_{11} - \epsilon_L)[SL]$$

By measuring the absorbance against a reference solution containing ligand at concentration L_t we get

$$A = \epsilon_s b s_0 + (\epsilon_{11} - \epsilon_L)[SL]$$

But the mass balance on substrate is $S_t = s_0 + [SL]$, so

$$A = \epsilon_s b s_0 + (\epsilon_{11} - \epsilon_L)(S_t - s_0) \tag{4.39}$$

Alternatively, we can write

$$A = \epsilon_s b s_0 + (\epsilon_{11} - \epsilon_L) K_{11} s_0 [L] \tag{4.40}$$

The difference $(S_t - s_0)$ at given L_t is available from solubility measurements, so Eq. (4.39) provides a means for the evaluation of ϵ_{11} by plotting A against $(S_t - s_0)$ at different L_t. According to Eq. (4.40), a plot of A against $[L]$ gives the product $(\epsilon_{11} - \epsilon_L) K_{11}$, so K_{11} is obtained by combining the results of these two plots.

Measurements of Acid–Base Strength

A weak acid can be viewed as a molecular complex, the conjugate base being the free substrate and the proton being the ligand. The strength of the acid is conventionally expressed as the acid dissociation constant K_a, usually in aqueous solution; this is therefore the reciprocal of a binding constant. The spectrophotometric determination of K_a is simple for monoprotic acids whose conjugate acid and base species have different absorption spectra. Let HA represent the weak acid and A^- the conjugate base (though the charge types are irrelevant to this treatment). At the analytical wavelength, selected to give a substantial difference in light absorption by the two species, we write

$$A_{HA} = \epsilon_{HA} b [HA]$$

$$A_{A^-} = \epsilon_{A^-} b [A^-]$$

For any solution containing both HA and A^- the observed absorbance is given by $A = A_{HA} + A_{A^-}$, or

$$A = \epsilon_{HA} b [HA] + \epsilon_{A^-} b [A^-]$$

Now define the apparent molar absorptivity ϵ_{obs} in terms of the total (formal) concentration of substrate, thus

$$A_{obs} = \epsilon_{obs} b S_t$$

where $S_t = [HA] + [A^-]$. Since $A = A_{obs}$, they can be set equal and combined with the expression for S_t to give

$$\frac{[A^-]}{[HA]} = \frac{\epsilon_{obs} - \epsilon_{HA}}{\epsilon_{A^-} - \epsilon_{obs}} \qquad (4.41)$$

for the case in which $\epsilon_{A^-} > \epsilon_{HA}$. If $\epsilon_{HA} > \epsilon_{A^-}$, the equivalent form is

$$\frac{[A^-]}{[HA]} = \frac{\epsilon_{HA} - \epsilon_{obs}}{\epsilon_{obs} - \epsilon_{A^-}}$$

The apparent acid dissociation constant $K_{a'}$ is defined

$$K_{a'} = \frac{a_{H^+}[A^-]}{[HA]} \qquad (4.42)$$

where a_{H^+} is the activity of the hydronium ion as measured potentiometrically with a pH meter. Combining the logarithmic form of Eq. (4.42) with Eq. (4.41),

$$pK_{a'} = pH - \log \frac{\epsilon_{obs} - \epsilon_{HA}}{\epsilon_{A^-} - \epsilon_{obs}} \qquad (4.43)$$

Equation (4.43) is the basis for the spectrophotometric determination of $pK_{a'}$. Correction to the thermodynamic pK_a requires consideration of the charge type and application of the Debye–Hückel equation.

To use Eq. (4.43) it is necessary to measure ϵ_{HA} and ϵ_{A^-} in solutions containing essentially pure HA and pure A^-, respectively; these are prepared by adjusting the pH to at least two units below (ϵ_{HA}) and above (ϵ_{A^-}) the pK'_a. Then ϵ_{obs} is measured in solutions buffered to pH values in the range $pK'_a \pm 1$, for in this range both species, HA and A^-, are present in comparable concentrations. If the total concentration S_t is held constant in all of these measurements, the absorbances A_{HA}, A_{A^-}, and A_{obs} may be substituted for the absorptivities ϵ_{HA}, ϵ_{A^-}, and ϵ_{obs}.

This procedure differs from the usual method of Section 4.1 in two ways; first, it requires that ϵ_{HA} (which corresponds to ϵ_{11}) be independently measurable; second, the free ligand concentration (activity) is directly measured.

A polyprotic acid whose pK_a values differ by at least four units can be handled in the same way, since each dissociation step is essentially independent, and the individual species' molar absorptivities can each be measured by appropriate adjustment of the pH. Often, however, the pK_a values are not sufficiently widely spaced to permit their independent measurement, and the solution may contain more than two species. Then the best approach is to use a curve-fitting procedure, essentially a nonlinear regression of ϵ_{obs} as a function of pH. Many programs have been described for this purpose (31).

The preceding treatment implicitly assumes that the molar absorptivities are unaffected by the medium, probably a good assumption for dilute aqueous solutions (say pH 2 to 12), where most pK_a's are measured. In more concentrated solutions, such as are required for the study of very weak bases, this is not a reasonable assumption, and the medium effect on absorptivities is a serious complication. Hammett (32) has reviewed this problem.

Almost all organic compounds (except for saturated hydrocarbons) are bases, but very acidic media may be required to measure the base strength of such very weakly basic substances as amides, nitriles, ethers, esters, and aromatic hydrocarbons. If the solvent is sufficiently acidic to react with the base, measurement of the equilibrium constant of this reaction is appropriate. By adding a reference acid to the system, a wider range of base strengths can be studied. When the solvent is essentially inert, the addition of a reference acid is essential, because the solvent cannot function as the reference. Thus an indicator acid HI might be added and the extent of reaction with base B measured spectrophotometrically.

$$B + HI \rightleftharpoons BHI$$

This stability constant may be called a salt formation constant. Davis and co-workers used this technique to measure base strengths in benzene (33). Another method is to add a reference acid selected for its strength. Perchloric acid has often been used for this purpose, especially for studies in acetic acid as the solvent. In this low dielectric constant solvent perchloric acid and salts are largely undissociated, and the reaction is written

$$B + HClO_4 \rightleftharpoons BHClO_4$$

the perchlorate formation constant being defined

$$K_f^{BHClO_4} = \frac{[BHClO_4]}{[B][HClO_4]}$$

This reaction can be studied by adding an indicator base I to the system and setting up this exchange reaction:

$$BHClO_4 + I \rightleftharpoons IHClO_4 + B$$

The equilibrium constant for this reaction is $K_f^{IHClO_4}/K_f^{BHClO_4}$. This quantity can be measured spectrophotometrically (34).

Spectrophotometric Titrations

In a spectrophotometric (photometric) titration a sample solution is titrated while its absorbance is monitored as a function of titrant volume. The technique has been developed by analytical chemists as a sensitive and accurate means for the detection of titration end points (35), but it has also been applied to measure the equilibrium constant of the titration reaction. There are two general classes of photometric titration, those in which the sample, titrant, or product (or some combination of these) produce the absorbance change (so-called self-indicating titrations), and those in which an indicator substance is added to the system. Self-indicating photometric titrations are, in the present context, simply an experimental variant of the direct spectral method for determining a stability constant, as discussed in Section 4.1, and need not be treated further. Indicator photometric titrations may be viewed as a variant of the competitive spectral method of Section 4.3, but the difference in data treatment and range of applications warrants a separate treatment.

In analytical uses of indicator photometric titrations the absorbance of the sample solution may be plotted against the titrant volume, or, in methods devised by Higuchi and his co-workers (36), the indicator ratio, that is, the ratio of concentrations of the indicator in its two forms (acid–base, oxidized–reduced, complexed–uncomplexed) is plotted as some function of titrant volume. It is these indicator ratio methods that are of interest for their potential as methods for stability constant estimation. The original developments were in the field of acid–base chemistry, but the following treatment will be general. We call the sample being titrated the substrate, and the titrant the ligand. Then the titration reaction is

$$S + L \rightleftharpoons SL$$

but the absorbance of the solution is governed by this exchange reaction

$$SL + I \rightleftharpoons IL + S$$

where I is the indicator, and a wavelength is selected where I and IL have different absorptivities and where S, L, and SL do not absorb. The equilibrium constant for this reaction is called an exchange constant, and it evidently is a ratio of binding constants (compare the preceding discussion on salt formation constants).

$$K_{ex} = \frac{[IL][S]}{[I][SL]} = \frac{K_{11}^{IL}}{K_{11}^{SL}} \tag{4.44}$$

The mass balance expressions are

$$S_t = [S] + [SL]$$
$$L_t = [L] + [SL] + [IL]$$

which, combined with Eq. (4.44), yield

$$K_{ex} = \frac{[IL]}{[I]} \frac{S_t - (L_t - r)}{(L_t - r)} \tag{4.45}$$

where $r = [L] + [IL]$. It is usually possible to ensure that the indicator concentration is so small that $[IL]$ makes a negligible contribution to r. A notable distinction between most analytical titrations and typical complex formation studies is that the titration reaction is essentially "quantitative," that is, K_{11}^{SL} in the preceding system is very large; moreover, before the end point $L_t < S_t$, and $[L]$ is therefore negligible. (In contrast, in typical binding studies $L_t \gg S_t$ and often $[L] \cong L_t$.) Thus we often can write Eq. (4.45) in the approximate form

$$K_{ex} = \frac{[IL]}{[I]} \frac{S_t - L_t}{L_t} \qquad (4.46)$$

The concentrations S_t and L_t (which are those in the sample solution) can be related to volumes of titrant solution added by $V_S = S_t V/M$ and $V_L = L_t V/M$, where M is the molarity of the titrant and V is the volume of sample solution; thus V_L is the volume of titrant solution added and V_S is the volume of titrant solution corresponding to the end point. Dilution effects can be rendered negligible by using very concentrated titrant or can be compensated for by multiplying each observed absorbance by $(V_0 + V_L)/V_0 = V/V_0$, where V_0 is the initial volume of the sample solution. The effect of dilution of the indicator can be overcome by incorporating indicator in the titrant in the same concentration it has in the sample solution.

Equation (4.46) can be written

$$\frac{[IL]}{[I]} = \frac{K_{ex} V_L}{V_S - V_L} \qquad (4.47)$$

which is the familiar form of a rectangular hyperbola, the indicator ratio being the dependent variable, the titrant volume the independent variable, and the exchange constant and end point volume the parameters. It can be placed in three linear plotting forms as usual, and the parameters thus evaluated (37). The indicator ratio is calculated with Eq. (4.48).

$$\frac{[IL]}{[I]} = \frac{A_I - A}{A - A_{IL}} = \frac{A - A_I}{A_{IL} - A} \qquad (4.48)$$

In Eq. (4.48), A_I and A_{IL} are the absorbances of the sample solution when the indicator is completely converted to the free and complexed forms, respectively, and A is the absorbance corresponding to titrant volume V_L; these all refer to the same total indicator concentration.

Since Eq. (4.47) is based on the assumption $L_t = [SL]$, it can be expected to fail when this assumption is not valid. This will be the case if K_{11}^{SL} is not very large; the titration reaction does not "go to completion." In aqueous acid–base chemistry this could be described as "hydrolysis of the salt." In this circumstance Eq. (4.45) is valid. An experimental procedure has been described (36b) to make use of Eq. (4.45). Two samples, identical except for size, are titrated identically; let the concentrations be S_t and L_t for one titration, and S'_t and L'_t for the other. Writing Eq. (4.45) for both titrations, and subtracting them at constant value of the indicator ratio gives

$$K_{ex} = \frac{[IL]}{[I]} \frac{(S_t - S'_t) - (L_t - L'_t)}{(L_t - L'_t)} \tag{4.49}$$

since r is the same in the two titrations. Equation (4.49) has the same form as Eq. (4.46), and can be treated in the same way. This is a form of corresponding solutions treatment.

According to Eq. (4.44), K_{ex} is a ratio of binding constants, so if one of them is known from an independent determination, the other can be calculated. As indicated previously, the titration method is most suitable for the study of very large binding constants, and it has been applied mainly to acid–base equilibria. The indicator should be selected so K_{ex} is of the order unity.

4.3. COMPETITIVE SPECTROPHOTOMETRY

If binding between substrate and ligand does not produce a significant spectral change, the direct spectrophotometric method cannot be used. An alternative is to employ competitive spectrophotometry, in which an equilibrium is established between the ligand and a surrogate substrate called the indicator, this equilibrium resulting in a spectral change. Then the substrate of interest is added, resulting in a competition between the indicator and the substrate for the ligand, with a displacement of some indicator from its complexed state, and a consequent spectral change. Competitive spectrophotometric methods have been applied to several kinds of systems, such as acid–base equilibria (38), protein binding (39), small organic complexes (40), metal ion equilibria (41), and inclusion complexes (42).

We first consider a system in which the indicator I and the substrate S form only 1:1 complexes with the ligand L. The competitive complexation equilibrium is

$$IL + S \rightleftharpoons SL + I$$

The mass balance expressions are

$$S_t = [S] + [SL]$$
$$L_t = [L] + [SL] + [IL]$$
$$I_t = [I] + [IL]$$

and the binding constants are defined

$$K_{11} = \frac{[SL]}{[S][L]}$$

$$K_I = \frac{[IL]}{[I][L]}$$

These equations are combined to give

$$L_t = [L] + \frac{K_{11}[L]S_t}{1 + K_{11}[L]} + \frac{K_I[L]I_t}{1 + K_I[L]} \tag{4.50}$$

Defining the indicator ratio $Q = [I]/[IL]$ allows us to write $K_I = 1/Q[L]$, which, substituted into Eq. (4.50), gives

$$L_t = \frac{1}{QK_I} + \frac{S_t K_{11}}{QK_I + K_{11}} + \frac{I_t}{Q + 1} \tag{4.51}$$

The quantity P is defined as

$$P = L_t - \frac{1}{QK_I} - \frac{I_t}{Q + 1} \tag{4.52}$$

Therefore Eq. (4.51) may be written

$$P = \frac{S_t K_{11}}{QK_I + K_{11}}$$

or

$$\frac{S_t}{P} = \frac{K_1}{K_{11}} Q + 1 \tag{4.53}$$

The indicator ratio Q can be obtained from Eq. (4.54),

$$Q = \frac{\epsilon - \epsilon_{IL}}{\epsilon_I - \epsilon} \tag{4.54}$$

where ϵ_I and ϵ_{IL} are the molar absorptivities of free and complexed indicator, respectively, and ϵ is the apparent molar absorptivity in any solution containing both forms. If the total indicator concentration is constant in all solutions, the absorptivities can be replaced by absorbances.

K_1 can be obtained by independent measurements on solutions of indicator and ligand by the conventional spectrophotometric method. Equation (4.53) therefore provides a graphical approach to determining K_{11} by plotting S_t/P against Q, where P is obtained using Eq. (4.52). It is of course assumed implicitly that only the indicator species I and IL absorb light at the analytical wavelength.

Figure 4.17 shows a plot of Eq. (4.53) in the competitive spectrophotometric determination of the binding between 4-nitrophenol and α-cyclodextrin (43); the indicator was methyl orange, and the value of K_1 was obtained from the data in Table 4.2 as described in Section 4.1. The value $K_{11} = 249\ M^{-1}$ found in this study agrees well with K_{11} as determined by other methods (42c).

It can be easily shown that if the substrate and ligand form isomeric 1:1 complexes, the K_{11} evaluated by competitive spectrophotometry is the sum of the isomeric binding constants.

We now expand the treatment to systems containing a 1:1 complex IL of indicator with ligand plus the 1:1 (SL) and 1:2 (SL_2) complexes of substrate with ligand (44). The mass balances on substrate and ligand are

$$S_t = [S] + [SL] + [SL_2]$$
$$L_t = [L] + [SL] + 2[SL_2] + [IL]$$

Define the quantity P by Eq. (4.55) [we shall see that this is the same quantity as that in Eq. (4.52)]:

$$P = [SL] + 2[SL_2] = L_t - [L] - [IL] \tag{4.55}$$

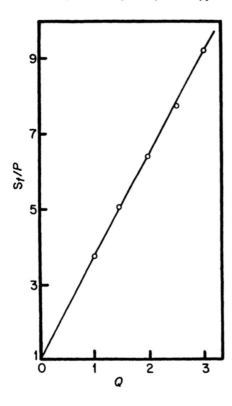

Figure 4.17. Plot according to Eq. (4.53) for the 4-nitrophenol (S):methyl orange (I):α-cyclodextrin (L) system (43).

Combining Eq. (4.55) with the definition of $\bar{i}$, the average number of ligand molecules bound per substrate molecule Eq. (2.46) shows that

$$\bar{i} = \frac{P}{S_t} \tag{4.56}$$

Combination of the mass balance on indicator with Eq. (4.55) gives Eq. (4.52). Thus P is experimentally accessible. For this system $\bar{i}$ [Eq. (2.48)] is

$$\bar{i} = \frac{K_{11}[L] + 2K_{11}K_{12}[L]^2}{1 + K_{11}[L] + K_{11}K_{12}[L]^2} \tag{4.57}$$

Combining Eqs. (4.56) and (4.57) with the earlier relationship $Q = 1/K_1[L]$ gives

$$\frac{PQK_1}{S_t - P} = K_{11} + \frac{K_{11}K_{12}}{K_1} \frac{1}{Q} \frac{(2S_t - P)}{(S_t - P)} \tag{4.58}$$

P and Q can be measured, S_t is known, and K_1 can be determined in a separate experiment, so the stability constants K_{11} and K_{12} can be obtained from the slope and intercept of a plot according to Eq. (4.58). If $K_{12} = 0$, the slope of this line will be zero, and Eq. (4.58) gives the simpler plotting form, Eq. (4.53). Equation (4.58) is the linearized form of Eq. (4.57) as seen in Eq. (2.102). This development shows that the competitive spectrophotometric method yields $\bar{i}$ as a measure of binding, and moreover it gives free ligand concentration in a direct experimental measurement through the expression $[L] = 1/K_1 Q$.

In order to evaluate Q it is necessary to know ϵ_I and ϵ_{IL}. Throughout a study the total indicator concentration I_t is held constant; hence absorptivities in Eq. (4.54) can be replaced by absorbances.

$$Q = \frac{A - A_{IL}}{A_I - A} \qquad (4.59)$$

A_I is readily measured in a solution of the indicator in the absence of ligand. A_{IL} is obtained as the product $\epsilon_{IL} I_t$; ϵ_{IL} is found as follows: In a solution containing only indicator and ligand, the mass balance on ligand is

$$L_t = [L] + [IL] \qquad (4.60)$$

This, combined with the mass balance on indicator, gives

$$[I] = \frac{I_t}{1 + K_1[L]} \qquad (4.61)$$

An iterative process is used. First, the assumption is made that $[L] = L_t$, and Eq. (4.61) is solved for $[I]$ (making use of an independently measured value of K_1). Then $[IL]$ is found from the mass balance $I_t = [I] + [IL]$, and this is used in Eq. (4.60) to produce a refined estimate of $[L]$. This process is repeated until no change in $[L]$ is noted in the fourth decimal place. The absorbance of the solution (path length = 1 cm) is $A = \epsilon_I[I] + \epsilon_{IL}[IL]$; hence

$$\epsilon_{IL} = \frac{A - \epsilon_I[I]}{[IL]}$$

The choice of concentrations is critical in the successful application of the competitive spectrophotometric method, particularly if the system contains both 1:1 and 1:2 complexes so that Eq. (4.58) is used. It is clear, in this case, that the quantity $(S_t - P)$ controls the reliability of the plotting variables. From the definitions of these quantities we find $S_t - P = [S] - [SL_2]$. Thus $(S_t - P)$ may be positive, zero, or negative, whereas $(2S_t - P)$ must be positive, as must the other variables and parameters in Eq. (4.58). Therefore data points plotted according to Eq. (4.58) must lie in either the first or third quadrants, with the point at $S_t = P$ being indeterminate. Points in the first and third quadrant correspond to the conditions $\bar{i} < 1$ and $\bar{i} > 1$, respectively.

The experimental data consist of absorbance values, which are bounded by the limits A_I and A_{IL}. With Eq. (4.59) these are converted to Q values, which may range from 0 to ∞. At large Q values, a small absorbance error results in a large error in Q; at small Q values, a small absorbance error results in a small absolute but large relative error in Q. For optimum results, Q should be of the order unity, with values in the range 0.3–3 being found usable (42c, 44).

According to the relationship $[L] = 1/K_I Q$, Q is uniquely related to free ligand concentration. Thus the limits placed on Q establish corresponding limits on $[L]$. To extend these limits in order to examine a greater range of the binding isotherm, it is necessary to have available a series of indicators with different K_I values.

For a given indicator, substrate, and ligand combination, each pair of plotting variables in Eq. (4.58) is associated with a unique $Q([L])$, which may be realized by innumerable combinations of S_t, I_t, and L_t. It is necessary to limit these quantities to values that will give acceptable uncertainties in P and in $S_t - P$. The following procedure is a typical optimization plan (44).

1. Choose I_t such that A_I or A_{IL} (whichever is larger) is approximately 2.
2. Select a series of Q values in the acceptable range (0.3–3) and for each Q calculate the corresponding $[L]$, using $[L] = 1/K_I Q$.
3. With preliminary estimates of K_{11} and K_{12}, calculate $\bar{i}$ values for each of these $[L]$ values, using Eq. (4.57). Reject any systems giving $\bar{i}$ values near unity.
4. For each acceptable $[L]$ value, choose L_t such that $L_t \geq 2[L]$. Calculate P with Eq. (4.52).
5. For each set of conditions, calculate S_t from Eq. (4.56).

Often it will be satisfactory to work with a fixed S_t and simply to calculate the desired L_t levels with Eq. (4.62), which is obtained from Eqs. (4.52) and (4.56).

$$L_t = \bar{i}S_t + [L] + \frac{I_t}{Q+1} \tag{4.62}$$

Table 4.4 lists spectral data obtained in a competitive study of the 4-cyanobenzoic acid:α-cyclodextrin system (45), with methyl orange serving as the indicator. Figure 4.18 is the plot according to Eq. (4.58), showing that a 1:2 complex is present. The binding constants can be evaluated from this plot, and can be refined, if desired, by nonlinear regression of $\bar{i}$ ($= P/S_t$) on $[L]$ ($= 1/K_1Q$).

This competitive spectral method has some severe limitations, including the requirements that the indicator form only a 1:1 IL complex and that no substrate–indicator complexes form. Moreover, as we have seen, it is difficult to investigate the full binding isotherm. Nevertheless, when other methods are inapplicable, it may provide useful results.

For substrates of low solubility it is difficult to establish accurate values of S_t for use in the preceding competitive spectral method, and the following modification has been devised (44). The experiment is carried out by varying the total ligand concentration L_t, but by maintaining the free substrate concentration $[S]$ constant at its saturation solubility s_0; this is accomplished by establishing the equilibria in the presence of solid substrate. Then Eq. (4.55) can be written

Table 4.4. Competitive Spectrophotometric Study of the 4-Cyanobenzoic Acid: α-Cyclodextrin System (45)[a]

$10^3 \, S_t/M$	$10^3 \, L_t/M$	Q	$\dfrac{PQK_1}{S_t - P}$	$\dfrac{1}{Q}\left(\dfrac{2S_t - P}{S_t - P}\right)$
4.675^b	2.518	1.644	562.9	1.530
2.330^b	2.568	1.024	589.9	2.000
5.981^b	4.997	0.849	651.0	3.718
4.429^b	4.982	0.673	714.8	5.351
5.849^c	3.663	1.214	564.4	2.224
5.771^c	4.640	0.901	626.6	3.383
6.348^c	6.636	0.605	746.8	6.379

[a] In 0.1 M HCl at 25°C with methyl orange indicator; $K_1 = 664 \, M^{-1}$.
[b] $I_t = 3.016 \times 10^{-5} \, M$.
[c] $I_t = 3.044 \times 10^{-5} \, M$.

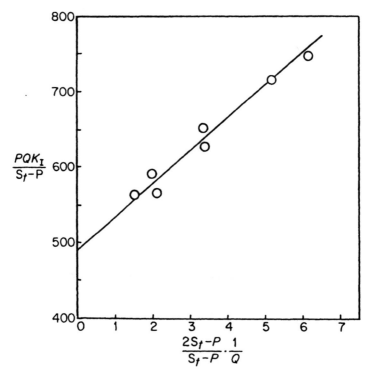

Figure 4.18. Plot according to Eq. (4.58) for the 4-cyanobenzoic acid (S):methyl orange (I): α-cyclodextrin (L) system. Data from Table 4.4.

$$P = K_{11}s_0[L] + 2K_{11}K_{12}s_0[L]^2$$

Using $[L] = 1/K_I Q$ gives

$$\frac{PQK_I}{s_0} = K_{11} + \frac{2K_{11}K_{12}}{K_I Q} \qquad (4.63)$$

Thus a plot of PQK_I/s_0 against $1/Q$ should be linear.

This combined solubility–competitive indicator method can be extended to include systems in which a 1:1 substrate–indicator complex SI is formed, provided its stability constant, K_{SI}, can be independently measured and a wavelength can be selected such that the molar absorptivities of SI and I are equal. The analysis (46) shows that Eq. (4.63) remains applicable but that Q and P must be evaluated by Eqs. (4.64) and (4.65), which may be compared with Eqs. (4.54) and (4.52).

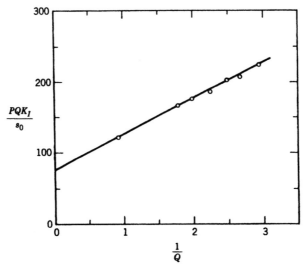

Figure 4.19. Plot according to Eq. (4.63), the solubility-competitive spectral method, for the 4-dimethoxybenzene (S): methyl orange (I):α-cyclodextrin system. $I_t = 4.64 \times 10^{-5}$ M, $s_0 = 5.74 \times 10^{-3}$ M.

$$Q = \frac{\epsilon - \epsilon_{IL}}{(\epsilon_I - \epsilon)(1 + K_{SI}s_0)} \qquad (4.64)$$

$$P = L_t - \frac{1}{K_I Q} - \frac{I_t}{1 + Q - QK_{SI}s_0} \qquad (4.65)$$

Figure 4.19 shows Eq. (4.63) plotted for the 4-dimethoxybenzene: α-cyclodextrin system; from this plot are obtained the parameter estimates $K_{11} = 75.4\ M^{-1}$ and $K_{12} = 221\ M^{-1}$.

4.4. INFRARED SPECTROSCOPY

Infrared (IR) spectroscopy has not been so widely applied as has ultraviolet–visible spectroscopy, except in the field of hydrogen bonding, where it is an important tool. Some IR measurements of stability constants have been made on metal ion complexes (47) and charge-transfer complexes (48).

The vibrational frequencies of O—H, N—H, and related bonds are sensitive to interaction of the hydrogen with a third atom to form a hydrogen bond, so IR spectroscopy is an excellent means for detecting and studying H-bonds (49). Most measurements of equilibrium constants have been based on the following simple formulation. K_{11} is defined in the usual way as $K_{11} = [SL]/[S][L]$, where SL is the hydrogen-bonded complex. Using the mass balances on substrate and ligand gives this equation:

$$K_{11} = \frac{S_t - [S]}{[S](L_t - S_t + [S])}$$

The uncomplexed substrate concentration [S] is determined spectroscopically by applying Beer's law at the frequency corresponding to the uncomplexed species. Self-association can be studied similarly.

The weakness of this approach is its assumption that only the unassociated monomeric substrate absorbs at the analytical wavelength. But it is unnecessary to make this assumption if account is taken of the contributions to the absorbance by other species. Then the formal development proceeds exactly as in Section 4.1, and all the equations and methods described there are applicable to IR data.

A typical system for study by the IR method consists of the proton donor (substrate), proton acceptor (ligand), and solvent. As in UV-visible studies, S_t is often held constant and L_t is varied. There is thus the possibility of a medium effect as the solution composition is altered with the incorporation of ligand. This change in solvent can affect the absorptivities and it can affect the binding constant. For example, Belobrov and Shurpach (50) studied the phenol–pyridine system in carbon tetrachloride, carrying out an error analysis of the instrumental and solution preparation phases and accounting for the medium effect on molar absorptivity; they found $K_{11} = 100 \pm 20\ M^{-1}$ when $L_t/S_t = 1$, whereas $K_{11} = 43 \pm 3\ M^{-1}$ where $L_t/S_t = 10$. This real difference in constants as a function of composition was attributed to a medium effect, a consequence of specific solvation resulting in a local solvent sheath of composition and properties different from the average macroscopic composition.

A serious limitation to the wider use of IR measurements is the difficulty caused by absorption by the solvent. Even more limiting has been the relatively low sensitivity (low compared with UV spectroscopy) of IR spectroscopy. With modern Fourier-transform IR spectrometers this experimental approach may develop into a more useful method for measuring binding constants.

REFERENCES

1. S.-F. Lin, Ph.D. Dissertation, University of Wisconsin-Madison, 1981, p. 158.

2. T. W. Rosanske, Ph.D. Dissertation, University of Wisconsin-Madison, 1979, pp. 56–64.

3. (a) J. Brynestad and G. P. Smith, *J. Phys. Chem.*, **72**, 296 (1968); (b) C. Chylewski, *Angew. Chem. Int. Ed.*, **10**, 195 (1971); (c) T. Nowicka-Jankowska, *J. Inorg. Nucl. Chem.*, **33**, 2043 (1971).

4. L. Skulski and A. Plucinski, *Pol. J. Chem.*, **54**, 999 (1980).

5. J. S. Coleman, L. P. Varga, and S. H. Martin, *Inorg. Chem.*, **9**, 1015 (1970).

6. D. D. Pendergast, Ph.D. Dissertation, University of Wisconsin-Madison, 1983, p. 63.

7. (a) G. Briegleb, *Elektronen-Donator-Acceptor-Komplexe*, Springer-Verlag, Berlin, 1961, Ch. XII; (b) C. W. Davies, *Ion Association*, Butterworths, Washington, 1962, Ch. 4; (c) L. J. Andrews and R. M. Keefer, *Molecular Complexes in Organic Chemistry*, Holden-Day, San Francisco, 1964, Ch. IV; (d) G. H. Nancollas, *Interactions in Electrolyte Solutions*, Elsevier, Amsterdam, 1966, Ch. 2; (e) J. Rose, *Molecular Complexes*, Pergamon, Oxford, 1967, Ch. 3; (f) R. W. Ramette, *J. Chem. Educ.*, **44**, 647 (1967); (g) R. Foster, *Organic Charge Transfer Complexes*, Academic, 1969, London, Ch. 6; (h) R. S. Mulliken and W. B. Person, *Molecular Complexes*, Wiley-Interscience, New York, 1969, Ch. 7; (i) M. A. Slifkin, *Charge Transfer Interactions of Biomolecules*, Academic, London, 1971, Ch. 1; (j) E. N. Gur'yanova, I. P. Gol'dshtein, and I. P. Romm, *Donor-Acceptor Bond*, Halsted, Wiley, New York, 1975, Ch. II; (k) W. A. E. McBryde, *Talanta*, **21**, 979 (1974); (l) F. R. Hartley, C. Burgess, and R. M. Alcock, *Solution Equilibria*, Ellis Horwood Ltd., Wiley, New York, 1980, Ch. 8; (m) B. K. Seal, H. Sil, and D. C. Mukherjee, *J. Indian Chem. Soc.*, **56**, 1081 (1979).

8. S.-F. Lin, Ph.D. Dissertation, University of Wisconsin-Madison, 1981, pp. 58–63.

9. J. A. Mollica, Ph.D. Dissertation, University of Wisconsin-Madison, 1966, p. 45.

10. H. Benesi and J. H. Hildebrand, *J. Am. Chem. Soc.*, **71**, 2703 (1949).

11. R. L. Scott, *Rec. Trav. Chim.*, **75**, 787 (1956).

12. R. Foster, D. L. Hammick, and A. A. Wardley, *J. Chem. Soc.*, 3817 (1953).

13. G. Scatchard, *Ann. N.Y. Acad. Sci.*, **51**, 660 (1949).

14. K. A. Connors, *Reaction Mechanisms in Organic Analytical Chemistry*, Wiley-Interscience, New York, 1973, pp. 76–85.

15. (a) N.J. Rose and R. S. Drago, *J. Am. Chem. Soc.*, **81**, 6138 (1959); (b) R. S. Drago and N. J. Rose, *J. Am. Chem. Soc.*, **81**, 6141 (1959).

16. (a) K. Conrow, G. D. Johnson, and R. E. Bowen, *J. Am. Chem. Soc.*, **86**,

1025 (1964); (b) B. K. Seal, A. K. Mukherjee, and D. C. Mukherjee, *Bull. Chem. Soc. Jpn.*, **52**, 2088 (1979).

17. W. Liptay, *Z. Elektrochem.*, **65**, 375 (1961).
18. (a) S. Nagakura, *J. Am. Chem. Soc.*, **76**, 3070 (1954); (b) S. Nagakura, *J. Am. Chem. Soc.*, **80**, 520 (1958).
19. B. K. Seal, H. Sil, M. Banerjee, and D. C. Mukherjee, *Bull. Chem. Soc. Jap.*, **54**, 639 (1981).
20. B. Grabarić, I. Piljac, and I. Filipović, *Anal. Chem.*, **45**, 1932 (1973).
21. M. Barigand, J. Orszagh, and J. J. Tondeur, *Bull. Soc. Chim. Fr.*, 4864 (1972).
22. B. K. Seal, H. Sil, and D. C. Mukherjee, *Spectrochim. Acta*, **38A**, 289 (1982).
23. G. D. Johnson and R. E. Bowen, *J. Am. Chem. Soc.*, **87**, 1655 (1965).
24. A. M. Halpern and K. Weiss, *J. Am. Chem. Soc.*, **90**, 6297 (1968).
25. (a) K. A. Connors and J. A. Mollica, *J. Pharm. Sci.*, **55**, 772 (1966); (b) H. J. G. Hayman, *J. Chem. Phys.*, **37**, 2290 (1962).
26. (a) B. Dodson, R. Foster, A. A. S. Bright, M. I. Forman, and J. Gorton, *J. Chem. Soc.*, B, 1283 (1971); (b) A. A. S. Bright, J. A. Chudek, and R. Foster, *J. Chem. Soc.*, *Perkin Trans.*, 2, 1256 (1975); (c) Y. E. Ho and C. C. Thompson, *Chem. Commun.*, 609 (1973).
27. (a) L. Newman and D. N. Hume, *J. Am. Chem. Soc.*, **79**, 4571 (1957); (b) G. Kortüm and G. Weber, *Z. Elektrochem.*, **64**, 642 (1960); (c) D. A. Deranleau, *J. Am. Chem. Soc.*, **97**, 1218 (1975); (d) M. T. Beck, *Chemistry of Complex Equilibria*, Van Nostrand Reinhold, London, 1970, pp. 93–97.
28. K. A. Connors and T. W. Rosanske, *J. Pharm. Sci.*, **69**, 173 (1980).
29. (a) J. A. Mollica, Ph.D. Dissertation, University of Wisconsin-Madison, 1966, p. 140; (b) J. A. Mollica and K. A. Connors, *J. Am. Chem. Soc.*, **89**, 308 (1967).
30. J. D. Childs, S. D. Christian, and J. Grundnes, *J. Am. Chem. Soc.*, **94**, 5657 (1972).
31. (a) C. Mongay, G. Ramis, and M. C. Garcia, *Spectrochim. Acta*, **38A**, 247 (1982); (b) D. J. Leggett, S. L. Kelly, L. R. Shiue, Y. T. Wu, D. Chang, and K. M. Kadish, *Talanta*, **30**, 579 (1983); (c) M. Meloun and J. Cermak, *Talanta*, **31**, 947 (1984); (d) A. Albert and E. P. Serjeant, *The Determination of Ionization Constants*, 3rd ed., Chapman and Hall, London, 1984, Ch. 4.
32. L.P. Hammet, *Physical Organic Chemistry*, 2nd ed., McGraw-Hill, New York, 1970, pp. 290–296.
33. (a) M. M. Davis and P. J. Schuhmann, *J. Res. Natl. Bur. Stand.*, **39**, 221 (1947); (b) M. M. Davis, P. J. Schuhmann, and M. E. Lovelace, *J. Res. Natl. Bur. Stand.*, **41**, 27 (1948).
34. T. Higuchi and K. A. Connors, *J. Phys. Chem.*, **64**, 179 (1960).
35. (a) A. L. Underwood, *J. Chem. Educ.*, **31**, 394 (1954); (b) R. F. Goddu and D. N. Hume, *Anal. Chem.*, **26**, 1740 (1954); (c) J. B. Headridge, *Talanta*, **1**,

293 (1958); (d) J. B. Headridge, *Photometric Titrations*, Pergamon, New York, 1961; (e) S. P. Eriksen and K. A. Connors, *J. Pharm. Sci.*, **53**, 465 (1964); (f) M. A. Leonard, *Comprehensive Analytical Chemistry*, Vol. III, Ch. III, G. Svehla, Ed., Elsevier, Amsterdam, 1977.

36. (a) T. Higuchi, C. Rehm, and C. Barnstein, *Anal. Chem.*, **28**, 1506 (1956); (b) K. A. Connors and T. Higuchi, *Anal. Chem.*, **32**, 93 (1960); (c) C. Rehm and T. Higuchi, *Anal. Chem.*, **29**, 367 (1957).

37. K. A. Connors and T. Higuchi, *Anal. Chim. Acta*, **25**, 509 (1961).

38. C. B. Monk, *Trans. Faraday Soc.*, **53**, 540 (1957).

39. (a) I. M. Klotz, *J. Am. Chem. Soc.*, **68**, 2299 (1946); (b) I. M. Klotz, H. Triwush, and F. M. Walker, *J. Am. Chem. Soc.*, **70**, 2935 (1948).

40. (a) R. Foster, *Nature (London)*, **173**, 222 (1954); (b) J. M. Corkill, R. Foster, and D. L. Hammick, *J. Chem. Soc.*, 1202 (1955).

41. (a) M. T. Beck, *Chemistry of Complex Equilibria*, Van Nostrand Reinhold, London, 1970, p. 100; (b) E. Ohyoshi, *Anal. Chem.*, **57**, 446 (1985).

42. (a) W. Lautsch, W. Bandel, and W. Broser, *Z. Naturforsch.*, **11b**, 282 (1956); (b) B. Casu and L. Ravá, *Ric. Sci.*, **36**, 733 (1966); (c) S.-F. Lin and K. A. Connors, *J. Pharm. Sci.*, **72**, 1333 (1983); (d) R. I. Gelb, L. M. Schwartz, B. Cardelino, and D. A. Laufer, *Anal. Biochem.*, **103**, 362 (1980).

43. S.-F. Lin, Ph.D. Dissertation, University of Wisconsin-Madison, 1981, p. 131.

44. D. D. Pendergast and K. A. Connors, *J. Pharm. Sci.*, **73**, 1779 (1984).

45. D. D. Pendergast, Ph.D. Dissertation, University of Wisconsin-Madison, 1983, p. 70.

46. D. D. Pendergast, Ph.D. Dissertation, University of Wisconsin-Madison, 1983, pp. 39–43.

47. (a) M. T. Beck, *Chemistry of Complex Equilibria*, Van Nostrand Reinhold, London, 1970, p. 101; (b) F. R. Hartley, C. Burgess, and R. M. Alcock, *Solution Equilibria*, Ellis Horwood, Chichester, 1980, p. 147.

48. R. Foster, *Organic Charge-Transfer Complexes*, Academic, London, 1969, p. 139.

49. (a) D. Hadži, Ed., *Hydrogen Bonding*, Pergamon Press, New York, 1959; (b) G. C. Pimentel and A. L. McClellan, *The Hydrogen Bond*, Freeman, San Francisco, 1960; (c) L. Pauling, *The Nature of the Chemical Bond*, 3rd ed., Ch. 12, Cornell University Press, Ithaca, N.Y., 1960; (d) M. D. Joesten and L. J. Schaad, *Hydrogen Bonding*, Dekker, New York, 1974, Ch. 3.

50. V. M. Belobrov and V. I. Shurpach, *Zh. Obshch. Khim.* (Engl. trans.), **52**, 632, 636 (1982).

5

MAGNETIC
RESONANCE
SPECTROSCOPY

This chapter treats nuclear magnetic resonance (NMR) and electron spin resonance (ESR, also called EPR for electron paramagnetic resonance) spectroscopy. Of these two techniques NMR has been more widely applied to binding studies. Valuable structural information, much of it unavailable by other methods, has been obtained about complexes in solution, primarily from chemical shift measurements of the complexed species. Our concern, however, is with the use of magnetic resonance methods for the determination of binding constants.

5.1. NMR CHEMICAL SHIFTS

Binding as Chemical Exchange

Consider a nucleus (observable by NMR) that can "partition" between two magnetically nonequivalent sites. Examples would be protons or carbon atoms involved in cis–trans isomerization, rotation about the carbon–nitrogen bond in amides, proton exchange between solvent and solute, and complex formation. In the NMR context the nucleus is said to undergo *chemical exchange* between the sites, and the nature of the observable signal depends on the rate of the exchange process. Let the system be the $1:1$ complex formation equilibrium

$$S + L \rightleftharpoons SL$$

where we observe a nucleus on the substrate, so the two sites are S and SL, and we simplify the description to

$$S \underset{k_S}{\overset{k_{SL}}{\rightleftharpoons}} SL$$

where k_{SL} is, at high concentrations of L relative to S, a pseudofirst-order rate constant for association and k_S is a first-order dissociation rate constant. From elementary kinetic theory we have

$$k = k_{SL} + k_S \tag{5.1}$$

where k is the observed first-order rate constant for the approach to equilibrium. In NMR nomenclature it is more common to speak of the lifetime of a nuclear state, where a lifetime τ is related to a rate constant by $\tau = 1/k$, so that Eq. (5.1) becomes

$$\frac{1}{\tau} = \frac{1}{\tau_{SL}} + \frac{1}{\tau_S} \tag{5.2}$$

or

$$\tau = \frac{\tau_S \tau_{SL}}{\tau_S + \tau_{SL}} \tag{5.3}$$

This symbolism is convenient for expressing the fractional occupancy of the two sites according to Eq. (5.4)

$$f_{10} = \frac{\tau_S}{\tau_S + \tau_{SL}} \tag{5.4a}$$

$$f_{11} = \frac{\tau_{SL}}{\tau_S + \tau_{SL}} \tag{5.4b}$$

In Eq. (5.4), f_{10} is the fraction of substrate present as S, and f_{11} is the fraction present as SL; this symbolism was introduced in Section 2.4. According to (5.4), $\tau = f_{10}\tau_{SL} = f_{11}\tau_S$. For the special case in which $\tau_S = \tau_{SL}$ (as in isotopic exchange), we have $2\tau = \tau_S = \tau_{SL}$.

Now we develop a qualitative view of the NMR experiment under conditions of chemical exchange (1). We adopt a coordinate system that is rotating about the applied field H_0 in the same direction as the precessing magnetization vector of the nuclei being observed. Let ν_S and

ν_{SL} be the Larmor precessional frequencies of the nucleus in sites S and SL, respectively. For present simplicity we set $f_{10} = f_{11}$, $\tau_S = \tau_{SL}$. As the frequency of the rotating frame of reference ν_0 we choose the average of ν_S and ν_{SL}, thus

$$\nu_0 = \tfrac{1}{2}(\nu_S + \nu_{SL}) \tag{5.5}$$

As a consequence, from the point of view of this rotating frame, a nucleus at site S precesses at frequency $(\nu_0 - \nu_S)$, whereas a nucleus at site SL precesses at frequency $(\nu_{SL} - \nu_0)$; that is, the two nuclei (actually their magnetization vectors) precess in opposite directions. We imagine several possible cases.

1. *Very slow exchange.* Slow exchange means that the lifetime $\tau_S = \tau_{SL}$ in each site is very long. Thus a nucleus in site S precesses many times, at frequency $(\nu_0 - \nu_S)$ in the rotating frame, before it leaves site S, and similarly for a nucleus in site SL. Thus there is time for absorption of energy from the radiofrequency field H_1, and resonance peaks appear at ν_S and ν_{SL} in the laboratory frame.

2. *Moderately slow exchange.* The state lifetime is 2τ; we ask how the absorption band is affected as this becomes smaller. According to the Heisenberg uncertainty principle, $(\delta E)(\delta t) \approx h$, where δE and δt are uncertainties associated with energy and time measurements. Writing $\delta E = h\delta\nu$ and identifying δt with the state lifetime, we get $\delta\nu \approx 1/2\tau$, which is interpreted to mean that the width of the absorption band $(\delta\nu)$ increases as the state lifetime decreases. This is a general phenomenon (lifetime broadening). Thus we expect resonance absorption at (or near) frequencies ν_S and ν_{SL}, but the bands will be broader than in the very slow exchange limit.

3. *Very fast exchange.* If the lifetimes are very short, a nucleus in site S cannot precess to a significant extent before it leaves S to enter SL, where it begins to precess in the opposite direction (in the rotating frame), again enters S, and so on. Therefore from the point of view of the rotating frame, the nucleus is essentially stationary. In the laboratory frame its frequency is ν_0, the frequency of the rotating frame. Thus, according to Eq. (5.5), a single absorption band will be seen at ν_0, the mean of ν_S and ν_{SL}.

A quantitative theory of chemical exchange describes the line shapes of the resonance absorption over the entire range of lifetimes (2). The behavior of a hypothetical system is shown in Figure 5.1. Of course, the lifetimes in the two states are generally different, and we replace Eq. (5.5) with

$$\nu_0 = f_{10}\nu_S + f_{11}\nu_{SL} \qquad (5.6)$$

The result is that in the slow exchange limit the areas of the absorption bands are proportional to the fractional occupancy of the sites, and in the fast exchange limit the observed resonance frequency is a weighted average of the site frequencies. Since rates are temperature dependent, it is often possible to control the chemical exchange regime of a system by changing its temperature.

The preceding discussion did not define the terms *slow* and *fast*, but their meaning is implicit in the imagined experiment with the rotating frame. In this frame the precessional frequencies of the nuclei are $(\nu_0 - \nu_S)$ and $(\nu_{SL} - \nu_0)$, or, using Eq. (5.5), each of these is equal to $(\nu_{SL} - \nu_S)/2$. It is the exchange rate $1/\tau$ relative to this frequency difference that establishes whether an exchange is slow or fast. For proton NMR typical resonance frequency differences would be of the order 10^2 Hz, corresponding to a lifetime of roughly 10^{-2} s. The slow exchange limit would require lifetimes much longer than this, the fast exchange limit lifetimes much shorter than this.

For a binding system in the slow exchange limit the NMR study yields the area of the absorption bands of the nuclei in the two sites S and SL; thus the ratio of the concentrations of these species is obtained, and the binding constant is calculated with the additional knowledge of the ligand concentration. Relatively few complex formation systems will meet the requirements for very slow exchange. However, Feeney et al. (3) point out that slow exchange may apply in very tightly bound complexes, whereas fast exchange usually applies in loosely bound complexes. These authors used simulated NMR data on hypothetical systems to show the extent of error when a system in the intermediate exchange region is erroneously treated as if it were in the fast exchange limit.

It is instructive to compare these results on the time scale in NMR with the situation in electronic absorption spectroscopy. In the UV–visible study of a binding equilibrium a typical wavelength shift is of the order 10^2 nm, corresponding to a frequency of about 10^{15} Hz, or a lifetime of 10^{-15} s. Since even a diffusion-controlled rate is slower than this by orders of magnitude, observations in the UV–visible region always

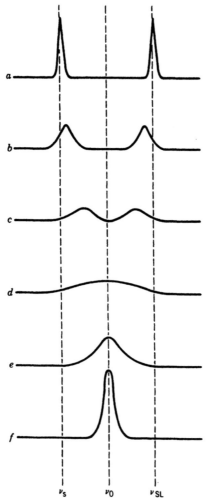

Figure 5.1. NMR absorption by a hypothetical two-identical site system with chemical exchange. (a) Slow exchange limit. (b) Moderately slow exchange. (d) Coalescence. (f) Fast exchange limit.

conform to the slow exchange condition. In contrast, the NMR observation usually is carried out in the fast exchange limit.

1:1 Binding in Fast Exchange

Suppose the system of substrate and ligand in a solvent forms a single 1:1 complex. The nucleus being observed is on the substrate. The quantitative treatment starts with the statement [which is equivalent to Eq. (5.6)] that the observed chemical shift δ is the average of the chemical shifts of the nucleus in S and in SL, weighted by the fractional occupancy of these states:

$$\delta = f_{10}\delta_S + f_{11}\delta_{SL} \tag{5.7}$$

Since $f_{10} + f_{11} = 1$, Eq. (5.8) can be written

$$\delta = f_{11}(\delta_{SL} - \delta_S) + \delta_S \tag{5.8}$$

Defining chemical shift differences

$$\Delta = \delta - \delta_S \tag{5.9a}$$

$$\Delta_{11} = \delta_{SL} - \delta_S \tag{5.9b}$$

allows Eq. (5.8) to be expressed as $\Delta = f_{11}\Delta_{11}$. Combining this with Eq. (2.57) for f_{11} gives

$$\Delta = \frac{\Delta_{11}K_{11}[L]}{1 + K_{11}[L]} \tag{5.10}$$

Equation (5.10) is the NMR 1:1 binding isotherm. It has the usual hyperbolic form, and comparison with Eq. (4.5) shows the formal identity of the NMR and optical spectroscopic methods. Linearization of Eq. (5.10) is accomplished as described generally in Section 2.5 and for optical data in Section 4.1. The double-reciprocal plot is made according to

$$\frac{1}{\Delta} = \frac{1}{\Delta_{11}K_{11}[L]} + \frac{1}{\Delta_{11}} \tag{5.11}$$

Equation (5.12) is the *y*-reciprocal plotting form.

$$\frac{[L]}{\Delta} = \frac{[L]}{\Delta_{11}} + \frac{1}{\Delta_{11}K_{11}} \tag{5.12}$$

The *x*-reciprocal form is

$$\frac{\Delta}{[L]} = -K_{11}\Delta + \Delta_{11}K_{11} \tag{5.13}$$

These plotting forms, and related ones discussed for optical data in Section 4.1, have been widely used to estimate the parameters K_{11} and Δ_{11} for molecular complexes (4).

Table 5.1 gives Δ values obtained by Carper et al. (4c) in a proton NMR study of 1,3,5-trinitrobenzene (S) and phenyl sulfide (L) in carbon tetrachloride. The Δ values are expressed in hertz (Hz); in the literature the units Hz (frequency difference) and ppm (chemical shift difference) are both found. In this study L_t was much larger than S_t, so the approximation $[L] = L_t$ is valid. Figures 5.2–5.4 show plots of these data according to Eqs. (5.11)–(5.13), respectively. Carper et al. (4c) report, for this system, $K_{11} = 0.95 \, M^{-1}$, $\Delta_{11} = 50.8 \, Hz$.

The functional similarity of the NMR and optical binding isotherms has been pointed out, but there is a difference as a consequence of the different physical quantities being measured. Equation (4.5) contains S_t, the substrate concentration, whereas Eq. (5.10) does not. [It is true that S_t could be divided out of Eq. (4.5), which could be expressed in terms of an apparent molar absorptivity, but this does not alter the argument.] This is because in optical spectroscopy an intensity is measured, and this is proportional to concentration, but in NMR spectroscopy a frequency is measured. Substrate concentration is therefore an experimental variable with which to manipulate the analytical signal in optical spectroscopy, but this is not available in NMR. Carta et al. (5) have examined the consequences of this difference for the optimization of the concentration range of the ligand in NMR studies. It remains valid to state that a substantial fraction of the binding curve must be examined if the model is to be tested, as discussed in Section 2.5; the optimization studies of Granot (6) are in agreement.

Table 5.1. Proton NMR Chemical Shift Differences for the 1,3,5-TNB : Phenyl Sulfide System (4c)[a]

L_t/M^{-1}	Δ/Hz^{-1}
0.110	5.0
0.204	8.6
0.496	17.1
0.753	22.6
1.002	26.6
1.33	30.5

[a] At 20°C; $S_t = 8.0 \times 10^{-3} \, M$.

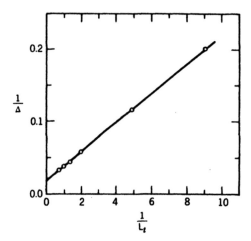

Figure 5.2. Double-reciprocal plot according to Eq. (5.11) of the NMR data in Table 5.1.

In early work by Huggins et al. (7) and Creswell and Allred (8) a curve-fitting treatment was used. We developed above the relationship $\Delta = f_{11}\Delta_{11}$, and from Eq. (2.57) we have $f_{11} = K_{11}[L]/(1 + K_{11}[L])$. A value is assumed for K_{11}, f_{11} is calculated, and Δ is plotted against the calculated f_{11} for a series of ligand concentrations. This calculation is repeated for other K_{11} values. Only the correct K_{11} should give a straight line.

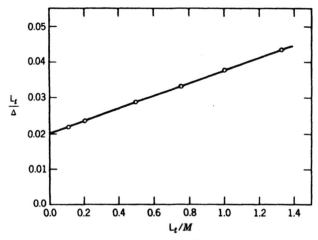

Figure 5.3. The y-reciprocal plot according to Eq. (5.12) of the data in Table 5.1.

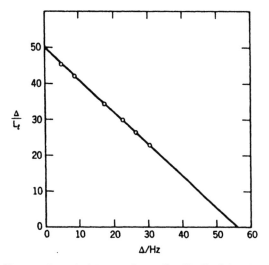

Figure 5.4. The x-reciprocal plot according to Eq. (5.13) of the data in Table 5.1.

We next consider the treatment of data when the approximation $[L] = L_t$ is not valid. From the mass balance on ligand, $L_t = [L] + [SL]$, we get

$$L_t = [L](1 + K_{11}[S]) \tag{5.14}$$

The assumption $[L] = L_t$ is therefore equivalent (9) to the assumption $K_{11}[S] \ll 1$. This can be satisfied if $S_t \ll L_t$, but this latter condition is not essential, as Sahai et al. have shown for NMR data (10). The very large majority of complex-forming systems investigated by NMR have had small K_{11} values, of the order $1\ M^{-1}$ or less, so $K_{11}[S]$ is usually very small. [A review of NMR in complex studies by Foster and Fyfe (11) lists many examples of K_{11} values.] When this condition is not met, it is necessary to use an iterative calculation; the general approach is outlined in Section 2.3. Nakano et al. (12) developed this graphical technique for NMR data. (The following derivation is different from theirs, but the result is the same.) Combine Eq. (5.10) with the mass balance $[L] = L_t - [SL]$, use the substitution $[SL] = S_t(\Delta/\Delta_{11})$ in the numerator (obtained from $f_{11} = \Delta/\Delta_{11}$), and rearrange to give

$$\frac{L_t}{\Delta} = \frac{L_t + S_t - [SL]}{\Delta_{11}} + \frac{1}{\Delta_{11}K_{11}} \tag{5.15}$$

Equation (5.15) is another way to write Eq. (5.12). The equation is used iteratively. First a plot is made of L_t/Δ against $(L_t + S_t)$. This gives an estimate of Δ_{11}, from which is calculated $[SL] = S_t(\Delta/\Delta_{11})$. Now the plot is made according to Eq. (5.15), and iteration is carried out until the plots converge. The method has been found to be successful when L_t and S_t are comparable. Difficulties with very strong complexing have been reported (13). Alternative plotting forms [corresponding to Eqs. (5.11) and (5.13)] can also be used. Stamm et al. (14) found for a very weak complex $(K_{11} = 0.1\ M^{-1})$ that several evaluation methods, including the Rose–Drago method and the Cresswell–Allred plot (8), gave the same results.

The most general way to make use of Eq. (5.10) is by nonlinear regression, treated in detail in Chapter 3, combined with the mass balance on ligand. The situation is just as described in Chapter 4 for optical data. Calculation of the free ligand concentration by the Taylor expansion method (15), as developed for optical spectroscopy in Section 2.3, should be effective, though it has not yet been applied to NMR data.

Let us consider the effect of isomerism on the NMR result. Suppose we have two 1:1 complexes, SL and LS, with corresponding parameters K_{SL}, K_{LS}, δ_{SL}, δ_{LS}. The observed chemical shift is given by a generalization of Eq. (5.7), namely,

$$\delta = f_S\delta_S + f_{SL}\delta_{SL} + f_{LS}\delta_{LS}$$

which becomes

$$\Delta = f_{SL}\Delta_{SL} + f_{LS}\Delta_{LS} \tag{5.16}$$

where $\Delta_{SL} = \delta_{SL} - \delta_S$ and $\Delta_{LS} = \delta_{LS} - \delta_S$. Equation (5.17) is easily obtained [see also Eq. (2.107)].

$$f_{SL} = \frac{K_{SL}[L]}{1 + (K_{SL} + K_{LS})[L]} \tag{5.17a}$$

$$f_{LS} = \frac{K_{LS}[L]}{1 + (K_{SL} + K_{LS})[L]} \tag{5.17b}$$

Combining Eqs. (5.16) and (5.17) gives

$$\Delta = \frac{(\Delta_{SL}K_{SL} + \Delta_{LS}K_{LS})[L]}{1 + (K_{SL} + K_{LS})[L]} \tag{5.18}$$

Equation (5.18) has the same ligand dependence as does Eq. (5.10). All of the comments made about Eq. (4.6) and the analogous situation in optical spectroscopy apply also to NMR. The observed K_{11} is the sum of the isomeric 1:1 binding constants; this is true even if one of the chemical shift differences is zero.

Presuming that the system is describable in terms of simple 1:1 binding, the preceding discussion provides relatively straightforward techniques for estimating the K_{11} value. Yet there are some potential difficulties in applying the NMR method. Most of these arise, or are most noticeable, in the study of very weak complexes ($K_{11} \approx 1\,M^{-1}$ or smaller). We have earlier (Section 2.6, "Random Association") commented on the unknown meaning of such numbers; nevertheless, these small constants can be determined quite precisely by NMR. A consequence of weak complexing is that high concentrations of ligand are usually required to drive the equilibrium, and this results in possible medium effects. These can be of several types.

One possibility is that the ligand undergoes self-association, and so it is advisable to work in a concentration range where self-association is negligible, if possible. The use of NMR to study self-association is treated subsequently. Another possibility is that Eq. (5.7) fails to describe the system because δ_S or δ_{SL} are not constants throughout the experimental concentration range. Such inconstancy could result from medium effects on S and SL [Stamm's (16) "additional unspecific shielding," or AUS], or from interaction with the internal reference (17). It has been suggested that an external reference may be preferable (18), but Chudek et al. (19), studying somewhat stronger complexes ($K_{11} = 5$ to $20\,M^{-1}$), found the same stability constant estimates using an internal reference and an external reference, the latter readings being corrected for bulk magnetic susceptibility differences. Stamm (16) accounted for curved Scatchard plots in terms of the AUS effect, but Chudek et al. (20) interpreted their curved plots in terms of 1:1 + 1:2 binding.

Besides these possible effects on chemical shifts there is the general medium effect leading to thermodynamic nonidealities. Much of the literature discussion on this topic (which was commented on in Section 2.6, "Medium Effects"), has been stimulated by NMR studies on very weak complexes (21). The matter of the "best" concentration scale for the processing of NMR data has been discussed (21) [see also ref. (28) in Chapter 2], because some estimates of Δ_{11} have been found to be dependent on the concentration scale (4c, 21), yet Δ_{11} contains no concentration units. Slejko and Drago (21c) argued on chemical grounds that the molar and mole fraction scales are appropriate, whereas the molal scale is not, for equilibrium systems. Bailey et al. (25c) found Δ_{11} values to be scale independent and the K_{11} values to be correctly related in the dilute solution range. Section 2.2 gives an extensive discussion of this problem.

The dependent variable in Eq. (5.10) is $\Delta = \delta - \delta_S$, which requires δ_S for its evaluation. This is obtained by measurement of a solution containing no ligand, or by extrapolation of δ to $L_t = 0$ of a series of measurements. It is possible also to estimate δ_S in solutions containing ligand by the Cresswell–Allred (8) type of curve fitting. For assumed values of K_{11}, the free ligand concentration is calculated (S_t and L_t being known quantities); see Eq. (2.39). Then f_{11} is calculated with Eq. (2.57), and δ is plotted against f_{11}. The best estimate of K_{11} will generate a straight line, and δ_S is its intercept, according to Eq. (5.8). Horman and Dreux (22) used this method to treat NMR data gathered under the condition $S_t = L_t$.

Multiple Equilibria

Suppose the system contains 1:1 (SL) and 1:2 (SL$_2$) complexes, in the fast exchange condition. The quantitative description follows the lines of the derivation for the 1:1 model. The weighted average expression for the observed chemical shift of a nucleus on the substrate is

$$\delta = f_{10}\delta_S + f_{11}\delta_{SL} + f_{12}\delta_{SL_2}$$

which is written

$$\Delta = f_{11}\Delta_{11} + f_{12}\Delta_{12}$$

where $\Delta = \delta - \delta_S$, $\Delta_{11} = \delta_{SL} - \delta_S$, $\Delta_{12} = \delta_{SL_2} - \delta_S$. Using Eq. (2.55) this becomes

$$\Delta = \frac{\Delta_{11}K_{11}[L] + \Delta_{12}K_{11}K_{12}[L]^2}{1 + K_{11}[L] + K_{11}K_{12}[L]^2} \tag{5.19}$$

Comparison of Eq. (5.19) with Eq. (4.28) for the same system as studied by optical spectroscopy shows their formal identity. The NMR method therefore shares the difficulty that each additional complex species adds two parameters to the model function. With so many parameters it is easy to obtain excellent curve fits to the experimental data points, but the parameters evaluated in this way are not necessarily reliable. It is important that the binding curve be examined over an extensive range; criteria have been suggested to assist in establishing the reliability of parameters in multiple equilibria systems (23). Nonlinear regression is the usual approach to extracting the parameters, and the calculational and extrapolation techniques described in discussing multiple equilibria in Sections 2.6 and 4.1 are helpful in providing initial estimates of the parameters.

The generalization of Eq. (5.19) is obvious, but it is not very useful because of the rapid increase in number of parameters. When additional equilibria are included in the model, it is usually necessary to invoke independent relationships among equilibrium constants in order to reduce the number of parameters, as was done by Davis and Schuster (24) in studying the interaction of trifluoroacetic acid with organic bases.

Most of the multiple equilibria systems that have been studied by NMR have been treated as $1:1 + 1:2$ systems, so that Eq. (5.19) is applicable. Foster and his co-workers (19, 20, 25) have studied many complexes of the charge-transfer type by NMR, finding for some of these that Scatchard plots [according to Eq. (5.13)] were curved (concave upward). Since these were moderately strong complexes, they could be studied in the dilute solution range where general nonideality effects were considered to be minor, and the curved plots were ascribed to the presence of $1:2$ complexes. The binding constants were evaluated by nonlinear regression. Gelb et al. (26) interpreted their C-13 NMR data on p-methylcinnamate anion (S): α-cyclodextrin (L), including observations of nuclei on both the substrate and ligand, in terms of $1:1 + 1:2$ complexes, using nonlinear regression.

When binding data clearly do not fit the $1:1$ binding model, the decision must be made whether to account for the deviations in terms of nonideal behavior or multiple equilibria. Foster's group has considered the question at length in the context of NMR data (25). A general comment is that it is usually more likely that deviations are due to nonideal behavior in weakly complexing systems, for in these the ligand concentration may extend out of the dilute range, whereas in strongly bound systems the multiple equilibria model is often more probable.

Self-Association

NMR has been widely used to study self-association equilibria, especially in hydrogen-bonded systems; Davis and Deb (27) have reviewed this subject. Most of the treatments have assumed that only the monomer and one associated species (which may be the dimer, trimer, tetramer, etc.) are present, this limitation being a result of the difficulty of analyzing a more general description. We use the symbolism of Section 2.4, "Examples of Models," thus we write

$$hS \rightleftharpoons S_h$$

$$\beta_h = \frac{[S]_h}{[S]^h} \qquad (5.20)$$

$$S_t = [S] + h[S_h] \qquad (5.21)$$

The fractions of S and S_h are f_1 and f_h, respectively; applying the usual fast exchange weighted-average assumption gives for the observed chemical shift

$$\delta = f_1 \delta_S + f_h \delta_h \qquad (5.22)$$

or, since $f_1 + f_h = 1$,

$$\delta = \delta_S + f_h(\delta_h - \delta_S) \qquad (5.23)$$

where δ_S and δ_h are the chemical shifts of S and S_h. If indeed S and S_h are the only species present, Eq. (5.23) shows that as $S_t \rightarrow 0$, $f_h \rightarrow 0$, and $\delta \rightarrow \delta_S$, whereas as $S_t \rightarrow \infty$, $f_h \rightarrow 1$, and $\delta \rightarrow \delta_h$. The extrapolation to zero S_t is often sound, but extrapolation to the presumed condition $f_h = 1$ is not usually valid because higher association products may form.

Saunders and Hyne (28) used a curve-fitting method to find h and β_h. Combining Eqs. (5.20) and (5.21) gives

$$S_t = [S] + h\beta_h[S]^h \qquad (5.24)$$

The fraction $f_h = h[S_h]/S_t$ is combined with Eqs. (5.20) and (5.23), giving

$$\delta = \delta_S + \frac{h\beta_h[S]^h}{S_t}(\delta_h - \delta_S) \qquad (5.25)$$

Arbitrarily chosen values of [S], in the experimental range, are inserted into Eq. (5.24) for a given h. With an assumed value for β_h the corresponding S_t are calculated. These quantities are used in Eq. (5.25) to calculate δ as a function of S_t, with δ_h also serving as an adjustable parameter. For a given value of h, the sigmoid curve of δ as a function of $\log S_t$ has a fixed shape, and different values of β_j correspond to horizontal displacements of the curve. The technique tends to emphasize data points in the central concentration region, and deviations are sometimes serious at the extreme ranges. Marcus and Miller have elaborated the method (29).

Most workers have taken $h = 2$ in the preceding equations and have investigated the monomer–dimer equilibrium, which is often the most important one, at least in the low concentration range. Then Eq. (5.25) becomes

$$\Delta = \frac{2\beta_2[S]^2}{S_t}\,\Delta_2 \tag{5.26}$$

where $\Delta = \delta - \delta_S$ and $\Delta_2 = \delta_2 - \delta_S$. For $h = 2$, Eq. (5.24) becomes a quadratic in [S], and with assumed values of β_h this is solved for [S]. The observed δ or Δ is plotted against $f_2 = 2\beta_2[S]^2/S_t$, as in Eq. (5.23), and β_h is adjusted until the best straight line is obtained (30). This is a variant of the Creswell–Allred method (8). In a related technique Bangerter and Chan (31) write f_2 as an explicit function of S_t and insert this in Eq. (5.23) to get

$$\Delta = \frac{1 + 4\beta_2 S_t - (1 + 8\beta_2 S_t)^{1/2}}{4\beta_2 S_t}\,\Delta_2 \tag{5.27}$$

Least-squares regression is then used to find β_2.

Huggins et al. (32) introduced a limiting-slope method. The observed chemical shift δ is plotted against S_t, and the initial slope $(d\delta/dS_t)_{S_t=0}$ is estimated. From Eqs. (5.24) and (5.26) are found $dS_t/d[S]$ and $d\delta/d[S]$, giving $d\delta/dS_t = (d\delta/d[S])(d[S]/dS_t)$.

$$\frac{d\delta}{dS_t} = \frac{2\beta_2\Delta_2}{(1 + 2\beta_2[S])^2(1 + 4\beta_2[S])} \tag{5.28}$$

The slope at $S_t = 0$ is therefore equal to $2\beta_2\Delta_2$. This method has the advantage that only the monomer–dimer equilibrium should be important at $S_t \to 0$, but it requires an estimate of Δ_2 in order that β_2 may be found. Davis and Debs (24) have discussed some approaches to estimating Δ_2 in self-associating systems.

Lippert (33) and Littlewood and Willmott (34) have developed extrapolation schemes to find δ_2. In Eq. (5.23) f_h is expressed in terms of chemical shift differences. Similarly, by eliminating f_h in Eq. (5.22) we obtain $(\delta - \delta_h) = f_1(\delta_s - \delta_h)$. For the case of dimerization we can write

$$\beta_2 = \frac{[S_2]}{[S]^2} = \frac{1 - f_1}{2 S_t f_1^2} \tag{5.29}$$

Substituting for f_1 gives

$$2 S_t \beta_2 = \frac{1 - (\delta - \delta_2)/(\delta_s - \delta_2)}{(\delta - \delta_2)^2/(\delta_s - \delta_2)^2} \tag{5.30}$$

When S_t is large, $(\delta - \delta_2) \approx 0$, the numerator is essentially unity, and Eq. (5.30) is rearranged to

$$\delta = \delta_2 + a S_t^{-1/2} \tag{5.31}$$

where $a = (\delta_s - \delta_2)/(2\beta_2)^{1/2}$. A plot of δ against $S_t^{-1/2}$ can be extrapolated from the high S_t region to obtain δ_2. With this quantity the right-hand side of Eq. (5.30) can be evaluated, and a plot of this against S_t gives a straight line whose slope yields β_2.

A linear graphical method for dimerization is based on a treatment of IR data by Spurr and Byers (35). The mass balance on S_t is combined with β_2 to obtain $[S_2]$ as a function of S_t and β_2, and at this stage a term in $[S_2]^2$ is neglected, which limits the applicability of the method. The expression for $[S_2]$ is used in $f_2 = 2[S_2]/S_t$, which is inserted in Eq. (5.23):

$$\Delta = \frac{2\Delta_2 \beta_2 S_t}{1 + 4\beta_2 S_t} \tag{5.32}$$

Equation (5.32) has the hyperbolic form, and it can be rearranged into the usual three linear forms, of which the double-reciprocal form is most frequently seen.

The inclusion of more than one self-associated species in the stoichiometric model is usually realistic, but the resulting functions are too complicated for general analysis. Simplifications are therefore introduced.

Thus, for example, Littlewood and Willmott (34) considered that successive linear polymers $S, S_2, S_3, \ldots, S_h$ might be formed in the 1-dodecanol system, and set stepwise binding constants $K_h = [S_h]/[S][S_{h-1}]$ all equal. Another form of simplification is to assume that all δ_h are equal, that is, $\delta_2 = \delta_3 = \delta_4$, and so on. Such assumptions reduce the number of parameters to a manageable set, but the assumptions may not be valid.

LaPlanche et al. (36) treated a continuous self-association system with these assumptions: (1) the monomer–dimer equilibrium has a unique binding constant K_2; (2) all subsequent equilibria, that is, dimer–trimer, trimer–tetramer, and so on, have the identical constant $\bar{K}$; (3) all associated species have the same proton chemical shift. They also included the possibility that the solvent may hydrogen-bond to the self-associated species. It was then possible to find parameter estimates by a curve-fitting method that is essentially a complicated version of the Creswell–Allred (8) technique. An even more complicated system was studied by Davis and Schuster (24), with self-association of substrate, dimerization of ligand, and complexing of ligand monomer with all substrate species being written into the model.

5.2. NMR RELAXATION TIMES

T_1, T_2, and the Correlation Time

A set of nuclei (the spin system) in a steady magnetic field H_0 will absorb energy from a radiofrequency (rf) field H_1 (normal to H_0) at the Larmor precessional angular frequency ω_0 of the nuclei. Upon the absorption of energy, the population of nuclear energy levels is altered from the equilibrium Boltzmann distribution appropriate to the sample temperature, so a process occurs of energy transfer from the spin system to the rest of the sample (collectively referred to as the lattice), the driving force being the return to thermal equilibrium. This process is called spin–lattice relaxation, and it takes place at a rate governed by a first-order rate constant $1/T_1$, where T_1 is the spin–lattice relaxation time. An equivalent viewpoint is that absorption of energy from the rf field generates a component of the spin system magnetization vector in the H_0 direction (the z direction) different from the equilibrium value M_0 in the absence of energy absorption. Relaxation of M_z back to M_0 occurs with a rate $(M_0 - M_z)/T_1$.

There can also be interactions within the spin system. Adjacent nuclei can interact through a dipole–dipole mechanism to cause a transition. The precessing moment of the nucleus **1** sets up an oscillating field at nucleus **2**; this field will, at some time, be in phase with the precession of nucleus **2**, which can absorb energy from **1**. The result is a transfer of energy by a spin–spin relaxation process. Because the precessional frequencies of nuclei differ by a small amount $\delta\omega_0$ because of local induced-field differences, the length of time for two nuclei to come into phase is $1/\delta\omega_0$. This is the approximate lifetime of a nuclear state; it is written $1/T_2$, where T_2 is the spin–spin relaxation time. Reverting to the description in terms of magnetization vectors, absorption of energy by the spin system generates components of the magnetization vector in the xy plane (the plane of the H_1 field). These components, M_x and M_y, decay to their equilibrium values of zero with the rate constant $1/T_2$. The spin–spin relaxation phenomenon is reflected in resonance peak widths (the lifetime broadening effect), narrow lines corresponding to long T_2 values.

Both spin–lattice and spin–spin relaxation depend on rates of molecular motion, for relaxation results from the interaction of fluctuating magnetic fields set up by nuclei in the spin system and in the lattice. A quantitative theory of this dependence was given by Bloembergen et al. (37), who obtained

$$\frac{1}{T_1} \propto \frac{\tau_c}{1 + 4\pi^2\nu_0^2\tau_c^2} \tag{5.33}$$

In Eq. (5.33) ν_0 is the Larmor precessional frequency in hertz ($\omega_0 = 2\pi\nu_0$), and τ_c is the *correlation time*, a measure of the rate of molecular motion. The reciprocal of the correlation time is a frequency, and $1/\tau_c$ may receive additive contributions from several sources, in particular $1/\tau_r$, where τ_r is the rotational correlation time. τ_r is, approximately, the time taken for the molecule to rotate through one radian.

According to Eq. (5.33), when $2\pi\nu_0\tau_c \ll 1$, T_1 is proportional to $1/\tau_c$, whereas when $2\pi\nu_0\tau_c \gg 1$, T_1 is proportional to τ_c. The theory also shows that for small τ_c, $T_1 = T_2$, whereas at large τ_c, $T_2 < T_1$. From the point of view of binding phenomena, these relationships are important because T_1 and T_2 can be measured by fairly straightforward techniques, and their connection with τ_c can be interpreted in terms of the effect of binding on molecular motion. For small molecules in solution, $T_1 = T_2$ and both are in the approximate range 0.1 to 10 s (38), corresponding to correlation times of the order 10^{-11}–10^{-12} s, but macromolecules may have much longer correlation times.

The magnetic dipole–dipole interaction is a general mechanism for relaxation, but there are some special mechanisms that can be more important in certain cases. If a nucleus has an electric quadrupole moment (i.e., if its spin quantum number $I > \frac{1}{2}$), molecular motion results in fluctuating electric fields that can induce relaxation. This can be very efficient, resulting in short relaxation times and broad absorption peaks. Another relaxation mechanism is provided by paramagnetic substances, which possess unpaired electrons. Then the dominant contributor to the relaxation is the large electron magnetic moment of the paramagnetic substance.

Relaxation Techniques

The qualitative, that is structural, information to be obtained from relaxation studies on complexing systems comes from the relationship between relaxation times and rotational correlation times. An increase in correlation time results in a decrease in relaxation time (i.e., an increase in relaxation rate), which is manifested as an increase in absorption line width. It is possible for relaxation times to differ for different nuclei in the same molecule, because of their different rotational times. Consider a substrate that is bound to a ligand. If relaxation times of several nuclei in the bound substrate are measured, that portion of the molecule having the shortest relaxation time (relative to the unbound species) is implicated as the binding site, because this is the group with the longest rotational correlation time; it is more immobilized by binding. Jardetzky and co-workers (39) used this approach to study the protein binding of sulfonamides and of penicillin. However, the correlation time may not be determined solely by the (intramolecular) rotational motion, but can also be influenced by an intermolecular translational correlation time; Pumpernik et al. (40) observed that the relaxation time of the hydrogen-bonded complex of trichloroacetic acid and dimethylsulfoxide was determined by the translational component. Examples in which relaxation is dominated by the quadrupolar mechanism include complexes of trialkylaluminum compounds with amines and other bases, [27]Al being the nucleus observed (41), and the binding of [23]Na$^+$ by ionophores such as monactin, valinomycin, and monensin (42).

For the extraction of quantitative information, namely, binding constants, the mechanism of the alteration in relaxation time is not a consideration; all that is needed is a sufficient change in this property. Most of the applications have been to biochemical systems; this subject has been reviewed by James (43). The fast exchange limit is usually assumed, and Fisher and Jardetzky (39b) justified this for the binding of penicillin to serum albumin. The study by James and Noggle (44) of the binding of $^{23}\text{Na}^+$ to RNA is an example of the quantitative formulation. Let $R = 1/T_1$. (This quantity is usually called a rate, although we identified it earlier as a rate constant.) Then the usual weighted-average assumption is made, Eq. (5.34), where R_{10} and R_{11} are the reciprocals of the relaxation times for free and bound sodium ions, respectively, and f_{10}, f_{11} are the fractions of free and bound ions.

$$R = f_{10}R_{10} + f_{11}R_{11} \tag{5.34}$$

Note the similarity of Eq. (5.34) to Eq. (5.7). The same subsequent development leads to Eq. (5.35) as the binding isotherm, where $\Delta R = R - R_{10}$ and $\Delta R_{11} = R_{11} - R_{10}$.

$$\Delta R = \frac{\Delta R_{11} \, K_{11}[L]}{1 + K_{11}[L]} \tag{5.35}$$

In this development the probe ion (the sodium ion) is the substrate and the macromolecule (RNA) is the ligand; the model is for 1:1 binding, but the interpretation is that [L] is the molar concentration of binding sites.

Equation (5.35) can be linearized in the usual ways or solved for K_{11} and ΔR_{11} by nonlinear regression. The study by James and Noggle was unusual in that the total ligand concentration L_t was held constant while the total substrate concentration S_t was varied. Algebraic manipulation of the quantity $f_{11} = [SL]/S_t$ yields

$$f_{11} = \frac{K_{11}L_t}{1 + K_{11}(L_t + S_t - [SL])}$$

This is combined with $\Delta R = f_{11} \Delta R_{11}$ to give an alternative expression for the binding isotherm.

$$\Delta R = \frac{\Delta R_{11} \, K_{11}L_t}{1 + K_{11}(L_t + S_t - [SL])} \tag{5.36}$$

[Note the formal identity of Eq. (5.36) with Eq. (5.15)]. Under the experimental condition that $L_t \ll S_t$, this gives the linear form

$$\frac{1}{\Delta R} = \frac{S_t}{\Delta R_{11} L_t} + \frac{1}{\Delta R_{11} K_{11} L_t} \tag{5.37}$$

Thus a plot of $1/\Delta R$ against S_t yields the binding constant. James and Noggle found for this system $R_{10} = 17.5 \text{ s}^{-1}$ and $R_{11} = 222 \text{ s}^{-1}$.

An indirect method involves the measurement of the proton relaxation time for water (the solvent) in a solution of a paramagnetic ion (the substrate) and a ligand, usually a macromolecule. The effectiveness of the paramagnetic ion in relaxing the water protons depends on the immediate environment of the ion, namely, on whether or not the ion is bound to the ligand and, if bound, the accessibility of the water to the bound ion. Following Eisinger et al. (45) we define the relaxation enhancement ε by

$$\epsilon = \frac{(1/T_1)_{s,L} - (1/T_1)_L}{(1/T_1)_s - (1/T_1)} \tag{5.38}$$

The T_1 values are proton relaxation times of water, the subscripts denoting the presence or absence of the paramagnetic ion S and the ligand L. The denominator of Eq. (5.38) is the change in relaxation rate when the paramagnetic ion is added to water; it is simply a normalizing factor, with the result that $\epsilon = 1$ in the absence of ligand or if the ion does not bind to the ligand. An enhancement less than 1 may signify binding of the ion at a site inaccessible to solvent, whereas $\epsilon > 1$ suggests binding at an external accessible site; Eisinger et al. (45b) have discussed these inferences in terms of the correlation times and hydration sphere compositions of the bound ions. The technique has been extensively applied by Cohn (46) to study protein binding. Consider the simple case of 1:1 binding between the paramagnetic ion and the ligand. Then in the familiar way we define the weighted-average relaxation enhancement by

$$\epsilon = f_{10}\epsilon_{10} + f_{11}\epsilon_{11} \tag{5.39}$$

But by definition $\epsilon_{10} = 1$. The further development is completely analogous to earlier treatments, for example, the derivation of Eq. (5.10). The result is

$$(\epsilon - 1) = \frac{(\epsilon_{11} - 1)K_{11}[L]}{1 + K_{11}[L]} \tag{5.40}$$

Equation (5.40) can be treated as we have seen for many other equations having this form.

If the ligand has m identical sites, the treatment is modified (47) by writing the mass balance on ligand as $L_t = ([L] + [SL])/m$, where $[L]$ is interpreted as the molar concentration of free sites. This expression for $[L]$ is combined with Eq. (5.40) to yield, after algebraic manipulation:

$$\frac{\epsilon - 1}{\epsilon_{11} - 1} = \frac{mK_{11}L_t f(\epsilon)}{1 + K_{11}S_t f(\epsilon)} \tag{5.41}$$

where

$$f(\epsilon) = \frac{\epsilon_{11} - \epsilon}{\epsilon_{11} - 1} \tag{5.42}$$

Equation (5.41) is of the hyperbolic form, but to use it an estimate of ϵ_{11} is needed. Mildvan and Cohn (47) have discussed this situation for the binding of Mn(II) to bovine serum albumin. Equation (5.43) is the x-reciprocal or Scatchard form of Eq. (5.42).

$$\frac{(\epsilon - 1)}{L_t(\epsilon_{11} - \epsilon)} = mK_{11} - \frac{K_{11}S_t(\epsilon - 1)}{L_t(\epsilon_{11} - 1)} \tag{5.43}$$

This water proton relaxation method has been extended to systems capable of forming a ternary (1:1:1) complex, for example, of the paramagnetic ion (S), an enzyme (L), and another ligand such as a coenzyme (C). Then the system may contain S, L, SL, SC, and SLC species, if we ignore the various acid–base states. The observed resonance enhancement is a weighted average of contributions from the four paramagnetic ion species. Mass balance expressions and equilibrium constant definitions are introduced; Cohn and co-workers have worked this out in detail (48). They have developed graphical methods for estimating the parameters, but find that these methods give results in poor agreement with those found by nonlinear least-squares regression analysis.

5.3. ELECTRON SPIN RESONANCE

Electron spin resonance (ESR) is formally analogous to NMR. It is applicable to paramagnetic substances (hence the synonymous term electron paramagnetic resonance, EPR), namely, those that possess an unpaired electron. ESR is therefore less widely applicable to the measurement of binding constants than is NMR. Its most important feature is its capability of revealing details of the molecular environment of the paramagnetic species, and for this purpose a paramagnetic "spin probe" may be introduced into the system, for example, to examine the details of membrane structure. In the context of binding equilibria, the most useful applications have been to the study of ion association phenomena in which at least one of the ions is paramagnetic; often this is a radical anion.

The ESR time scale is shorter than is that of NMR. Microwave frequency radiation, of the order 10^9 Hz, generates the field normal to the applied magnetic field, and typical resonance frequency differences are of the order 10 MHz, corresponding to a characteristic time of roughly 10^{-7} s. In Section 5.1 we noted that the corresponding quantity in NMR is about 10^{-2} s, with the result that NMR observations of binding phenomena are usually in the fast exchange limit. In UV spectroscopy, on the other hand, with a time scale of about 10^{-15} s, we observe binding in the slow exchange limit. Thus we see that the ESR experiment provides a probe on a time scale intermediate to NMR and UV spectroscopy. As a consequence, some equilibria are in the slow exchange limit and some are in the fast exchange limit when investigated by ESR. The appearance of the spectrum distinguishes between these limiting possibilities, with slow exchange being characterized by the additive presence of lines from both species participating in the equilibrium; in fast exchange, however, coalescence to a time-averaged spectrum is seen.

Many systems have been studied in the slow exchange limit, the equilibrium constants being estimated from the line intensities. (In ESR the derivative of the absorption line is observed, and the peak area is taken as the measure of intensity.) Atherton and Weissman (49) observed the system sodium ion–naphthalene anion in tetrahydrofuran and similar solvents and were able to calculate equilibrium constants on the assumption of ion pair formation. Much of the interest in such systems is in establishing the existence and relative importance of more than one form of the ion pair, as described in the following equilibria for a metal ion M^+ in aqueous solution in the presence of an anion L^-.

$$M^+(aq) + L^-(aq) \underset{a}{\overset{K_{ab}}{\rightleftharpoons}} \underset{b}{M^+(aq)L^-} \overset{K_{bc}}{\rightleftharpoons} \underset{c}{M^+L^-(aq)}$$

State *a* consists of the dissociated free ions, the symbol (aq) signifying the hydration shell of an ion [its cosphere in Gurney's (50) terminology]. State *b* is called an outer-sphere complex, loose ion pair, or solvent-separated ion pair, and state *c* is an inner-sphere complex, tight ion-pair, or intimate ion pair. Either M^+ or L^- or both of these may be paramagnetic, and the ESR signal is sensitive to the environment of the paramagnetic species, so in the slow exchange case it may be possible to distinguish among the states and thus to evaluate the equilibrium constants. When only one equilibrium need be considered, an analytical solution is easily obtained (49, 51), but the more general case requires curve fitting. Among the ion-pair systems that have been studied by ESR intensity measurements are Na^+–naphthalene anion (49, 52), alkali metal ion–durosemiquinone (51), Mn^{2+} with both diamagnetic and paramagnetic anions (53), and K^+–benzoquinone anion (54).

In the fast exchange condition the ESR parameters that are observed are weighted averages of the quantities characteristic of the states contributing to the signal. The situation is exactly analogous to the NMR example. Corresponding to the chemical shift difference in NMR is the *g*-factor difference; in fact Allendoerfer and Papez (51) have recommended that a *g* shift be defined on a ppm basis just as the NMR δ shift is. The observed *g*-shift difference is written as a weighted average, and the development proceeds as in the treatment of Eq. (5.7), so further elaboration seems unnecessary. Linear plotting forms are obtained as usual. Stevenson and Alegria (55) studied the K^+-2,6-di-*tert*-butylbenzoquinone anion system in this way. The weighted-average description of fast exchange observables. has also been employed to interpret ESR relaxation times (56) and hyperfine splitting constants (57).

REFERENCES

1. E. D. Becker, *High Resolution NMR*, Academic, New York, 1969, Ch. 10.
2. (a) H. S. Gutowsky, D. W. McCall, and C. P. Slichter, *J. Chem. Phys.*, **21**, 279 (1953); (b) H. S. Gutowsky and A. Saika, *J. Chem. Phys.*, **21**, 1688 (1953).
3. J. Feeney, J. G. Batchelor, J. P. Albrand, and G. C. K. Roberts, *J. Magn. Reson.*, **33**, 519 (1979).
4. (a) M. W. Hanna and A. L. Ashbaugh, *J. Phys. Chem.*, **68**, 811 (1964); (b) R. Foster and C. A. Fyfe, *J. Chem. Soc.*, B, 926 (1966); (c) W. R. Carper, C. M. Buess, and G. R. Hipp, *J. Phys. Chem.*, **74**, 4229 (1970); (d) H. N.

Wachter and V. Fried, *J. Chem. Educ.*, **51**, 798 (1974); (e) K. F. Wong and S. Ng, *Spectrochim. Acta*, **32A**, 455 (1976); (f) B. K. Seal, A. K. Mukherjee, D. C. Mukherjee, P. G. Farrell, and J. V. Westwood, *J. Magn. Reson.*, **51**, 318 (1983); (g) R. Mathur, E. D. Becker, R. B. Bradley, and N. C. Li, *J. Phys. Chem.*, **67**, 2190 (1963).

5. G. Carta, G. Crisponi, and A. Lai, *J. Magn. Reson.*, **48**, 341 (1982).

6. J. Granot, *J. Magn. Reson.*, **55**, 216 (1983).

7. C. M. Huggins, G. C. Pimentel, and J. N. Shoolery, *J. Chem. Phys.*, **23**, 1244 (1955).

8. C. J. Creswell and A. L. Allred, *J. Phys. Chem.*, **66**, 1469 (1962).

9. K. A. Connors and J. A. Mollica, *J. Pharm. Sci.*, **55**, 772 (1966).

10. R. Sahai, G. L. Loper, S. H. Lin, and H. Eyring, *Proc. Nat. Acad. Sci. US*, **71**, 1499 (1974).

11. R. Foster and C. A. Fyfe, *Prog. NMR Spectrosc.*, **4**, 1 (1969).

12. M. Nakano, N. I. Nakano, and T. Higuchi, *J. Phys. Chem.*, **71**, 3954 (1967).

13. T. Gramstad and O. Mundheim, *Spectrochim. Acta*, **28A**, 1405 (1972).

14. H. Stamm, W. Lamberty, and J. Stafe, *Tetrahedron*, **32**, 2045 (1976).

15. K. A. Connors and D. D. Pendergast, *Anal. Chem.*, **56**, 1549 (1984).

16. (a) H. Stamm, W. Lamberty, and J. Stafe, *J. Am. Chem. Soc.*, **102**, 1529 (1980); (b) H. Stamm and W. Lamberty, *Tetrahedron*, **37**, 565 (1981).

17. J. Homer, *J. Magn. Reson.*, **34**, 31 (1979).

18. J. Homer, E. J. Hartland, and C. J. Jackson, *J. Chem. Soc.*, *A*, 931 (1970).

19. J. A. Chudek, R. Foster, and D. J. Livingstone, *J. Chem. Soc.*, *Faraday Trans.*, *I*, 1222 (1979).

20. J. A. Chudek, R. Foster, and F. M. Jarrett, *J. Chem. Soc.*, *Faraday Trans.*, *I*, 2729 (1983).

21. (a) Y. Y. Lim and R. S. Drago, *J. Am. Chem. Soc.*, **94**, 84 (1972); (b) M. W. Hanna and D. G. Rose, *J. Am. Chem. Soc.*, **94**, 2601 (1972); (c) F. L. Slejko and R. S. Drago, *J. Am. Chem. Soc.*, **94**, 6546 (1972); (d) J. Homer, M. H. Everdell, C. J. Jackson, and P. M. Whitney, *J. Chem. Soc.*, *Faraday Trans.*, *II*, 874 (1972).

22. I. Horman and S. Dreux, *Anal. Chem.*, **56**, 299 (1984).

23. (a) T. O. Maier and R. S. Drago, *Inorg. Chem.*, **11**, 1861 (1972); (b) R. E. Lenkiski, G. A. Elgavish, and J. Reuben, *J. Magn. Reson.*, **32**, 367 (1978).

24. J. P. Davis and I. I. Schuster, *J. Solution Chem.*, **13**, 167 (1984).

25. (a) B. Dodson, R. Foster, A. A. S. Bright, M. I. Foreman, and J. Gorton, *J. Chem. Soc.*, *B*, 1283 (1971); (b) A. A. S. Bright, J. A. Chudek, and R. Foster, *J. Chem. Soc.*, *Perkin Trans.*, *2*, 1256 (1975); (c) R. J. Bailey, J. A. Chudek, and R. Foster, *J. Chem. Soc.*, *Perkin Trans.*, *2*, 1590 (1976).

26. R. I. Gelb, L. M. Schwartz, and D. A. Laufer, *J. Am. Chem. Soc.*, **100**, 5875 (1978).

27. J. C. Davis, Jr., and K. K. Deb, *Adv. Magn. Reson.*, **4**, 201 (1970).

28. M. Saunders and J. B. Hyne, *J. Chem. Phys.*, **29**, 253, 1319 (1958).
29. S. H. Marcus and S. I. Miller, *J. Am. Chem. Soc.*, **88**, 3719 (1966).
30. (a) C. F. Jumper, M. T. Emerson, and B. B. Howard, *J. Chem. Phys.*, **35**, 1911 (1961); (b) I. Horman and B. Dreux, *Helv. Chim. Acta*, **67**, 754 (1984).
31. B. W. Bangerter and S. I. Chan, *J. Am. Chem. Soc.*, **91**, 3910 (1969).
32. C. M. Huggins, G. C. Pimentel, and J. N. Shoolery, *J. Phys. Chem.*, **60**, 1311 (1956).
33. (a) E. Lippert, *Ber. Bunsenges. Phys. Chem.*, **67**, 267 (1963); (b) U. Jentschura and E. Lippert, *Ber. Bunsenges. Phys. Chem.*, **75**, 556 (1971).
34. A. B. Littlewood and F. W. Willmott, *Trans. Faraday Soc.*, **62**, 3287 (1966).
35. R. A. Spurr and H. F. Byers, *J. Phys. Chem.*, **62**, 425 (1958).
36. L. A. LaPlanche, H. B. Thompson, and M. T. Rogers, *J. Phys. Chem.*, **69**, 1482 (1965).
37. N. Bloembergen, E. M. Purcell, and R. V. Pound, *Phys. Rev.*, **73**, 679 (1948).
38. E. D. Becker, *High Resolution NMR*, Academic, New York, 1969, p. 205.
39. (a) O. Jardetzky and N. G. Wade-Jardetzky, *Mol. Pharmacol.*, **1**, 214 (1965); (b) J. J. Fisher and O. Jardetzky, *J. Am. Chem. Soc.*, **87**, 3237 (1965).
40. D. Pumpernik, G. Lahajnar, D. Hadži, and A. Ažman, *Chem. Phys. Lett.*, **26**, 53 (1974).
41. (a) H. E. Swift, C. P. Poole, and J. F. Itzel, *J. Phys. Chem.*, **68**, 2509 (1964); (b) L. Petrakis and H. E. Swift, *J. Phys. Chem.*, **72**, 546 (1968).
42. D. H. Haynes, B. C. Pressman, and A. Kowalsky, *Biochemistry*, **10**, 852 (1971).
43. T. L. James, *Nuclear Magnetic Resonance in Biochemistry*, Academic, New York, 1975, Ch. 6.
44. T. L. James and J. H. Noggle, *Proc. Natl. Acad. Sci., U.S.*, **62**, 644 (1969).
45. (a) J. Eisinger, R. G. Schulman, and W. E. Blumberg, *Nature (London)*, **192**, 963 (1961); (b) J. Eisinger, R. G. Schulman, and B. M. Szymanski, *J. Chem. Phys.*, **36**, 1721 (1962).
46. M. Cohn, *Biochemistry*, **2**, 623 (1963).
47. A. S. Mildvan and M. Cohn, *Biochemistry*, **2**, 910 (1963).
48. (a) A. S. Mildvan and M. Cohn, *J. Biol. Chem.*, **241**, 1178 (1966); (b) W. J. O'Sullivan and M. Cohn, *J. Biol. Chem.*, **243**, 2737 (1968); (c) G. H. Reed, M. Cohn, and W. J. O'Sullivan, *J. Biol. Chem.*, **245**, 6547 (1970).
49. N. M. Atherton and S. I. Weissman, *J. Am. Chem. Soc.*, **83**, 1330 (1961).
50. R. W. Gurney, *Ionic Processes in Solution*, McGraw-Hill, New York, 1953; reprinted by Dover Publications, New York, 1962, p. 4.
51. R. D. Allendoerfer and R. J. Papez, *J. Phys. Chem.*, **76**, 1012 (1972).
52. K. Höfelmann, J. Jagur-Grodzinski, and M. Szwarc, *J. Am. Chem. Soc.*, **91**, 4645 (1969).

53. (a) L. Burlamacchi and E. Tiezzi, *J. Mol. Struct.*, **2**, 261 (1968); (b) L. Burlamacchi, G. Martini, and E. Tiezzi, *J. Phys. Chem.*, **74**, 3980 (1970).

54. G. R. Stevenson and A. E. Alegria, *J. Phys. Chem.*, **77**, 3100 (1973).

55. G. R. Stevenson and A. E. Alegria, *J. Phys. Chem.*, **79**, 1042 (1975).

56. G. R. Stevenson, R. Concepción, and I. Ocasio, *J. Phys. Chem.*, **80**, 861 (1976).

57. G. R. Stevenson and H. Hidalgo, *J. Phys. Chem.*, **77**, 1027 (1973).

6

REACTION
KINETICS

6.1. COMPLEX FORMATION AND CHEMICAL REACTIVITY

Any property of a substrate that is altered upon complex formation with a ligand may serve as a basis for measuring the binding constant of the complexation equilibrium. When this property is chemical reactivity, as expressed quantitatively in a rate or a rate constant, we have the subject of the present chapter. Thus we are not concerned here with measurements of rates of association and dissociation of complexes [although such measurements themselves can be useful in estimating binding constants (1)]. We shall therefore usually make the assumption that the system is at equilibrium with respect to the binding process; that is, the rate of attainment of the binding equilibrium is considered to be much greater than the rate of the "probe" reaction. Exceptions to this condition will be noted.

There are three kinds of experimental situations in which chemical reactivity and complex formation must both be considered in analyzing the system. In one of these the complex is a possible intermediate in a reaction pathway, and the essential goal of the study is the elucidation of the mechanism of the reaction. Kosower (2) has reviewed the chemistry of reactions proceeding by means of charge-transfer complexes. The entire field of enzyme chemistry is an example of this class of systems. A second situation is a more-or-less contrived one in which the system is designed to exhibit kinetic behavior analogous to that in the preceding class of system, the usual goal being to achieve better understanding of the complicated system by studies of the simple analog. In this way "enzyme models" have been extensively studied, as reviewed for micellar systems by Fendler and Fendler (3) and for cyclodextrins by Bender and Komiyama (4). Morawetz (5) has reviewed catalysis by polymers and has expressed reservations about the value of such model studies.

In the third type of situation the experimentalist is primarily interested in the complex, and the chemical reaction is selected for convenience and effectiveness. The more complicated the kinetics of the probe reaction, the less useful the method as a tool for studying the binding. Because of the many possibilities for kinetic complications, it is not feasible to attempt to develop a general description of the method. Instead the simplest cases will be described, with some extensions to indicate how more complicated systems might be handled. The (partial) freedom to choose the reaction makes this approach useful.

Let the substrate S be the interactant that undergoes structural change in the chemical reaction; the concentration of S is monitored in the kinetic study. The ligand L is the substance whose binding to S is the feature of interest; usually, the concentration of L is the independent variable. The system may also include a reagent R, which is chosen with the knowledge or hope that it reacts with S but not with L, and that it complexes with neither. We designate reaction products as P.

There are two types of systems to consider. In one of these the only initial solute species are S and L. Complex formation between these can take place, but in addition it is possible for L either to react with S in a "chemical" (covalent) reaction or to catalyze the solvolysis of S. The interpretation of the role of the complex is subject to ambiguity, however (6–8). One possibility is that S and L react in a bimolecular process to form P, but that S and L also interact to form the (assumed 1:1) complex SL, which is unreactive; this is shown as

$$S + L \underset{\text{fast}}{\rightleftharpoons} SL$$

$$S + L \xrightarrow[\text{slow}]{k} P$$

Scheme I

(In this scheme "fast" and "slow" are relative terms, implying only that the system is essentially at equilibrium with respect to complex formation.) The rate equation is $v = k[S][L]$, or, since $f_{10} = [S]/S_t = 1/(1 + K_{11}[L])$ from Eq. (2.51),

$$v = \frac{kS_t[L]}{1 + K_{11}[L]} \qquad (6.1)$$

Thus the rate is a hyperbolic function of ligand concentration. However, Scheme II could also be written for this system:

$$S + L \underset{}{\overset{\text{fast}}{\rightleftharpoons}} SL$$

$$SL \underset{\text{slow}}{\overset{k'}{\longrightarrow}} P$$

Scheme II

In Scheme II the ligand is functioning as a catalyst through the complex SL, which is said to be "on the reaction path." The rate equation is $v = k'[SL] = k'K_{11}[S][L]$, or

$$v = \frac{k'K_{11}S_t[L]}{1 + K_{11}[L]} \qquad (6.2)$$

Equations (6.1) and (6.2) are kinetically indistinguishable, so the applicable mechanism cannot be determined on the basis of the rate data alone. Independent evidence, in the form of a reasonable estimate of the expected bimolecular rate constant for the reaction, is needed to decide between the possibilities. Note, however, that the binding constant can be estimated from either form of the kinetic description.

The second type of system is composed of substrate S, ligand L, and reagent R. There are many possibilities for mutual interactions with complicating kinetic consequences, but this experimental design can be very useful if S and L do not undergo "chemical" reaction; then the addition of R to the system provides a reaction between S and R that constitutes the observational tool. This kind of system is examined in the following sections.

6.2. 1:1 BINDING SYSTEMS

Inhibition by Complexation

For a system containing only S and L, kinetic schemes I and II can be written. The kinetic indistinguishability of these schemes has been pointed out. There is an associated ambiguity, namely, that on the basis of the kinetic data alone, we do not know whether the rate is *decreased* by complexation (Scheme I) or whether it is *increased* by complexation (Scheme II) (7). Let us suppose that, with the aid of an independent estimate of k for the reaction, it has been concluded that Scheme I is applicable. A convenient experimental approach is to evaluate the pseudofirst-order rate constant defined by $v = k_{obs}S_t$, where we take S_t to represent the total concentration of substrate at the time of measurement. From Eq. (6.1) we get

$$k_{obs} = \frac{k[L]}{1 + K_{11}[L]} \qquad (6.3)$$

Thus the parameters k and K_{11} can be evaluated from measurements of k_{obs} as a function of ligand concentration.

We now turn to a treatment of a system containing S, L, and R subject to

$$S + L \overset{K_{11}}{\rightleftharpoons} SL$$

$$S + R \overset{k_S}{\longrightarrow} P \qquad \qquad \textit{Scheme III}$$

$$SL + R \overset{k_{11}}{\longrightarrow} P$$

In this scheme, k_S and k_{11} are second-order rate constants for the reaction of R with S and SL, respectively. We assume that R does not form complexes with S or L, and that the complexation process is essentially at equilibrium. The hypothetical rate equation is

$$v = k_S[S][R] + k_{11}[SL][R]$$

If the study is carried out under pseudofirst-order conditions ([R] essentially constant), the experimental rate equation is

$$v = k_{obs}S_t$$

(Note that S_t, the total substrate concentration, decreases with time.) Setting these equal gives

$$k_S' = k_S f_{10} + k_{11} f_{11} \qquad (6.4)$$

where $k_S' = k_{obs}/[R]$ is the apparent second-order rate constant and f_{10}, f_{11}

are the fractions of substrate present as S and SL, respectively. Equation (6.4) can be written $\Delta k_S = f_{11}(k_S - k_{11})$, where $\Delta k_S = k_S - k_S'$. Introducing the definition

$$q_{11} = 1 - \frac{k_{11}}{k_S} \tag{6.5}$$

allows the binding isotherm to be expressed as in Eq. (6.6).

$$\frac{\Delta k_S}{k_S} = \frac{q_{11}K_{11}[L]}{1 + K_{11}[L]} \tag{6.6}$$

Equation (6.6) is identical in form with the isotherms derived for optical spectroscopy [Eq. (4.5)] and NMR [Eq. (5.10)]. The three linear plotting forms are obtained as usual. Equations (6.7), (6.8), and (6.9) are the double-reciprocal, y-reciprocal, and x-reciprocal forms, respectively.

$$\frac{k_S}{\Delta k_S} = \frac{1}{q_{11}K_{11}[L]} + \frac{1}{q_{11}} \tag{6.7}$$

$$\frac{k_S[L]}{\Delta k_S} = \frac{[L]}{q_{11}} + \frac{1}{q_{11}K_{11}} \tag{6.8}$$

$$\frac{\Delta k_S}{k_S[L]} = -K_{11}\left(\frac{\Delta k_S}{k_S}\right) + q_{11}K_{11} \tag{6.9}$$

Several points are of interest. As these equations are written, the rate variable is the ratio $k_S/\Delta k_S$ or its reciprocal, so it is not necessary actually to calculate the second-order rate constants; the corresponding function of the observed pseudofirst-order rate constants has the same value. The analytical procedure used in the kinetic studies will determine the initial value of S_t, and with a sufficiently sensitive method it is often feasible to set S_t much smaller than L_t, thus permitting the approximation $[L] = L_t$ to be used. This is also advantageous if L complexes with P, for then this phenomenon (which has no direct effect on the rate if the reaction is irreversible) may be neglected. If optical spectroscopy is used to follow the kinetics, the absorbances extrapolated back to zero time give a set of data for the spectroscopic estimation of K_{11}, and likewise the "infinity time" absorbance values provide an estimate of complexing between P and L.

The inhibition of ester hydrolysis by complex formation has been extensively studied by the kinetic method (9, 10). Table 6.1 gives rate

Table 6.1. Pseudofirst-Order Rate Constants for the Alkaline Hydrolysis of Methyl *trans*-Cinnamate in the Presence of Theobromine Anion[a,b]

L_t/M	$10^3 \, k_{obs}/s^{-1}$	$\dfrac{\Delta k_s}{k_s}$
0.000	6.45	0.000
0.022	5.20	0.194
0.027	4.89	0.242
0.033	4.72	0.268
0.044	4.33	0.329
0.062	3.82	0.408
0.091	3.17	0.509

[a]pH 12.88, 25°C. Initial ester concentration = 4.6×10^{-4} M.
[b]From ref. 11.

data for the alkaline hydrolysis of methyl *trans*-cinnamate (S) in the presence of theobromine anion (L); in this system the hydroxide ion plays the role of R. Figure 6.1 is a plot of k_{obs} against L_t, and Figure 6.2 shows the double-reciprocal plot according to Eq. (6.7). From this plot the parameters $K_{11} = 11.5 \, M^{-1}$ and $q_{11} = 1.00$ were evaluated (11).

If two 1:1 complexes, SL and LS, are formed, with second-order rate constants k_{SL} and k_{LS}, the same type of development leads to

$$\frac{\Delta k_S}{k_S} = q_{SL} f_{SL} + q_{LS} f_{LS} \tag{6.10}$$

Using Eq. (5.17) for the fractions gives Eq. (6.11).

$$\frac{\Delta k_S}{k_S} = \frac{(q_{SL} K_{SL} + q_{LS} K_{LS})[L]}{1 + (K_{SL} + K_{LS})[L]} \tag{6.11}$$

With the aid of the linear forms of this equation the following identities are readily found:

$$K_{11} = K_{SL} + K_{LS} \tag{6.12}$$

$$q_{11} = \frac{q_{SL} K_{SL} + q_{LS} K_{LS}}{K_{SL} + K_{LS}} \tag{6.13}$$

Thus, as we have seen for optical and NMR spectroscopy, the experimentally evaluated K_{11} is the sum of all 1:1 binding constants, and q_{11}

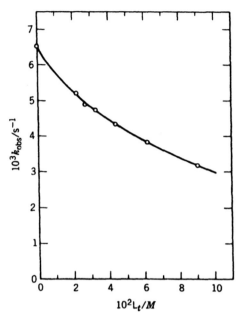

Figure 6.1. Pseudo-first-order rate constants for the hydrolysis of methyl cinnamate as a function of ligand concentration. Data are from Table 6.1.

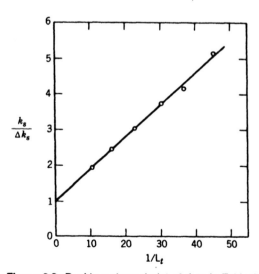

Figure 6.2. Double-reciprocal plot of data in Table 6.1.

223

is a weighted average. These conclusions are correct even if the reactivity of one of the complexes is the same as that of the uncomplexed substrate (12, 13).

According to Eq. (6.5), the parameter q_{11} can be interpreted as the fractional decrease in reactivity of the substrate upon complexation. A relationship between q_{11} and the difference in free energy of activation between S and SL is obtained from the transition state theory expression (6.14) for k_S and the corresponding equation for k_{11}.

$$k_S = \frac{kT}{h} e^{-\Delta G_S^{\ddagger}/RT} \tag{6.14}$$

Combining these with Eq. (6.5) gives

$$\delta_R \Delta G^{\ddagger} = -2.3RT \log(1 - q_{11}) \tag{6.15}$$

where $\delta_R \Delta G^{\ddagger} = \Delta G_{11}^{\ddagger} - \Delta G_S^{\ddagger}$. Figure 6.3 is a plot of $\delta_R \Delta G^{\ddagger}$ as a function of q_{11}. It is evident that small values of q_{11} correspond to very minor perturbations of the reactivity of the substrate, whereas values of q_{11} in excess of 0.9 represent major effects. Typical uncertainties in q_{11} estimates are 0.05 to 0.10, which do not result in large energy errors at low q_{11} values but can lead to difficulties of interpretation at high q_{11} values.

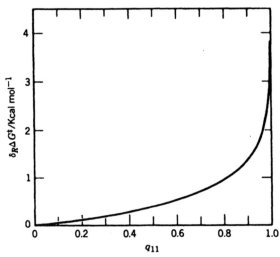

Figure 6.3. Plot of Eq. (6.15), the vertical axis giving the difference in free energy of activation (kcal mol^{-1}) between SL and S.

Because of the implications of Eq. (6.15), q_{11} values can be useful indicators of complex structure. For example, complexes of cinnamates (S) with xanthines (L) yield, for attack at the sites shown in the following structure, $q_{11}(a) = 0.4$, $q_{11}(b) = 1.0$, $q_{11}(c) = 0.9 - 1.0$.

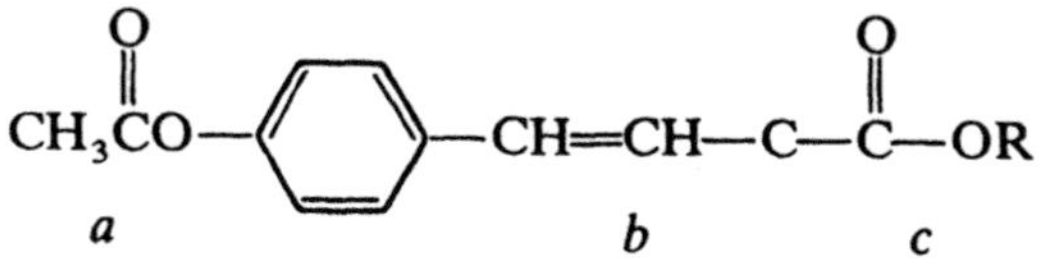

$$CH_3\overset{\overset{O}{\|}}{C}O-\langle\;\rangle-CH{=}CH-\overset{\overset{O}{\|}}{C}-\overset{}{C}-OR$$

$$\qquad a \qquad\qquad\qquad\qquad b \qquad\qquad c$$

These results strongly suggest that, in the complex, the ligand lies in the immediate vicinity of sites b and c, with only a minor polar effect being exerted on the reactivity of the acetoxy group (10, 14).

Catalysis by Complexation

If the ligand functions according to Scheme II, namely, by forming a complex SL that is more reactive than is S, the reaction rate is described by Eq. (6.2); it is assumed here that the uncatalyzed reaction of S is too slow to make an appreciable contribution. Enzyme-catalyzed reactions are of this type, and we illustrate with the simplest possible case. Scheme II is redrawn as

$$E + S \underset{k_{-1}}{\overset{k_1}{\rightleftharpoons}} ES$$

Scheme IV

$$ES \overset{k_2}{\longrightarrow} P + E$$

It has been found that it is not always valid to assume that an enzyme-catalyzed system is essentially at equilibrium with respect to formation of the enzyme–substrate complex. We therefore replace the equilibrium assumption with the steady-state approximation, and write Eqs. (6.16) and (6.17) for the complex.

$$\frac{d[ES]}{dt} = k_1[E][S] - k_{-1}[ES] - k_2[ES] = 0 \qquad (6.16)$$

$$[ES] = \frac{k_1[E][S]}{k_{-1} + k_2} \qquad (6.17)$$

Since enzyme studies are carried out with the condition $S_t \gg E_t$, we set

$[S] = S_t$. Equation (6.17) is combined with the mass balance on enzyme, $E_t = [E] + [ES]$, to give

$$[ES] = \frac{E_t S_t}{K_m + S_t} \tag{6.18}$$

where

$$K_m = \frac{k_{-1} + k_2}{k_1} \tag{6.19}$$

Combining Eq. (6.18) with the rate equation $v = k_2[ES]$ gives

$$v = \frac{V_{max} S_t}{K_m + S_t} \tag{6.20}$$

where $V_{max} = k_2 E_t$. Equation (6.20) is the Michaelis–Menten equation.

Because of complications such as product inhibition, enzyme kinetic studies usually are carried out by measuring initial rates rather than rate constants, so Eq. (6.20) is applied directly, S_t being the substrate concentration at zero time. The usual linear plotting forms are widely used; the double-reciprocal form is called the Lineweaver–Burk plot (15), and the x-reciprocal form is the Eadie plot (16). Further discussion of the data treatment in enzyme kinetics is given in Chapter 3. The field of enzyme kinetics is a very large one that cannot be pursued here.

It is important to observe that K_m, the Michaelis constant, is not, in general, an equilibrium constant. Only when $k_{-1} \gg k_2$ can K_m be interpreted as an equilibrium constant; in such cases it is the dissociation constant of the enzyme–substrate complex. Since it is the universal practice to define K_m according to Eq. (6.19), that is, as a dissociation constant (when $k_{-1} \gg k_2$), many workers studying enzyme models report their equilibrium constants as dissociation constants.

In a system containing S, L, and R, Scheme III is applicable, but now $k_{11} > k_S$. The quantitative development is similar to that given for the inhibition case but it is convenient to define $\Delta k_S = k_S' - k_S$ and

$$p_{11} = \frac{k_{11}}{k_S} - 1 \tag{6.21}$$

Then the isotherm is written

$$\frac{\Delta k_S}{k_S} = \frac{p_{11} K_{11}[L]}{1 + K_{11}[L]} \tag{6.22}$$

Although this has the same form as Eq. (6.6) for inhibition, note that p_{11} does not have the character of a fraction (unlike q_{11}, which possesses the limits of 0 and 1). Colter et al. (12) used the double-reciprocal form of Eq. (6.22) in studying the catalysis of the acetolysis of 2,4,7-trinitro-9-fluorenyl p-toluenesulfonate by phenanthrene. Van Etten et al. (17) preferred the x-reciprocal (Eadie) form in their enzyme model studies of the acceleration of phenol release from phenyl esters by cyclodextrins.

Concurrent Inhibition and Catalysis

It is possible that the ligand may produce concurrent opposing effects of inhibition and acceleration, but if only 1:1 complexes are present, the dependence of rate on ligand concentration will not reveal the presence of two processes. Consider kinetic scheme V, which is the same scheme that led earlier to Eq. (6.11), but we now suppose that $k_{SL} < k_S$ (inhibition) and $k_{LS} > k_S$ (catalysis).

$$S + L \xrightleftharpoons{K_{SL}} SL$$

$$S + L \xrightleftharpoons{K_{LS}} LS$$

$$S + R \xrightarrow{k_S} P \qquad\qquad Scheme\ V$$

$$SL + R \xrightarrow{k_{SL}} P$$

$$LS + R \xrightleftharpoons{k_{LS}} P$$

Carrying out the development as before, defining $\Delta k_S = k_S' - k_S$, gives the isotherm, Eq. (6.23), which is clearly equivalent to Eq. (6.11).

$$\frac{\Delta k_S}{k_S} = \frac{(-q_{SL}K_{SL} + p_{LS}K_{LS})[L]}{1 + (K_{SL} + K_{LS})[L]} \qquad (6.23)$$

Equation (6.23) has the same dependence on ligand concentration as is seen for the simple systems described by Eqs. (6.6) (inhibition) and (6.22) (catalysis), so the rate behavior does not by itself show that opposing effects are occurring. In the special case that $q_{SL}K_{SL} = p_{LS}K_{LS}$, evidently there will be no observed net effect of ligand on the rate, the inhibitory and catalytic effects exactly balancing each other.

Scheme VI is another, possibly more likely, scheme leading to opposing rate effects.

$$S + L \underset{K_{11}}{\overset{K_{11}}{\rightleftharpoons}} SL$$

$$S + R \xrightarrow{k_S} P$$

$$S + L + R \xrightarrow{k_L} P \qquad \text{\textit{Scheme VI}}$$

$$SL + R \xrightarrow{k'_{11}} P$$

We set $k_{11} < k_S$ (inhibition by complexation) and $k_L[L] > k_S$ (concurrent catalysis by ligand). Development as before gives Eq. (6.24), where $\Delta k_S = k'_S - k_S$.

$$\frac{\Delta k_S}{k_S} = \frac{(-q_{11}K_{11} + k_L/k_S)[L]}{1 + K_{11}[L]} \qquad (6.24)$$

Again we obtain the hyperbolic relationship characteristic of 1:1 binding. It is clear that the unsuspected presence of either the catalytic or the inhibitory effect would compromise interpretations of data in terms of Eq. (6.6) or Eq. (6.22) (10a).

6.3 NONHYPERBOLIC ISOTHERMS

We have seen, in Section 6.2, that 1:1 complexes yield the usual hyperbolic dependence of observed property (rate or rate constant) on ligand concentration, whether the ligand is responsible for a rate increase or a decrease, or both. When a nonhyperbolic isotherm is observed, a stoichiometric relationship other than 1:1 must be suspected. Suppose, for example, that SL and SL_2 complexes are formed; then, defining $\Delta k_S = k_S - k'_S$, the relationship Eq. (6.25) is obtained (13).

$$\frac{\Delta k_S}{k_S} = q_{11}f_{11} + q_{12}f_{12} \qquad (6.25)$$

Using Eqs. (2.55) this becomes

$$\frac{\Delta k_S}{k_S} = \frac{q_{11}K_{11}[L] + q_{12}K_{11}K_{12}[L]^2}{1 + K_{11}[L] + K_{11}K_{12}[L]^2} \qquad (6.26)$$

Equation (6.26) is of the same functional form as Eq. (4.28) for optical spectroscopy. The kinetic version does possess a potential advantage, however, in that experience on related systems may permit reasonable initial estimates of the parameters q_{11} and q_{12} to be made.

The most obvious deviation from a hyperbolic relationship is the appearance of a maximum in the isotherm. It is then evident that the ligand is exerting both acceleratory and inhibitory effects, and that some stoichiometry other than 1:1 must be operating. Hamilton et al. (18) show an example in the hydroxylation of anisole by H_2O_2 catalyzed by Fe(III) and catechol; it was postulated that hydroxylation occurs through formation of a 1:1:1 complex of H_2O_2, Fe(III), and catechol. A maximum is also seen in the dependence of the rate of reaction of t-butyllithium with ethylene in the presence of tetrahydrofuran (19). Figure 6.4 shows a maximum in the rate of hydrolysis of cinnamic anhydride as a function of the concentration of theophylline (20); this experiment will now be analyzed in illustration of an approach to this type of problem.

The rate constants in Figure 6.4 are pseudo-first-order constants for alkaline hydrolysis at pH 10.00; at this pH about 95% of the ligand (theophylline) is in the anion form. On the basis of observations on related systems, such as that shown in Figure 6.1, it is reasonable to suggest that both 1:1 (SL) and 1:2 (SL_2) complexes can form between cinnamic anhydride and theophylline anion. Moreover, we expect these complexes to be less susceptible to attack by hydroxide than is the uncomplexed substrate. The catalytic effect of the ligand is most reasonably attributed to nucleophilic attack of theophylline anion on the

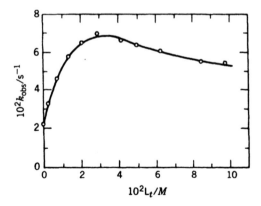

Figure 6.4. Effect of theophylline concentration on the rate of hydrolysis of cinnamic anhydride at pH 10 and 25°C. The smooth line was drawn with Eq. (6.34).

anhydride, forming the reactive intermediate 7-cinnamoyltheophylline, and there is kinetic evidence for this species. Let HT represent theophylline, T^- theophylline anion, and I the intermediate. We neglect the involvement of HT at pH 10; then Scheme VII constitutes the model for this system.

$$S + T^- \underset{}{\overset{K_{11}}{\rightleftharpoons}} ST^-$$

$$ST^- + T^- \underset{}{\overset{K_{12}}{\rightleftharpoons}} ST_2^{-2}$$

$$S + OH^- \xrightarrow{k_{OH}} P$$

$$ST^- + OH^- \xrightarrow{k_{11}} P \qquad\qquad \textit{Scheme VII}$$

$$S + T^- \underset{k_{-1}}{\overset{k_1}{\rightleftharpoons}} I + \text{cinnamate}$$

$$ST^- + T^- \underset{k'_{-1}}{\overset{k'_1}{\rightleftharpoons}} I + \text{cinnamate} + T^-$$

$$I + OH^- \xrightarrow{k_I} P$$

These equations, or rather the omission of several other equations, imply the following assumptions:

1. The 1:2 complex is completely unreactive.
2. "Water" reactions of S and I are negligible relative to hydroxide reactions (at high pH).
3. Since the hydrolysis of I is assumed to be faster than its formation, complexation of I with T^- (and consequent decrease in its reactivity) is ignored.
4. All involvement of HT is neglected.

The steady-state approximation is now applied to I:

$$\frac{d[I]}{dt} = k_1[S][T^-] + k'_1[ST^-][T^-] - k_{-1}[I][\text{cinnamate}]$$

$$- k'_{-1}[I][\text{cinnamate}][T^-] - k_I[I][OH^-] = 0$$

Because [cinnamate] is very small, the k_{-1} and k'_{-1} terms are neglected, giving

$$[I] = \frac{k_1[S][T^-] + k_1'[ST^-][T^-]}{k_1[OH^-]} \quad (6.27)$$

The rate equation for product formation is

$$\frac{d[P]}{dt} = k_{OH}[S][OH^-] + k_{11}[ST^-][OH^-] + k_1[I][OH^-] \quad (6.28)$$

Combining Eqs. (6.27) and (6.28) gives

$$\frac{d[P]}{dt} = k_{OH}[S][OH^-] + k_{11}[ST^-][OH^-]$$
$$+ k_1[S][T^-] + k_1'[ST^-][T^-] \quad (6.29)$$

The mass balance expressions are

$$L_t = [HT] + [T^-] + [ST^-] + [IT^-] + \cdots$$
$$S_t = [S] + [HTS] + [ST^-] + [ST_2^{-2}] + \cdots$$

Since $L_t \ggg S_t$ and HT is being neglected, these become

$$L_t = [T^-]$$
$$S_t = [S] + [ST^-] + [ST_2^{-2}]$$

Combining these with the stability constants K_{11} and K_{12}:

$$[S] = \frac{S_t}{1 + K_{11}L_t + K_{11}K_{12}L_t^2} \quad (6.30)$$

$$[ST^-] = \frac{K_{11}S_tL_t}{1 + K_{11}L_t + K_{11}K_{12}L_t^2} \quad (6.31)$$

The experimental rate equation is $d[P]/dt = k_{obs}S_t$. This is combined with Eqs. (6.29), (6.30), and (6.31).

$$k_{obs} = \frac{k_{OH}[OH^-] + k_1L_t + K_{11}L_t(k_{11}[OH^-] + k_1'L_t)}{1 + K_{11}L_t + K_{11}K_{12}L_t^2} \quad (6.32)$$

Equation (6.32) is the desired relationship between k_{obs} and L_t, incorporating the model scheme together with the several simplifying assumptions. It can be further simplified with the additional assumption that the effect of complexing on the reactivity of the substrate is independent of the nature of the attacking nucleophile, which provides

$$r_{11} = \frac{k_{11}}{k_{OH}} = \frac{k_1'}{k_1} \tag{6.33}$$

Then Eq. (6.32) becomes

$$k_{obs} = \frac{(k_{OH}[OH^-] + k_1 L_t)(1 + r_{11} K_{11} L_t)}{1 + K_{11} L_t + K_{11} K_{12} L_t^2} \tag{6.34}$$

The rate constant k_{OH} can be measured in the absence of ligand. When L_t is very small, Eq. (6.34) becomes $k_{obs} \approx k_{OH}[OH^-] + k_1 L_t$, so the slope of a plot of k_{obs} against L_t at very small L_t gives an initial estimate of k_1. At very large L_t it is possible that k_{obs} will approach the limiting value $r_{11} k_1 / K_{12}$.

For the data in Figure 6.4, it was found that $k_{OH} = 225\ M^{-1} s^{-1}$ and $k_1 = 4.25\ M^{-1} s^{-1}$. The first estimate of r_{11} is reasonably taken as 0.5, because in the 1:1 complex one of the two reactive sites is blocked by ligand. The smooth line in Figure 6.4 is drawn with Eq. (6.34), the preceding numerical estimates, and the values $K_{11} = 20\ M^{-1}$ and $K_{12} = 70\ M^{-1}$. Because of the many approximations that were made in reaching Eq. (6.34) these estimates are not highly reliable, but it is evident that the model is a reasonable one.

In the preceding example the nonhyperbolic isotherm was the result of ternary complex formation of the SL_2 type. We now consider a scheme in which a maximum in the isotherm is generated by a ternary complex of composition SLR. Laidler and Hoare (21) accounted in this way for a maximum in the rate of hydrolysis of urea catalyzed by urease, the complex being a 1:1:1 association of urea, water, and the enzyme. We consider the micellar catalysis of a bimolecular reaction in illustration of the problem. Berezin and co-workers (22) accounted for a maximum in the isotherm describing the acceleration of the aminolysis of 1-fluoro-2,4-dinitrobenzene (R) with N-benzoyl-L-histidine methyl ester (S) by micelles of cetyltrimethylammonium bromide (L). In the Berezin model both R and S are distributed between the micellar (M) and aqueous (W) phases in accordance with the simple distribution law. We shall call this the partitioning–partitioning model, since it postulates a partitioning mechanism for the uptake of both solutes.

If V_T, V_M, and V_W represent the total volume of the system, the volume of the micellar phase, and the volume of the aqueous phase, respectively, then $V_T = V_M + V_W$, and

$$\frac{V_W}{V_T} = 1 - \frac{V_M}{V_T}$$

A mass balance on R yields

$$[R]_T V_T = [R]_M V_M + [R]_W V_W$$

and similarly for S. Combining these gives

$$[R]_T = [R]_M \frac{V_M}{V_T} + [R]_W \left(1 - \frac{V_M}{V_T}\right) \tag{6.35}$$

The molar concentration of surfactant present as micelles, C, is defined:

$$C = C_T - (\text{CMC})$$

where (CMC) is the critical micelle concentration. Letting V be the molar volume of surfactant, it follows that the product CV is the liters of surfactant present as micelles per liter of solution, or the volume fraction of the micellar phase, hence:

$$CV = \frac{V_M}{V_T}$$

which, combined with Eq. (6.35), gives

$$[R]_T = [R]_M CV + [R]_W (1 - CV) \tag{6.36}$$

The distribution equilibria, according to this model, are

$$R_W \rightleftharpoons R_M$$
$$S_W \rightleftharpoons S_M$$

The partition coefficient for R is defined

$$P_R = \frac{[R]_M}{[R]_W} \tag{6.37}$$

and similarly for S. Equations (6.36) and (6.37) are combined to yield:

$$\frac{[R]_T}{[R]_w} = 1 + (P_R - 1)CV$$

The quantity K_R is defined

$$K_R = (P_R - 1)V \tag{6.38}$$

Then Eq. (6.39) is written for R, and Eq. (6.40) is the analogous equation for S:

$$\frac{[R]_T}{[R]_w} = 1 + K_R C \tag{6.39}$$

$$\frac{[S]_T}{[S]_w} = 1 + K_S C \tag{6.40}$$

In these equations K_R and K_S have the units of M^{-1}; that is, they have the character of 1:1 binding constants.

The bimolecular reaction between R and S can occur in both phases, with corresponding rate equations

$$v_M = k_M[R]_M[S]_M$$
$$v_w = k_w[R]_w[S]_w$$

The observed velocity (v) is equal to the sum of the products of the velocities in the individual phases and the volume fractions of the phases:

$$v = v_M CV + v_w(1 - CV) \tag{6.41}$$

An experimentally observed second-order rate constant can be defined by

$$v = k_{exp}[R]_T[S]_T \tag{6.42}$$

Equations (6.41) and (6.42) are combined with Eqs. (6.39) and (6.40) and the partition coefficient definitions as follows:

$$k_{exp} = \frac{k_M P_R P_S CV + k_w(1 - CV)}{(1 + K_R C)(1 + K_S C)} \tag{6.43}$$

Equation (6.43) can be simplified for the present purpose. When partitioning into the micelle is favored, as with anionic and hydrophobic species, P is much larger than unity, and the binding constants from Eq. (6.39) can be written as $K_R = P_R V$ and $K_S = P_S V$. Moreover, when $CV \ll 1$, Eq. (6.43) becomes

$$k_{exp} = \frac{k_M' K_R K_S C + k_w}{(1 + K_R C)(1 + K_S C)} \tag{6.44}$$

where $k_M' = k_M / V$.

This model can account for the appearance of a maximum in the dependence of k_{exp} on C. As Berezin and co-workers point out (22), to account for this maximum it is necessary to describe the uptake of both solutes, making the model, and the corresponding equation, qualitatively different from the hyperbolic model.

We now show that a result formally equivalent to the partitioning–partitioning model equation can be obtained with a different physical picture of the process. It is postulated that one of the reactants, that in large excess (S in this example), partitions between the micellar and aqueous phases, whereas the other reactant binds to the micelle with 1:1 stoichiometry. These equilibria are

$$S_w \rightleftharpoons S_M$$

$$R + MS \rightleftharpoons MSR$$

In this formulation, $[S]_M$ is the concentration of S in the micellar phase, and $[MS]$ is the solution concentration of S-containing micelles. There are no micelles without S, but there can be micelles without R. This is physically reasonable when $[S]_T > [\text{micelles}] > [R]_T$. By arguments given earlier, Eq. (6.40) applies to S, whereas Eq. (6.45) defines the binding constant for R.

$$K_M' = \frac{[MSR]}{[R]_w [MS]} \tag{6.45}$$

The rate equation for loss of R is given by

$$-\frac{d[R]}{dt} = k_w [R]_w [S]_w + k_M [MSR][S]_M \tag{6.46}$$

The mass balance on surfactant is

$$C_T = [CMC] + n[MS] + n[MSR] \tag{6.47}$$

where n is the micelle aggregation number. Combining Eqs. (6.45) and (6.47),

$$[MS] = \frac{C}{n + nK_M'[R]_w} \tag{6.48}$$

The mass balance on R is

$$[R]_T = [R]_w + [MSR] \tag{6.49}$$

since [MSR] is the concentration of micelles each containing an R. This leads to

$$[R]_w = \frac{[R]_T}{1 + K_M'[MS]} \tag{6.50}$$

Combining the partition coefficient $P_S = [S]_M/[S]_M$ with Eqs. (6.45), (6.46), and (6.50) gives the rate equation

$$-\frac{d[R]_T}{dt} = \left[\frac{k_M K_M' P_S[MS] + k_w}{1 + K_M'[MS]} \right] [S]_w[R]_T \tag{6.51}$$

This is first-order in $[R]_T$, with the apparent first-order rate constant k_{obs}:

$$k_{obs} = \left[\frac{k_M K_M' P_S[MS] + k_w}{1 + K_M'[MS]} \right] [S]_w \tag{6.52}$$

Equation (6.48) is combined with Eq. (6.52) to give

$$k_{obs} = \left[\frac{(k_M K_M' P_S C)/(n + nK_M'[R]_w) + k_w}{1 + (K_M'C)/(n + nK_M'[R]_w)} \right] [S]_w \tag{6.53}$$

Equation (6.53) is simplified by using the reasonable assumption $K_M'[R]_w \ll 1$, substituting $K_S = P_S V$, and replacing $[S]_w$ from Eq. (6.40) to give

$$k_{exp} = \frac{k_M' K_M K_S C + k_w}{(1 + K_M C)(1 + K_S C)} \tag{6.54}$$

where $k'_M = k_M/V$, $K_M = K'_M/n$, and $k_{exp} = k_{obs}/[S]_T$. Equation (6.54) for the binding–partitioning model is identical in form with Eq. (6.44) for the partitioning–partitioning model. In these equations the quantity K_S has the same significance, but K_R in Eq. (6.44) is interpreted as K'_M/n in Eq. (6.54) as a consequence of the different assumptions concerning the uptake of R into the micelle.

Evaluation of the parameters is a curve-fitting problem, and preliminary estimates of or constraints on the parameters can be obtained as follows. Equations (6.44) and (6.54) can be written in the equivalent form

$$k_{obs} = \left[\frac{k'_M K_M K_S C + k_W}{(1 + K_M C)(1 + K_S C)} \right] [S]_T \qquad (6.55)$$

A plot of k_{obs} against C_T, at very low C_T, will give $k_W[S]_T$ as the value of k_{obs} below the CMC. Above the CMC, the slope of the straight line is equal to $k'_M K_M K_S[S]_T$, from which the product $k'_M K_M K_S$ is found. The intersection of the straight line segments marks the CMC.

A plot of k_{obs} against C over the full range of C exhibits a maximum if Eq. (6.55) describes the system. The concentration C_{max} corresponding to this maximum is found by setting the derivative dk_{obs}/dC equal to zero:

$$C_{max} = (K_M K_S)^{-1/2} \qquad (6.56)$$

This result yields the product $K_M K_S$, and k'_M is obtained from this and the estimate of $k'_M K_M K_S$.

Inserting Eq. (6.56) into Eq. (6.54), and neglecting k_W, leads to Eq. (6.57), where k_{exp}^{max} is the value of k_{exp} when $C = C_{max}$:

$$\left[\frac{k_{exp}^{max}}{k'_M} \right]^{-1/2} = K_M^{-1/2} + K_S^{-1/2} \qquad (6.57)$$

With these relationships the four parameters k_W, k'_M, K_M, and K_S can be evaluated. Because Eq. (6.55) is symmetrical in K_M and K_S, it is not possible to assign these quantities unambiguously.

Another approach is a modification of a method described by Yatsimirskii et al. (22d). Equation (6.54) can be rearranged to

$$k_{exp} - k_W = \left[\frac{K_M C}{1 + K_M C} \right] \left[\frac{k'_M K_S}{1 + K_S C} - k_W \right]$$

From this, Eq. (6.58) is obtained,

$$\frac{C}{k_{exp} - k_W} = \frac{1}{A} + \frac{(K_M + K_S)}{A} C + \frac{K_M K_S}{A} C^2 \tag{6.58}$$

where $A = k'_M K_M K_S - k_W K_M (1 + K_S C)$. A plot of $C/(k_{exp} - k_W)$ against C is a curve, whose intercept at $C = 0$ is $1/(k'_M K_M K_S - k_W K_M)$. If the k_W term is negligible, this gives an estimate of $k'_M K_M K_S$.

Equation (6.54) is then rearranged to

$$\frac{k'_M K_M K_S C + k_W - k_{exp}}{k_{exp} C} = K_M K_S C + (K_M + K_S) \tag{6.59}$$

A plot of the left side against C yields $K_M K_S$ and $(K_M + K_S)$ from the slope and intercept.

By either of these methods K_M and K_S are found by solving the quadratic formula. In some instances the solution is indeterminate—that is, the estimates $K_M K_S$ and $(K_M + K_S)$ are mutually inconsistent, because of experimental error. The individual constants are then evaluated by curve fitting, in which Eq. (6.55) is fitted to the experimental points. This is a straightforward process, because the quantities $k'_M K_M K_S$ and $K_M K_S$ serve as constraints.

Figure 6.5 is a plot of k_{obs} against C for the reaction of 1-fluoro-2,4-dinitrobenzene with 4-methylaniline in the presence of cetyltrimethylammonium bromide (23). The smooth curve was drawn with Eq. (6.55) and the parameters $K_S = 27.0 \ M^{-1}$, $K_M = 55.5 \ M^{-1}$, $k'_M = 0.20 \ s^{-1}$, $k_W = 0.143 \ M^{-1} s^{-1}$; and k_M was calculated to be $0.57 \ M^{-1} s^{-1}$. The ratio $k_M / k_{exp}^{max} = 0.30$, where k_{exp}^{max} is the maximum value of the observed second-order rate constant. This ratio is a measure of the fraction of true micellar catalysis contributing to the total micellar acceleration effect, as suggested by Eq. (6.60):

$$\frac{k_M}{k_{exp}^{max}} = \frac{k_M / k_W}{k_{exp}^{max} / k_W} \tag{6.60}$$

That is, about 30% of the maximum observed acceleration is a consequence of catalysis by the micelle, the remainder being a concentration effect, through which the micelle increases the local concentrations of the reactants.

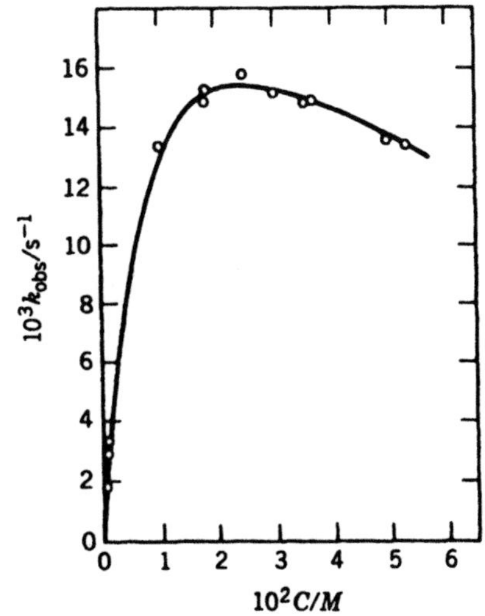

Figure 6.5. Dependence of first-order rate constant on surfactant concentration for the reaction of 4-methylaniline with fluorodinitrobenzene (23). The smooth curve was drawn with Eq. (6.55).

REFERENCES

1. W. Spencer and J. R. Sutter, *J. Phys. Chem.*, **83**, 1573 (1979).
2. E. M. Kosower, *Prog. Phys. Org. Chem.*, **3**, 81 (1965).
3. J. H. Fendler and E. J. Fendler, *Catalysis in Micellar and Macromolecular Systems*, Academic, New York, 1975.
4. M. L. Bender and M. Komiyama, *Cyclodextrin Chemistry*, Springer-Verlag, Berlin, 1978.
5. H. Morawetz, *Advan. Catal. Relat. Subj.*, **20**, 341 (1969).
6. S. D. Ross and I. Kuntz, *J. Am. Chem. Soc.*, **76**, 3000 (1954).
7. W. P. Jencks, *Catalysis in Chemistry and Enzymology*, McGraw-Hill, New York, 1969, p. 440.
8. F. M. Menger, *J. Am. Chem. Soc.*, **90**, 4387 (1968).
9. (a) T. Higuchi and L. Lachman, *J. Am. Pharm. Assoc., Sci. Ed.*, **44**, 521 (1955); (b) L. Lachman, L. J. Ravin, and T. Higuchi, *J. Am. Pharm. Assoc., Sci. Ed.*, **45**, 290 (1956); (c) L. Lachman, D. Guttman, and T. Higuchi, *J. Am. Pharm. Assoc., Sci. Ed.*, **46**, 36 (1957).
10. (a) J. A. Mollica and K. A. Connors, *J. Am. Chem. Soc.*, **89**, 308 (1967); (b) H. Stelmach and K. A. Connors, *J. Am. Chem. Soc.*, **92**, 863 (1970).
11. J. L. Cohen, Ph.D. Dissertation, University of Wisconsin-Madison, 1969, p. 63.
12. A. K. Colter, S. S. Wang, G. H. Mergerle, and P. S. Ossip, *J. Am. Chem. Soc.*, **86**, 3106 (1964).

13. K. A. Connors and J. A. Mollica, *J. Pharm. Sci.*, **55**, 772 (1966).

14. (a) P. A. Kramer and K. A. Connors, *J. Am. Chem. Soc.*, **91**, 2600 (1969); (b) J. L. Cohen and K. A. Connors, *J. Pharm. Sci.*, **59**, 1271 (1970).

15. H. Lineweaver and D. Burk, *J. Am. Chem. Soc.*, **56**, 658 (1934).

16. G. S. Eadie, *J. Biol. Chem.*, **146**, 85 (1942).

17. R. L. Van Etten, J. F. Sebastian, G. A. Clowes, and M. L. Bender, *J. Am. Chem. Soc.*, **89**, 3242 (1967).

18. G. A. Hamilton, J. P. Friedman, and P. M. Campbell, *J. Am. Chem. Soc.*, **88**, 5266 (1966).

19. P. D. Bartlett, C. V. Goebel, and W. P. Weber, *J. Am. Chem. Soc.*, **91**, 7425 (1969).

20. M. H. Infeld, Ph.D. Dissertation, University of Wisconsin-Madison, 1969, p. 133.

21. K. J. Laidler and J. P. Hoare, *J. Am. Chem. Soc.*, **71**, 2699 (1949).

22. (a) A. K. Yatsimirskii, K. Martinek, and I. V. Berezin, *Dokl. Akad. Nauk SSSR*, **194**, 840 (1970); (b) A. K. Yatsimirskii, K. Martinek, and I. V. Berezin, *Tetrahedron*, **27**, 2855 (1971); (c) A. K. Yatsimirskii, K. Martinek, and I. V. Berezin, *Russ. Chem. Revs.*, **42**, 787 (1973); (d) A. K. Yatsimirskii, Z. A. Strel'tsova, K. Martinek, and I. V. Berezin, *Kinet. Katal.*, **15**, 354 (1974).

23. M. P. Wong and K. A. Connors, *J. Pharm. Sci.*, **72**, 146 (1983).

7

POTENTIOMETRY

7.1. TYPES OF SYSTEMS

The potential E of an electrochemical cell is related to the activity a of an ion taking part in a reversible electrode process by Eq. (7.1), the Nernst equation.

$$E = \text{constant} + \frac{RT}{nF} \ln a \qquad (7.1)$$

In this equation F is the Faraday and n is the number of electrons involved in the (actual or formal) process. Redox electrodes are based on oxidation–reduction reactions, and membrane electrodes (which include glass electrodes) are based on ion concentration gradients across a membrane phase.

If the activity (i.e., the concentration) of an ionic species undergoes a change as a consequence of a complexation equilibrium, and if an electrode is available that is reversible with respect to this ion, so that Eq. (7.1) applies, then measurement of the potential can be used to study the complexation. In fact, potentiometry has been the most widely used technique for the determination of acid–base dissociation constants and of stability constants of metal ion coordination complexes.

Note that it is not essential that the ion whose activity controls the potential actually be part of the complex, but only that its activity be a function of the complex formation; the monitored ion may be involved in a secondary equilibrium. Several applications of potentiometry are indirect competitive techniques.

When potentiometry is applicable, ionic species are necessarily present, and in this circumstance we may, in a very general sense, view these systems as containing acids and bases. The following scheme includes the important features of the systems that are discussed in Sections 7.2 and 7.3:

$$\text{Acid substrate} + \text{ligand} \rightleftharpoons \text{acid complex(es)}$$
$$\big\updownarrow \qquad\qquad\qquad \big\updownarrow \qquad\qquad \textit{Scheme I}$$
$$\text{Base substrate} + \text{ligand} \rightleftharpoons \text{base complex(es)}$$

Section 7.2 deals with systems for which only the top of this scheme need be considered. The acid substrates are Lewis acids, namely, electron-pair acceptors, most notably metal ions. The ligand is then a base, usually a neutral or anionic species. If the ligand does not undergo significant protonation under the experimental conditions, as when it is the conjugate base of a strong acid (such as Cl^-) or when its pK_a is much lower than the pH, Scheme I becomes simply

$$S + L \rightleftharpoons \text{complexes} \qquad\qquad \textit{Scheme II}$$

and the equilibria can be studied potentiometrically by measuring pS (i.e., pM) or pL. When the ligand is sufficiently basic to be protonated, the scheme is modified as follows:

$$S + L \rightleftharpoons \text{complexes}$$
$$\textit{Scheme III}$$
$$H^+ + L \rightleftharpoons HL^+$$

This can be described as a competition between S and H^+ for L, where S is a metal ion; the system can be studied by measuring pH.

When the acid substrate in Scheme I is a Brønsted acid (a proton donor), the scheme must include protons as appropriate. Scheme IV shows the simplest example.

$$HA + L \rightleftharpoons HAL$$
$$\big\updownarrow \qquad\qquad \big\updownarrow$$
$$A^- + L \rightleftharpoons AL^- \qquad\qquad \textit{Scheme IV}$$
$$+ \qquad\qquad +$$
$$H^+ \qquad\qquad H^+$$

This can be viewed as a competition between the two substrates HA and A⁻ for L, with the substrates being related by a protonic equilibrium. Because of this connection between HA and A⁻, the complex formation can be studied by measuring the pH. Such systems are treated in Section 7.3.

Potentiometric studies of binding can be carried out in the usual manner in which separate solutions are prepared each to contain the same total substrate concentration S_t but varying total ligand concentrations L_t. It is very common, however, to conduct the experiment as a titration. Usually, the ionic strength is held constant so as to minimize changes in ionic activity coefficients and liquid junction potentials. Temperature control is particularly important in the potentiometric method because of the temperature dependence of the potential in Eq. (7.1). Hartley et al. (1) have discussed the practical experimental problems associated with potentiometric studies of complex formation.

7.2. LEWIS ACID SUBSTRATES

Since potentiometry has been widely applied to the study of metal ion equilibria, many detailed reviews of this technique are readily available, and it is not necessary to repeat this material here. Consequently, in this section we briefly show how, from potentiometric data, measures of binding can be obtained. The subsequent analysis to extract stability constants is treated in Sections 2.5 and 2.6 and in Chapter 3. Many computer programs have been described for carrying out nonlinear least-squares regression analysis for potentiometric studies of metal ion complexation (2).

Measurement of pL

If an electrode is available that is reversible toward the ligand L, the free ligand concentration [L] is accessible, and the appropriate measure of binding is $\bar{i}$, the mean number of ligand molecules bound per molecule of substrate. This quantity is discussed in Section 2.4, "Measures of Binding." It is calculated from the experimental data by means of

$$\bar{i} = \frac{L_t - [L]}{S_t} \tag{7.2}$$

Thus $\bar{i}$ can be evaluated as a function of [L]. For systems in which $m = 1$

(mononuclear complexes in the terminology of the metal complex field), $\bar{i}$ is related to the stability constants by

$$\bar{i} = \frac{\beta_{11}[L] + 2\beta_{12}[L]^2 + \cdots}{1 + \beta_{11}[L] + \beta_{12}[L]^2 + \cdots} \tag{7.3}$$

Thus the β_{1i} can be estimated from the $\bar{i}$, [L] data set. It is evident, from Eq. (7.2), that L_t and [L] must differ significantly.

Measurement of pM

If the substrate (metal ion) concentration is measured potentiometrically, we have two ways to proceed. One of these is simply to redefine the interactants, calling the metal ion the ligand; then the preceding discussion and Eqs. (7.2) and (7.3) are applicable.

The second approach is to calculate f_{10}, the fraction of substrate in the uncomplexed form, by

$$f_{10} = \frac{[S]}{S_t} \tag{7.4}$$

We then calculate $F_{\bar{s}} = 1 - f_{10}$ (see Section 2.4). This quantity, for mononuclear complexes, is related to the stability constants [Eq. (2.56c)] by

$$F_{\bar{s}} = \frac{\beta_{11}[L] + \beta_{12}[L]^2 + \cdots}{1 + \beta_{11}[L] + \beta_{12}[L]^2 + \cdots} \tag{7.5}$$

which may be compared with Eq. (7.3). Estimation of the β_{1i} from the $F_{\bar{s}}$ values requires corresponding [L] concentrations, which are obtained by iterative calculation. Hartley et al. (3) show a numerical example of this type of calculation.

Measurement of pH

Let us make Scheme III slightly more general by supposing that the ligand L is a diacidic base subject to these protonation equilibria (charges omitted for convenience):

$$H_2L \underset{}{\overset{K_{a1}}{\rightleftharpoons}} HL \underset{}{\overset{K_{a2}}{\rightleftharpoons}} L$$

It is assumed that only the L form can complex with the substrate. The experiment is carried out such that total substrate and ligand concentrations, S_t and L_t, are known, as well as the total strong acid concentration H_t, which is the source of the protons. The pH is measured. It is also presumed that the acid dissociation constants K_{a1} and K_{a2} are known. H_t is defined by Eq. (7.6).

$$H_t = [H^+] + [HL] + 2[H_2L] \qquad (7.6)$$

Letting the concentration of protons bound to ligand be denoted H_L gives

$$H_t = [H^+] + H_L \qquad (7.7)$$

The total concentration of uncomplexed ligand is L_f:

$$L_f = [L] + [HL] + [H_2L] \qquad (7.8)$$

Accordingly the binding measure $\bar{i}$ is defined by Eq. (7.9).

$$\bar{i} = \frac{L_t - L_f}{S_t} \qquad (7.9)$$

Define the mean number of protons per ligand molecule, $\bar{j}$, by Eq. (7.10), and the fraction of uncomplexed ligand that is unprotonated, α, by Eq. (7.11).

$$\bar{j} = \frac{H_L}{L_f} \qquad (7.10)$$

$$\alpha = \frac{[L]}{L_f} \qquad (7.11)$$

Algebraic combination of Eqs. (7.10) and (7.11) with the expressions for K_{a1} and K_{a2} yields

$$\bar{j} = \frac{2[H^+]^2 + K_{a1}[H^+]}{[H^+]^2 + K_{a1}[H^+] + K_{a1}K_{a2}} \qquad (7.12)$$

$$\alpha = \frac{K_{a1}K_{a2}}{[H^+]^2 + K_{a1}[H^+] + K_{a1}K_{a2}} \qquad (7.13)$$

From Eqs. (7.10) and (7.11) we find

$$[L] = \frac{\alpha H_L}{\bar{j}} \qquad (7.14)$$

One way in which to conduct the experiment is to titrate a solution having fixed S_t and L_t with standard strong acid, measuring the pH. The calculation is made with the following steps:

1. Calculate $\bar{j}$ and α with Eqs. (7.12) and (7.13).
2. Calculate H_L with Eq. (7.7).
3. Calculate $[L]$ with Eq. (7.14).
4. Calculate L_f with Eq. (7.11).
5. Calculate $\bar{i}$ with Eq. (7.9).

Hartley et al. (4) give experimental data and calculations for the nickel (II)–ethylenediamine system, in which the ligand is diacidic, as in the preceding treatment.

7.3. BRØNSTED ACID SUBSTRATES

1:1 Binding

Scheme IV describes the equilibria in a system containing the two substrates HA and A⁻ (a conjugate Brønsted acid–base pair) and a ligand L, which is considered not to take part in acid–base equilibria under the experimental conditions. Several laboratories (5–8) developed potentiometric methods for the study of such systems in the 1970s.

The four equilibria in Scheme V can be written, and the corresponding equilibrium constants are defined in Eqs. (7.15)–(7.18).

$$\text{HA} + \text{L} \rightleftharpoons \text{HAL}$$

$$\text{A}^- + \text{L} \rightleftharpoons \text{AL}^-$$

Scheme V

$$\text{HA} \rightleftharpoons \text{H}^+ + \text{A}^-$$

$$\text{HAL} \rightleftharpoons \text{H}^+ + \text{AL}^-$$

$$K_{11a} = \frac{[HAL]}{[HA][L]} \tag{7.15}$$

$$K_{11b} = \frac{[AL^-]}{[A^-][L]} \tag{7.16}$$

$$K_a = \frac{[H^+][A^-]}{[HA]} \tag{7.17}$$

$$K_{a11} = \frac{[H^+][AL^-]}{[HAL]} \tag{7.18}$$

Combination of these equations gives (7.19).

$$\frac{K_a}{K_{a11}} = \frac{K_{11a}}{K_{11b}} \tag{7.19}$$

Consequently, only three of these constants are independent quantities. These inequalities follow from Eq. (7.19): when $K_{11a} > K_{11b}$, $K_a > K_{a11}$ (and the reverse condition). It is a matter of taste which of these inequalities is regarded as the cause and which as the effect.

The experiment consists of measuring the pH as a function of total ligand concentration L_t in a solution containing comparable concentrations of HA and A^-. For the present (an alternative treatment is given subsequently) let us define an apparent dissociation constant K_a' by

$$K_a' = \frac{[H^+]([A^-] + [AL^-])}{[HA] + [HAL]} \tag{7.20}$$

Combination of Eqs. (7.15)–(7.17) and (7.20) gives

$$K_a' = K_a \frac{(1 + K_{11b}[L])}{(1 + K_{11a}[L])} \tag{7.21}$$

which may also be written

$$\Delta pK_a' = \log \frac{(1 + K_{11a}[L])}{(1 + K_{11b}[L])} \tag{7.22}$$

where

$$\Delta pK_a' = pK_a' - pK_a \tag{7.23}$$

We denote the argument of the logarithm as C, thus $\Delta pK_a' = \log C$, and

$$C = \frac{1 + K_{11a}[L]}{1 + K_{11b}[L]} \tag{7.24}$$

Equation (7.24) is the binding isotherm. The quantity C is obtained from $\Delta pK_a'$, which in turn is given by Eq. (7.23). Experimentally, one measures the apparent dissociation constant, or pK_a', potentiometrically as a function of L_t.

In this procedure it is necessary that the substrate (acid–base pair) concentration be high enough to provide adequate buffer capacity, so the approximation $[L] = L_t$ is seldom valid. An expression for $[L]$ is found by combining the mass balances on S_t and L_t. The procedure is described subsequently in the treatment of multiple equilibria, the 1:1 system emerging as a special case. From that treatment we get

$$L_t = [L] + S_t \left[\frac{K_{11a}[L]}{1 + K_{11a}[L]} + \frac{K_{11b}[L]}{1 + K_{11b}[L]} \right] \tag{7.25}$$

Equations (7.24) and (7.25) describe the system.

Equation (7.24) is a rectangular hyperbola, as shown in Appendix A. The binding measure C is related to the measures $\bar{i}$ and $F_{\bar{s}}$ (see Section 2.4), but it is a simpler function than these in this system. Equation (7.24) can be put into the usual three linear forms, namely, the double-reciprocal [Eq. (7.26)], y-reciprocal (7.27), and x-reciprocal (7.28) forms,

$$\frac{1}{C - 1} = \frac{1}{(K_{11a} - K_{11b})[L]} + \frac{K_{11b}}{K_{11a} - K_{11b}} \tag{7.26}$$

$$\frac{[L]}{C - 1} = \frac{K_{11b}}{K_{11a} - K_{11b}} [L] + \frac{1}{K_{11a} - K_{11b}} \tag{7.27}$$

$$\frac{C - 1}{[L]} = K_{11a} - K_{11b} C \tag{7.28}$$

A value of $C > 1$ implies $K_{11a} > K_{11b}$, and vice versa. When $C < 1$ it is convenient to define $C' = 1/C$ and to write the isotherm as

$$C' = \frac{1 + K_{11b}[L]}{1 + K_{11a}[L]} \tag{7.29}$$

Then Eqs. (7.26)–(7.27) are applicable by interchanging the subscripts.

Figure 7.1 shows potentiometric data, as a plot of $\Delta pK'_a$ against L_t, for the substrates benzoic acid and 4-nitrophenol, the ligand being α-cyclodextrin (9). The stability constants evaluated from these data (10, 11) are, for benzoic acid, $K_{11a} = 722\ M^{-1}$, $K_{11b} = 11.2\ M^{-1}$; for 4-nitrophenol, $K_{11a} = 245\ M^{-1}$, $K_{11b} = 2408\ M^{-1}$.

Table 7.1 gives potentiometric data for the system 4-cyanophenol: α-cyclodextrin (12). Since $\Delta pK'_a$ is negative, C is smaller than unity, and form (7.29) of the isotherm is used. Figure 7.2 is a plot of Eq. (7.28) modified by replacing C by C' and by interchanging K_{11a} and K_{11b}. The slope of the plot is therefore $-K_{11a}$, the intercept on the ordinate is K_{11b}, and the intercept on the abscissa is K_{11b}/K_{11a}. (For this experiment it would have been helpful to collect some data points at lower L_t). The stability constants are found to be $K_{11b} = 662\ M^{-1}$, $K_{11a} = 158\ M^{-1}$.

The free ligand concentration can be obtained by means of the Taylor's series approximation of Section 2.3. For this case Eq. (7.25) corresponds to Eq. (2.36). The derivative $g'(L_t)$ is

$$g'(L_t) = 1 + \frac{S_t K_{11a}}{2(1 + K_{11a}L_t)^2} + \frac{S_t K_{11b}}{2(1 + K_{11b}L_t)^2} \qquad (7.30)$$

Substituting Eq. (7.30) into Eq. (2.37), and truncating at the linear term, gives Eq. (7.31) for [L].

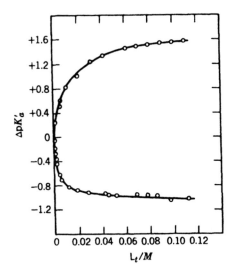

Figure 7.1. Plots of $\Delta pK'_a$ against ligand concentration for substrates benzoic acid (upper curve) and 4-nitrophenol (lower). The ligand is α-cyclodextrin.

Table 7.1. Potentiometric Data for 4-Cyanophenol (S): α-Cyclodextrin (L)[a]

L_t/M	$\Delta pK_a'$	C'	$[L]/M$
0.0075	−0.383	2.415	0.0051
0.0101	−0.431	2.698	0.0073
0.0152	−0.493	3.112	0.0120
0.0200	−0.517	3.289	0.0167
0.0249	−0.541	3.475	0:0214
0.0300	−0.554	3.581	0.0265
0.0399	−0.568	3.698	0.0363
0.0501	−0.578	3.784	0.0464
0.0600	−0.586	3.855	0.0562

[a]At 25.0°C, $S_t = 0.00401 \; M$ and $pK_a = 7.814$.

$$[L] = L_t - \frac{1}{g'(L_t)}\left[\frac{S_t K_{11a} L_t}{2(1 + K_{11a}L_t)} + \frac{S_t K_{11b} L_t}{2(1 + K_{11b}L_t)}\right] \quad (7.31)$$

An alternative calculation of [L] is given in the later treatment.

It is easy to show that the experimentally measured K_{11a} and K_{11b} are sums of binding constants of isomeric 1:1 complexes. Suppose that two complexes HAL and LAH, with stability constants K_{11a1} and K_{11a2}, are formed from the conjugate acid form of the substrate, and similarly for the two basic complexes AL⁻ and LA⁻, with constants K_{11b1} and K_{11b2}. The apparent dissociation constant is defined

$$K_a' = \frac{[H^+]([A^-] + [AL^-] + [LA^-])}{[HA] + [HAL] + [LAH]}$$

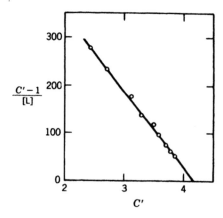

Figure 7.2. Plot of the Eq. (7.28) form for the system 4-cyanophenol: α-cyclodextrin. For this system $K_{11b} > K_{11a}$, so Eq. (7.29) is applicable.

Making substitutions leads to the isotherm:

$$C = \frac{K_a}{K'_a} = \frac{1 + (K_{11a1} + K_{11a2})[L]}{1 + (K_{11b1} + K_{11b2})[L]} \qquad (7.32)$$

Comparison of Eqs. (7.24) and (7.32) shows that $K_{11a} = K_{11a1} + K_{11a2}$ and $K_{11b} = K_{11b1} + K_{11b2}$.

The K_a value defined by Eq. (7.17) is itself an apparent dissociation constant (although it is referred to a solution in which $L_t = 0$), because it is a concentration quotient. If desired, activity coefficient corrections can be introduced for all ionic species, but in many systems such corrections are not justified because of the complexity of the medium. Instead the ionic strength may be held constant, with the hope that activity coefficients remain essentially constant over the experimentally studied range of ligand concentrations. There is a further medium effect associated with the observational technique; this is the possible variation in electrode response in solutions of different ligand concentrations. For example, Gelb et al. (8) studied the effect of cyclodextrin on pH electrode response by measuring the pH of strong acid solutions containing cyclodextrin and comparing the observed pH with the known pH. They concluded that the principal medium effect was on electrode response (through the liquid junction potential), rather than on activity coefficients. Note, incidentally, that this technique-specific medium effect is not explicitly apparent in the binding isotherm, Eq. (7.24), in contrast to its obvious appearance by way of a proportionality factor, such as a molar absorptivity, in other methods.

Multiple Equilibria

Let us consider the potentiometric study of a Brønsted acid–base pair HA and A$^-$ with ligand L in which 1:1 and 1:2 complexes may form with both of the substrate species (10). The constants K_{11a}, K_{11b}, K_a, and K_{a11} have been defined in Eqs. (7.15)–(7.18); we add now the corresponding constants for the 1:2 complex species:

$$K_{12a} = \frac{[HAL_2]}{[HAL][L]} \qquad (7.33)$$

$$K_{12b} = \frac{[AL_2^-]}{[AL^-][L]} \qquad (7.34)$$

$$K_{a12} = \frac{[\text{H}^+][\text{AL}_2^-]}{[\text{HAL}_2]} \quad (7.35)$$

Equation (7.19) holds in this system, and we also find

$$K_{12a}K_{a12} = K_{a11}K_{12b} \quad (7.36)$$

Letting S_t and L_t be the total molar concentrations of substrate and ligand, respectively, the mass balance equations are

$$S_t = [\text{HA}] + [\text{A}^-] + [\text{HAL}] + [\text{AL}^-] + [\text{HAL}_2] + [\text{AL}_2^-] \quad (7.37)$$

$$L_t = [\text{L}] + [\text{HAL}] + [\text{AL}^-] + 2[\text{HAL}_2] + 2[\text{AL}_2^-] \quad (7.38)$$

For convenience the quantities A, B, M, and N are defined.

$$A = 1 + K_{11a}[\text{L}] + K_{11a}K_{12a}[\text{L}]^2 \quad (7.39)$$

$$B = 1 + K_{11b}[\text{L}] + K_{11b}K_{12b}[\text{L}]^2 \quad (7.40)$$

$$M = K_{11a}[\text{L}] + 2K_{11a}K_{12a}[\text{L}]^2 \quad (7.41)$$

$$N = K_{11b}[\text{L}] + 2K_{11b}K_{12b}[\text{L}]^2 \quad (7.42)$$

Algebraic combination leads to

$$S_t = [\text{HA}]\left[\frac{A[\text{H}^+] + BK_a}{[\text{H}^+]}\right] \quad (7.43)$$

$$L_t = [\text{L}] + [\text{HA}]\left[\frac{M[\text{H}^+] + NK_a}{[\text{H}^+]}\right] \quad (7.44)$$

Eliminating [HA] from Eqs. (7.43) and (7.44) gives a general relationship between [L] and L_t:

$$L_t = [\text{L}] + S_t\left[\frac{M[\text{H}^+] + NK_a}{A[\text{H}^+] + BK_a}\right]. \quad (7.45)$$

The electroneutrality equation for this system is

$$[Na^+] + [H^+] = [OH^-] + [A^-] + [AL^-] + [AL_2^-] \qquad (7.46)$$

where it is supposed that sodium is the counterion to the substrate. Combining Eq. (7.46) with the preceding expressions gives

$$[Na^+] + [H^+] - [OH^-] = \frac{S_t B K_a}{A[H^+] + B K_a} \qquad (7.47)$$

The quantity $[Na^+]/S_t$ is the analytical fraction of substrate in the conjugate base form. Equation (7.47) is the general equation relating hydrogen ion concentration to free ligand concentration $[L]$. By means of Eqs. (7.45) and (7.47), $[H^+]$ is related to total ligand concentration L_t. In these equations, $[Na^+]$, S_t, and L_t are independent variables; $[H^+]$ and $[L]$ are dependent variables.

Equation (7.47) can be cast in a more usual form. If $[Na^+] \gg ([H^+] - [OH^-])$, then Eq. (7.47) becomes

$$[Na^+] = \frac{S_t B K_a}{A[H^+] + B K_a} \qquad (7.48)$$

For a solution in which L_t is 0, it follows that $[L] = 0$, $A = 1$, $B = 1$, and Eq. (7.48) becomes

$$[Na^+]_0 = \frac{S_t K_a}{[H^+]_0 + K_a} \qquad (7.49)$$

A series of measurements of $[H^+]$ as a function of L_t is made at constant S_t and $[Na^+]$; hence $[Na^+] = [Na^+]_0$, and these equations give

$$\frac{[H^+]_0}{[H^+]} = \frac{A}{B} = C \qquad (7.50)$$

Defining a quantity $\Delta pH = pH - pH_0$:

$$\Delta pH = \log C \qquad (7.51)$$

Equation (7.51) describes the change in the solution pH as a function of free ligand concentration for fixed $[Na^+]/S_t$. If $[Na^+]/S_t = \frac{1}{2}$, then the pH can be interpreted as pK_a' (apparent dissociation constant), and Eq. (7.51) becomes $\Delta pK_a' = \log C$, or

$$\Delta p K_a' = \log \left[\frac{1 + K_{11a}[\text{L}] + K_{11a}K_{12a}[\text{L}]^2}{1 + K_{11b}[\text{L}] + K_{11b}K_{12b}[\text{L}]^2} \right] \tag{7.52}$$

This derivation shows the level of approximation involved in interpreting ΔpH as $\Delta p K_a'$.

More generally, the exact Eq. (7.47) must be used. Defining the operational dissociation constant K_a' by

$$K_a' = \frac{[\text{H}^+]([\text{Na}^+] + [\text{H}^+] - [\text{OH}^-])}{S_t - ([\text{Na}^+] + [\text{H}^+] - [\text{OH}^-])} \tag{7.53}$$

Substitution into Eq. (7.53) from Eq. (7.47) leads to $K_a/K_a' = A/B = C$, or $\Delta p K_a' = \log C$. That is, Eq. (7.52) is general, provided the dissociation constants are evaluated by Eq. (7.53), which takes into account solvent dissociation.

The binding isotherm is, from Eq. (7.52),

$$C = \frac{1 + K_{11a}[\text{L}] + K_{11a}K_{12a}[\text{L}]^2}{1 + K_{11b}[\text{L}] + K_{11b}K_{12b}[\text{L}]^2} \tag{7.54}$$

As in the case of 1:1 complexing, the experimental data consist of $\Delta p K_a'$ values measured as a function of L_t, at constant S_t and reaction conditions. Throughout the following, we consider that $C > 1$; if, however, $C < 1$, we define $C' = 1/C$, and then Eq. (7.54) is written for C' with all subscripts interchanged.

Equation (7.54) is a ratio of polynomials. It is very easy to curve-fit data to this form, and thus to estimate the parameters. However, in the absence of restrictions on the reasonable ranges of these parameters, the curve-fit estimates may be unreliable. Many complex systems have been found to be special cases of Eq. (7.54), and a useful approach is to treat these special cases. The method is to make a plot of C versus L_t. From the shape of this curve, the system is tentatively assigned to one of the special cases. The free ligand concentration [L] is calculated, and stability constants are evaluated from the appropriate linear plot. Iterations are carried out until the stability constants are essentially unchanged.

To find [L], Eq. (7.45) is used. For this purpose, we set the ratio $[\text{Na}^+]/S_t = \frac{1}{2}$, so $[\text{H}^+] = K_a'$, to a level of accuracy acceptable for the present purpose. Since $C = K_a/K_a'$, this gives $[\text{H}^+] = K_a/C$. Substitution into Eq. (7.45) gives

$$L_t = [L] + S_t \left[\frac{M}{2A} + \frac{N}{2B} \right] \tag{7.55}$$

The value of [L] is estimated by combining Eq. (7.55) with the appropriate special case of Eq. (7.54). In this development the quantity C is incorporated into the expression for [L] to the greatest extent possible. This has two advantages: First, since the stability constants are needed in the calculation, their introduction in the form of experimental C values ensures that the correct values are being used; second, the solution of the equation for [L] is usually simplified, since much of the algebraic complexity is possessed by the numerical C values. The particular equations for [L] will be given when the special cases are discussed. This approach is an example of technique 3 in Section 2.3.

   Case I. $K_{12a} = 0$, $K_{12b} = 0$. Equation (7.54) becomes Eq. (7.24). A plot of C against L_t will approach a limiting value at high ligand concentration. Diagnosis of Case I behavior is tentative because Eq. (7.54) also results in this type of curve. The ligand concentration can be obtained iteratively through Eq. (7.31), or alternatively by the approach given in the preceding paragraph. The result is

$$[L] = L_t - \frac{S_t}{X + 1} \tag{7.56}$$

where

$$X = \frac{(C + 1)(R - C)}{(C - 1)(R + C)} \tag{7.57}$$

and $R = K_{11a}/K_{11b}$. The parameter R is estimated by extrapolating a plot of $\log C$ versus $1/L_t$ to $1/L_t = 0$, since C approaches R as L_t approaches infinity. The constants K_{11a} and K_{11b} are then obtained by linear plotting as described earlier. The calculated R is compared with the initial estimate; if they differ significantly, the process is repeated.

   Case II. $K_{11b} = 0$, $K_{12a} = 0$, $K_{12b} = 0$. From Eq. (7.54),

$$C = 1 + K_{11a}[L] \tag{7.58}$$

where C is a linear function of ligand concentration. Of course, any system will approach linearity at sufficiently low ligand concentration, so L_t must be made large enough to determine if curvature exists. The free ligand concentration is obtained with

$$[L] = L_t - \frac{(C-1)S_t}{2C} \tag{7.59}$$

Case III. $K_{11b} = 0$, $K_{12b} = 0$. Equation (7.54) becomes

$$C = 1 + K_{11a}[L] + K_{11a}K_{12a}[L]^2 \tag{7.60}$$

The plot of C versus L_t will reveal curvature that is concave upward; $[L]$ is calculated with

$$[L] = \frac{2C(L_t - S_t) + 2S_t}{2C - S_t K_{11a}} \tag{7.61}$$

The linear form of Eq. (7.60) is

$$\frac{C-1}{[L]} = K_{11a} + K_{11a}K_{12a}[L] \tag{7.62}$$

Case IV. $K_{12b} = 0$. Then Eq. (7.54) becomes

$$C = \frac{1 + K_{11a}[L] + K_{11a}K_{12a}[L]^2}{1 + K_{11b}[L]} \tag{7.63}$$

The plot of C versus L_t will approach a linear segment of positive slope at high L_t. The equation of this line is

$$C = R + RK_{12a}[L] \tag{7.64}$$

where $R = K_{11a}/K_{11b}$. This gives part of the desired information. The concentration $[L]$ is given by

$$K_{11b}[L]^2 + \left[1 + S_t K_{11b}\left(\frac{3}{2} - \frac{R}{2C}\right) - L_t K_{11b}\right][L] - \left[L_t - \frac{S_t(C-1)}{C}\right] = 0 \tag{7.65}$$

One effective way to evaluate the constants is to measure K_{11b} by a different, independent, experimental method, such as spectroscopy. This will usually be straightforward, the Case IV diagnosis having shown that $K_{12b} = 0$. The linear plot according to Eq. (7.66) is made:

$$\frac{C-1}{[L]} + CK_{11b} = K_{11a} + K_{11a}K_{12a}[L] \tag{7.66}$$

This plot is also useful for confirming Case I systems, which should yield a slope equal to zero.

Other methods have been suggested (10) for analyzing Case IV systems. Experience with simulated data shows that the slope of the plot according to Eq. (7.64) is a fairly good estimate of $K_{11a}K_{12a}/K_{11b}$, but that the intercept of this plot may not give a reliable estimate of K_{11a}/K_{11b}. The following procedure can be effective. From data at high ligand concentration, estimate $K_{11a}K_{12a}/K_{11b} = D$ from the plot according to Eq. (7.64). Then plot data at low ligand concentration according to Eq. (7.67), which is obtained from (7.63).

$$\frac{C-1}{[L]} = K_{11a} - K_{11b}(C - D[L]) \qquad (7.67)$$

Iterations are then carried out. The free ligand concentration is calculated with the quadratic equation (7.65), or alternatively, a linear Taylor's series approximation can be used (13). The expression for the Case IV system is

$$[L] = L_t - \frac{S_t L_t}{2g'(L_t)} \left[\frac{K_{11a} + 2K_{11a}K_{12a}L_t}{1 + K_{11a}L_t + K_{11a}K_{12a}L_t^2} + \frac{K_{11b}}{1 + K_{11b}L_t} \right]$$

$$(7.68)$$

where

$$g'(L_t) = 1 + \frac{S_t K_{11a}(1 + 4K_{12a}L_t + K_{11a}K_{12a}L_t^2)}{2(1 + K_{11a}L_t + K_{11a}K_{12a}L_t^2)^2} + \frac{S_t K_{11b}}{2(1 + K_{11b}L_t)^2}$$

$$(7.69)$$

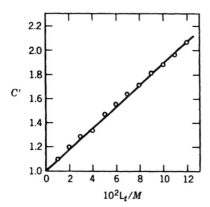

Figure 7.3. Plot of C' against L_t for aniline: α-cyclodextrin, a Case II system; $S_t = 0.00405\ M$, $pK_a = 4.646$. From the plot according to Eq. (7.58), $K_{11b} = 8.8\ M^{-1}$.

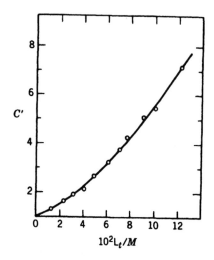

Figure 7.4. Plot of C' against L_t for isoquinoline : α-cyclodextrin, a Case III system; $S_t = 0.00407\ M$, $pK_a = 5.480$. From the plot according to Eq. (7.62), $K_{11b} = 22.7\ M^{-1}$, $K_{12b} = 10.8\ M^{-1}$.

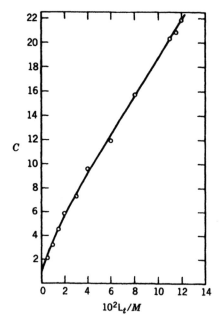

Figure 7.5. Plot of C against L_t for 4-cyanobenzoic acid : α-cyclodextrin; $S_t = 0.00323\ M$, $pK_a = 3.468$. From the plot according to Eq. (7.66), using $K_{11b} = 71.4\ M^{-1}$, are found $K_{11a} = 441\ M^{-1}$, $K_{12a} = 26.2\ M^{-1}$.

By these graphical techniques it is possible to obtain parameter estimates for these special cases of Eq. (7.54). An alternative procedure is to use these estimates as initial values in nonlinear regression analysis (in Cases I, III, and IV). When the full Eq. (7.54) is required to account for the data, it can be helpful to study the system by other experimental techniques and in this way to place constraints on the possible values the parameters may take. Thus the potentiometric study of the *trans*-cinnamic acid: α-cyclodextrin system was interpreted with the aid of spectrophotometric and solubility data (14).

All of these special cases have been encountered in practice (10, 11, 15). Figures 7.1 and 7.2 show data for Case I systems. Figures 7.3–7.5 show diagnostic plots of C or C' (as appropriate) against L, for Case II, Case III, and Case IV systems. Note the small range of C' values in Figure 7.3, a Case II system; it is to be expected that diagnosis of Case II is more likely for weakly complexing systems.

Other multiple binding systems can be treated similarly to this description of $1:1 + 1:2$ binding. If a $2:1$ complex is present, the binding isotherm is a function of substrate concentration. This can be a useful qualitative observation, but such a system is not conveniently analyzed quantitatively.

REFERENCES

1. F. R. Hartley, C. Burgess, and R. M. Alcock, *Solution Equilibria*, Ellis Horwood, Chichester, 1980, pp. 117–131.

2. (a) F. R. Hartley, C. Burgess, and R. M. Alcock, *Solution Equilibria*, Ellis Horwood, Chichester, 1980, pp. 319–345; (b) R. J. Motekaitis and A. E. Martell, *Can. J. Chem.*, **60**, 2403 (1982); (c) M. Cromer-Morin, J. P. Scharff, and R. P. Martin, *Analusis*, **10**, 92 (1982); (d) G. H. Nancollas and M. B. Tomson, *Pure Appl. Chem.*, **54**, 2675 (1982); (e) D. J. Leggett, Ed., *Computational Methods for the Determination of Formation Constants*, Plenum, New York, 1985.

3. F. R. Hartley, C. Burgess, and R. M. Alcock, *Solution Equilibria*, Ellis Horwood, Chichester, 1980, pp. 193–206.

4. F. R. Hartley, C. Burgess, and R. M. Alcock, *Solution Equilibria*, Ellis Horwood, Chichester, 1980, pp. 179–192.

5. A. K. Colter and D. Buben, *Can. J. Chem.*, **54**, 2141 (1976).

6. K. A. Connors and J. M. Lipari, *J. Pharm. Sci.*, **65**, 379 (1976).

7. T. Miyaji, Y. Kurono, K. Uekama, and K. Ikeda, *Chem. Pharm. Bull.*, **24**, 1155 (1976).

8. R. I. Gelb, L. M. Schwartz, R. F. Johnson, and D. A. Laufer, *J. Am. Chem. Soc.*, **101**, 1869 (1979).

9. A. B. Wong, Ph.D. Dissertation, University of Wisconsin-Madison, 1980, pp. 80, 109.

10. K. A. Connors, S.-F. Lin, and A. B. Wong, *J. Pharm. Sci.*, **71**, 217 (1982).

11. S.-F. Lin and K. A. Connors, *J. Pharm. Sci.*, **72**, 1333 (1983).

12. S.-F. Lin, Ph.D. Dissertation, University of Wisconsin-Madison, 1981, p. 105.

13. K. A. Connors and D. D. Pendergast, *Anal. Chem.*, **56**, 1549 (1984).

14. K. A. Connors and T. W. Rosanske, *J. Pharm. Sci.*, **69**, 173 (1980).

15. A. B. Wong, S.-F. Lin, and K. A. Connors, *J. Pharm. Sci.*, **72**, 388 (1983).

8

SOLUBILITY
MEASUREMENT

8.1. MOLECULAR INTERACTIONS AND SOLUBILITY

The solubility of a substance depends on the composition of the solvent, and this dependence (like the dependence of any other property of the system) can be interpreted in terms of nonideality effects, as discussed in Sections 1.3 and 2.6, "Medium Effects," or as evidence of a mass action interaction. Since solubility measurement is a classical technique for the estimation of activity coefficients, we briefly consider this subject before devoting the rest of this chapter to the solubility technique for measuring binding constants.

Determination of Activity Coefficients

The effect of salts on the activity coefficient of a nonelectrolyte can be studied by measuring the solubility of the nonelectrolyte in the presence and the absence of the salt. The chemical potential of the nonelectrolyte solute in its saturated solutions is equal to its chemical potential in the solid state, or

$$\mu(\text{solid}) = \mu(\text{sat. soln.}, c_s = 0) = \mu(\text{sat. soln.}, c_s)$$

where c_s represents the molar concentration of salt. It follows that the activity of the nonelectrolyte is identical in both solutions, or

$$s_0 \gamma_0 = s\gamma \qquad (8.1)$$

where s_0 and s are the molar solubilities in the absence and presence of salt, respectively, and γ_0 and γ are the corresponding molar activity coefficients. It is found experimentally that the logarithm of the nonelectrolyte solubility is very often a linear function of c_s over quite wide ranges in salt concentration. This observation combined with Eq. (8.1) gives

$$\log \frac{\gamma}{\gamma_0} = \log \frac{s_0}{s} = k_s c_s \qquad (8.2)$$

where k_s, the empirical proportionality constant, is called the salting-in (or salting-out) coefficient or the Setschenow constant. If s_0 is very small it is reasonable to set $\gamma_0 = 1$, and then Eq. (8.1) provides a measure of the activity coefficient ($\gamma = s_0/s$) or, equivalently, Eq. (8.2) allows k_s to be estimated. If $s_0/s > 1$, k_s is positive, and the solute is said to be salted out; if $s_0/s < 1$, k_s is negative, and the solute is salted in. Gordon (1) and Schneider (2) have reviewed salting-out and salting-in data and theories.

The activity coefficient of a sparingly soluble electrolyte can be measured by a solubility technique (3). In illustration we consider the univalent electrolyte MA, which dissociates as follows:

$$MA(s) \rightleftharpoons M^+ + A^-$$

The equilibrium constant for this reaction is K_s, the solubility product, given by Eq. (8.3) because the activity of the pure solid is unity.

$$K_s = a_{M^+} a_{A^-} = \gamma_\pm^2 c_\pm^2 \qquad (8.3)$$

In Eq. (8.3), $\gamma_\pm$ is the mean ionic activity coefficient and $c_\pm$ is the mean ionic molar concentration, which is equal to the individual ion concentrations $[M^+]$ and $[A^-]$ for this $1:1$ electrolyte. If the dissolved electrolyte is completely dissociated, $c_\pm = s$, the molar solubility, and we get

$$\gamma_\pm = \frac{K_s^{1/2}}{s} \qquad (8.4)$$

The solubility product can itself be determined by solubility measurements, the solubility s being measured as a function of ionic strength (established with other electrolytes) and extrapolated to zero ionic strength by plotting against the square root of ionic strength in accordance with the Debye–Hückel limiting law. This extrapolation limit is the reference state, hence $\gamma_\pm = 1$ and, by Eq. (8.4), the extrapolated solubility is equal to $K_s^{1/2}$. If the salt is incompletely dissociated, an iterative calculation must be carried out, as described by Davis (4).

Interpretation as Binding

If the solubility of one substance is altered by the presence of a second substance, this alteration may be interpreted as evidence of complex formation between the two substances, rather than as (or in addition to) an activity coefficient effect. For example, Srivastava and Gupta (5) measured the solubilities of aromatic hydrocarbons in nitrobenzene and compared them with the ideal (thermodynamic) solubilities. Discrepancies were attributed to molecular complex formation. Anthracene shows cosolvency in binary mixtures of an iodo compound and a hydrocarbon, and this effect was interpreted as due to complex formation (6). The aqueous solubility of caffeine is reduced by the presence of sodium salts of low-molecular-weight fatty acids, a salting-out effect, whereas sodium salts of high-molecular-weight fatty acids increase the solubility of caffeine, and this was attributed to aggregation (micelle formation) of these amphiphilic substances (7). In this example the caffeine serves as a probe of the aggregation phenomenon. Mukerjee and Cardinal studied the solubilization of naphthalene in solutions of sodium cholate and emphasized the difficulties in establishing the aggregation pattern from such observations (8).

In better-defined systems than these it is possible to evaluate equilibrium constants from the solubility data. A method for the measurement of acid dissociation constants of slightly soluble acids and bases makes use of the compound's solubility as a function of pH (9). The apparent dissociation constant of the slightly soluble weak acid HA is defined

$$K_a' = \frac{a_{H^+}[A^-]}{[HA]} \tag{8.5}$$

The observed solubility of the acid (in solutions of buffered pH and constant ionic strength) is

$$s = [HA] + [A^-] \tag{8.6}$$

and we define the intrinsic solubility

$$s_0 = [\text{HA}] \tag{8.7}$$

The intrinsic solubility is measured in a solution whose pH is several units below pK_a', so that all of the solute is in the HA form. Combining these equations gives Eq. (8.8).

$$\log\left(\frac{s}{s_0} - 1\right) = \text{pH} - pK_a' \tag{8.8}$$

Thus a plot of $\log(s/s_0 - 1)$ against pH should be linear with slope $= +1$, and pK_a' is found as the intercept on the pH axis when $\log(s/s_0 - 1) = 0$. If the conjugate base is a neutral, slightly soluble, species (as with an amine), a similar development gives Eq. (8.9).

$$\log\left(\frac{s}{s_0} - 1\right) = pK_a' - \text{pH} \tag{8.9}$$

This plot has a negative slope. Krebs and Speakman (9a) determined both pK_a' values of sulfadiazine by this method. Zimmermann (9b) analyzed solubility–pH data by least-squares regression, obtaining estimates of both pK_a' and s_0 as parameters of the system.

Evidently, this solubility method is limited in its applicability, and it is laborious compared with potentiometry or spectrophotometry, but for slightly soluble acids or bases it can be very useful, provided an analytical method is available for the solubility measurements. In many practical situations the reverse problem, that of predicting the total solubility s at a given pH when pK_a' is known, must be solved. This problem may be complicated by the presence of slightly soluble salts of the acid (or base), for which solubility products must be known. A general solution of this problem has been described (10).

Several workers have developed solubility methods for studying stepwise complex formation in metal ion coordination systems (11). One of the complexes must be sparingly soluble, and the composition of this solid phase must not change throughout the range of solution compositions examined. A typical application is to the study of the halide complexes of silver halides (12). The principle of the method can be demonstrated with a hypothetical example similar to the silver halide case. Let the cation M^+

form $1:1$, $1:2$, and $1:3$ complexes with the anion (ligand) L^-, the neutral species ML being slightly soluble.

$$M^+ + L^- \rightleftharpoons ML(s)$$

$$M^+ + 2L^- \rightleftharpoons ML_2^-$$

$$M^+ + 3L^- \rightleftharpoons ML_3^{2-}$$

The equilibrium constants are

$$K_s = [M^+][L^-] \tag{8.10}$$

$$\beta_{12} = \frac{[ML_2^-]}{[M^+][L^-]^2} \tag{8.11}$$

$$\beta_{13} = \frac{[ML_3^{2-}]}{[M^+][L^-]^3} \tag{8.12}$$

The measured solubility s is given by Eq. (8.13), and combination with the equilibrium constant expressions gives the isotherm, Eq. (8.14).

$$s = [M^+] + [ML_2^-] + [ML_3^{2-}] \tag{8.13}$$

$$s = \frac{K_s}{[L^-]} (1 + \beta_{12}[L^-]^2 + \beta_{13}[L^-]^3) \tag{8.14}$$

Let L_t be the total concentration of ligand in solution.

$$L_t = [L^-] + 2[ML_2^-] + 3[ML_3^{2-}] \tag{8.15}$$

Combination of (8.15) with the equilibrium constants yields Eq. (8.16).

$$L_t = [L^-] + 2\beta_{12}K_s[L^-] + 3\beta_{13}K_s[L^-]^2 \tag{8.16}$$

The experiment is carried out by measuring s as a function of L_t. The stability constants are estimated by iterative treatment of Eqs. (8.14) and (8.16). Johansson (13) has discussed the iterative calculation of free ligand concentration and advises that $[L^-]$ be measured experimentally when possible.

8.2. INTERPRETATION OF SOLUBILITY DIAGRAMS

In this section we consider in detail the solubility method for studying the interaction between a slightly soluble substrate S and a soluble ligand L. This includes the metal ion coordination systems described in the preceding section. Andrews and Keefer (14) used this experimental method in 1949 to study the formation of complexes between silver ion and aromatic hydrocarbons, and Higuchi and his co-workers further developed and applied the technique in the 1950s. The subject was reviewed in 1965 (15).

The experimental procedure is to determine the total molar solubility of substrate, S_t, as a function of total molar concentration of ligand, L_t, at constant temperature and ionic strength. This is accomplished by preparing several vials or ampules to contain equal amounts (in considerable excess of its normal solubility) of the slightly soluble substrate, and adding a fixed volume of solvent containing increasing concentrations of ligand. The systems are brought to solubility equilibrium by tumbling or shaking in a constant-temperature bath; this commonly takes 1 to 2 days, but equilibration times of 1 to 2 weeks are sometimes required. The solution phase is then analyzed (after separation from solid phase by filtration or centrifugation) for total substrate concentration, no matter what its molecular state may be. Usually, absorption spectroscopy is a suitable analytical method, but prior experimental study must show that in the (usually) diluted analytical solutions, there is no spectral shift caused by the presence of complexes. The phase solubility diagram is then constructed by plotting S_t against L_t. From this diagram it is possible to learn something about the complex stoichiometry and to estimate binding constants.

Types of Diagrams and Determination of Stoichiometry

The simplest type of solubility diagram shows a linear increase of S_t as a function of L_t, as in Figure 8.1 for the system methyl 2-naphthoate (S): 8-nitrotheophyllinate (L) (16). The basic premise is that the increase in solubility is a consequence of the formation of one or more soluble complexes by interaction of S with L. The presence of pure solid S throughout the range of the solubility diagram ensures that the chemical potential, and therefore the activity, of free S in solution is constant, and since the study is carried out at constant ionic strength we usually assume

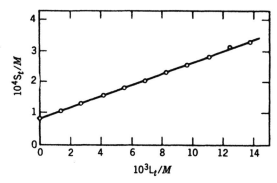

Figure 8.1. Solubility diagram for the system methyl 2-naphthoate (S):8-nitrotheophylli-nate (L) at 25°C.

that activity coefficients remain constant, hence that the concentration of free S is constant. Since [S] is fixed, we cannot vary it so as to learn something about the stoichiometry with respect to S.

As will be shown subsequently, if all complexes present are first-order in L (i.e., if $n = 1$ in all species S_mL_n), then the solubility diagram will be linear. The converse is not necessarily true, but a linear diagram is often taken as evidence for $n = 1$. We also note that if the slope of a linear diagram is greater than unity, then at least one complex must be present for which m is greater than 1, for it is obviously impossible for one mole of L to take more than one mole of S into solution if the complex has 1:1 stoichiometry. On the other hand, a slope of less than unity does not necessarily mean that only a 1:1 complex is formed.

Figure 8.2 is the solubility diagram for 1,4-dimethylterephthalate (S): α-cyclodextrin (L) (17). The nonlinear plot with concave-upward curvature means that at least one complex is present having $n > 1$. A nonlinear plot with concave-downward curvature may be evidence of nonideality effects (nonconstancy of activity coefficients) or of self-association by the ligand.

A solubility diagram may exhibit a plateau region representing a maximum value of S_t that additional quantities of ligand do not alter. A trivial origin of this discontinuity is the complete solubilization of the solid substrate that had been added to the system. A more interesting cause is that the break in the curve represents saturation with respect to L, if L is not highly soluble. Just as complex formation of S with L increases the solubility of S, so it also increases the solubility of L. If the

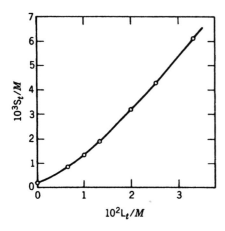

Figure 8.2. Solubility diagram for the system 1,4-dimethylterephthalate (S):α-cyclodextrin (L) at 25°C.

discontinuity represents saturation with respect to L, a stoichiometric ratio can be calculated. Figure 8.3 shows such a system. The normal solubility of the ligand, 8-nitrotheophyllinate, is $13.9 \times 10^{-3} M$, and its concentration at the discontinuity is $15.4 \times 10^3 M$, so the difference, $1.5 \times 10^3 M$, is the amount of ligand that has been taken into solution by complex formation. The amount of substrate taken into solution is $3.11 \times 10^{-3} M$ (the plateau concentration) minus $1.97 \times 10^3 M$ (its normal solubility), or $1.14 \times 10^3 M$. The ratio is 1.3/1, consistent with 1:1 stoichiometry within the accuracy of estimation of the point of discontinuity.

A third, and most interesting, cause of a discontinuity and plateau in a solubility diagram is the presence of a complex whose solubility is low enough to be exceeded in the range of ligand concentrations studied. Figure 8.4 shows this behavior for the system salicylic acid (S):caffeine (L) (18). A qualitative interpretation of this diagram is based on Gibb's phase rule and the complex formation hypothesis (15). For the present purpose let us suppose that only a single complex is formed. The initial rising portion of the graph, from the solubility at $L_t = 0$ to the point of discontinuity at a, is accounted for just as in the treatment of Figures 8.1 to 8.3—namely, the increase in substrate concentration is a consequence of the formation of a complex species. At point a, however, the solution phase has become saturated with respect to the complex. Addition of more L results in the formation of more complex, which must precipitate. Throughout the plateau region the concentration of unbound S is held constant by dissolution of solid S, and the essential process in this region

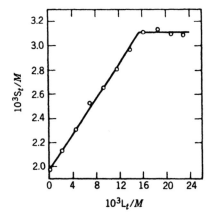

Figure 8.3. Solubility diagram for the system methyl *trans*-cinnamate (S):8-nitro-theophyllinate (L) at 25°C.

is the transformation of solid S into solid complex. The only degrees of freedom in the system are temperature and pressure throughout the plateau region.

At point *b* all of the solid substrate has been consumed by this process, and further addition of L results in depletion of S in the solution phase by complex formation and precipitation. Since from point *b* there is no remaining solid S, the system contains one less phase and therefore one more degree of freedom, the concentration of S. If only a single complex is present throughout the entire diagram, the final value of S_t near point *c* should be very similar to the rise in S_t from the initial point to point *a*, that is, equal to the solubility of the complex (modified by the different

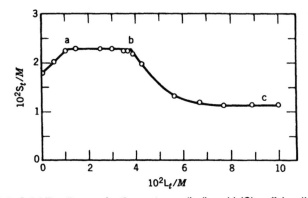

Figure 8.4. Solubility diagram for the system salicylic acid (S):caffeine (L) at 30°C.

solution compositions in these two regions). A substantial difference in these two concentrations is indicative of the presence of two complexes.

A calculation of stoichiometry can be made from the plateau data. The amount of L entering the complex in the plateau region is equal to the amount added between points a and b, or, from Figure 8.4.

$$\text{L in complex} = 3.80 \times 10^{-2} \, M - 1.10 \times 10^{-2} \, M$$

$$= 2.70 \times 10^{-2} \, M$$

The amount of S entering the complex in this same region is equal to the total amount of S in the system minus the S in solution at point a. In this study 0.70 g of salicyclic acid per 100 ml was taken in each sample, corresponding to a total initial concentration of $5.07 \times 10^{-2} \, M$, and the value of S_t at a is $2.29 \times 10^{-2} \, M$, giving a difference of $2.78 \times 10^{-2} \, M$. The ratio S/L is therefore 1.03/1, showing that this is probably a 1:1 complex.

Another graphical technique is available for soluble complexes. Let us write the overall formation of complex SL_n as

$$S + nL \rightleftharpoons SL_n$$

with overall stability constant

$$\beta_{1n} = \frac{[SL_n]}{[S][L]^n}$$

Rearranging and taking logarithms gives Eq. (8.17).

$$\log \frac{[SL_n]}{[S]} = \log \beta_{1n} + n \log [L] \tag{8.17}$$

The concentration of free substrate [S] is fixed by the presence of solid S; we denote this quantity s_0. The concentration of complexed substate is equal to the difference between S_t and s_0. Hence Eq. (8.18) is obtained.

$$\log \frac{S_t - s_0}{s_0} = \log \beta_{1n} + n \log [L] \tag{8.18}$$

A log–log plot according to Eq. (8.18) will give a straight line for the system, the slope being equal to n. Of course, the system may not be as simple as this, and multiple complexes will give rise to curves, but these still may be helpful in suggesting reasonable stoichiometric ratios (11a,19). In such instances a plot of the slope, that is $d \log [(S_t - s_0)/s_0]/d \log [L]$, against $d \log [L]$ can be useful in identifying dominant stoichiometric numbers (19).

The chemical analysis of solid phases collected from the solubility study can be valuable. Referring to Figure 8.4, the solid phase from s_0 to point a is pure S; the solid phase from a to b is a mixture of S and complex, its composition changing continuously from pure S to pure complex; and the solid phase beyond point b is pure complex.

As pointed out above, the solubility method does not permit the substrate concentration to be varied, and this limitation prevents us from investigating the stoichiometry with respect to S. However, if the solubilities of S and L are comparable, the system can be reversed, treating S as L and vice versa.

1:1 Binding

If a single 1:1 complex is formed, the mass balance expressions in the solubility experiment are

$$S_t = [S] + [SL]$$
$$L_t = [L] + [SL] \tag{8.19}$$

As we have seen, however, [S] is constant because pure solid substrate is present in the system. Calling this quantity s_0, the material balance on substrate becomes

$$S_t = s_0 + [SL] \tag{8.20}$$

Combining Eqs. (8.19) and (8.20) with the definition of the stability constant K_{11} gives Eq. (8.21) as the solubility isotherm.

$$S_t = s_0 + \frac{K_{11}s_0L_t}{1 + K_{11}s_0} \tag{8.21}$$

Thus a plot of S_t against L_t is linear with intercept s_0 and slope $K_{11}s_0/(1 - K_{11}s_0)$. The stability constant is given by Eq. (8.22).

$$K_{11} = \frac{\text{slope}}{s_0(1 - \text{slope})} \tag{8.22}$$

Notice that slope ≤ 1 if only $1:1$ complexation occurs. Figures 8.1 and 8.3 are plots of systems that have been interpreted according to Eq. (8.21); from Figure 8.1 we find $K_{11} = 215\,M^{-1}$, and Figure 8.3 yields $K_{11} = 40\,M^{-1}$. The initial rising portion of Figure 8.4 is also this type of plot.

Equation (8.21), the $1:1$ solubility isotherm, provides an interesting contrast with the isotherms derived in earlier chapters. In the solubility method the measure of binding $(S_t - s_0)$ is a linear function of total ligand concentration L_t, whereas in other methods (such as spectroscopy) the measure of binding is a hyperbolic function of free ligand concentration $[L]$. These differences arise because in the solubility method the free substrate concentration $[S]$ is constant and total substrate concentration S_t varies, but in other methods S_t is held constant and $[S]$ varies. As shown in Section 4.2, it is possible to combine the spectral and solubility conditions and to make spectral observations at constant $[S]$.

If two (or more) isomeric $1:1$ complexes are formed, a similar development (20) shows that the experimental K_{11} value is the sum of the isomeric $1:1$ binding constants, as with other methods.

It is a basic assumption of the solubility method that the composition of the pure undissolved substrate is unchanged throughout the region in which data are taken for stability constant measurements. This assumption leads to the condition that the free substrate concentration is constant and known. The formation of a solution of ligand or complex in the substrate phase would vitiate the binding constant calculation. With solid substrates such solution formation is, though possible, not expected to occur widely, but with liquid substrates it seems much more probable.

Multiple Equlibria

We first consider a system containing complexes $SL, S_2L, S_3L, \ldots, S_mL$. Defining overall stability constants β_{h1}, where $h = 1$ to m, and setting $[S] = s_0$, we obtain, from the mass balance expressions on substrate and ligand:

$$S_t = s_0 + \sum_{h=1}^{m} h\beta_{h1}s_0^h[L] \qquad (8.23)$$

$$L_t = [L]\left(1 + \sum_{h=1}^{m} \beta_{h1}s_0^h\right) \qquad (8.24)$$

Combining these equations gives the isotherm, Eq. (8.25), where the summations are from 1 to m.

$$S_t = s_0 + \frac{L_t \Sigma h\beta_{h1}s_0^h}{1 + \Sigma \beta_{h1}s_0^h} \qquad (8.25)$$

Equation (8.25) shows that the solubility diagram is a linear function of L_t as long as the system contains no species $S_m L_n$ for which n is greater than 1. If the experimentally measured slope has a value greater than some integer, then at least one complex $S_h L$ must be present for which h is greater than this integer; for example, a slope greater than unity shows that there cannot be only 1:1 complexes present. A binding constant can be calculated only if it is known (or assumed) that a single stoichiometric formula $S_m L$ is present; then Eq. (8.25) becomes

$$S_t = s_0 + \frac{m\beta_{m1}s_0^m L_t}{1 + \beta_{m1}s_0^m} \qquad (8.26)$$

Equation (8.21) is a special case of (8.26).

Next consider a system in which the species $SL, SL_2, SL_3, \ldots, SL_n$ may be present. Substituting into the mass balance expression for substrate in the usual manner gives Eq. (8.27).

$$S_t = s_0 + s_0 \sum_{i=1}^{n} i\beta_{1i}[L]^i \qquad (8.27)$$

The mass balance on ligand leads to Eq. (8.28).

$$L_t = [L] + s_0 \sum_{i=1}^{n} i\beta_{1i}[L]^i \qquad (8.28)$$

Equations (8.27) and (8.28) together describe the system. Unlike the earlier treatments in this section, however, the free ligand concentration cannot in general be eliminated by combining these expressions. Equation (8.27) therefore serves as the isotherm for the system. Equation (8.14) is a special application of Eq. (8.27).

The most important case of this system is the $n = 2$ case, for which Eqs. (8.27) and (8.28) become

$$S_t = s_0 + K_{11}s_0[L] + K_{11}K_{12}s_0[L]^2 \qquad (8.29)$$

$$L_t = [L] + K_{11}s_0[L] + 2K_{11}K_{12}s_0[L]^2 \qquad (8.30)$$

Several ways have been devised to obtain the stability constants from the

S_t, L_t, data for this case. Equation (8.29) is rearranged to the linear plotting form, Eq. (8.31).

$$\frac{S_t - s_0}{[L]} = K_{11}s_0 + K_{11}K_{12}s_0[L] \qquad (8.31)$$

It may be acceptable to make the approximation $[L] \approx L_t$, and then to plot $(S_t - s_0)/L_t$ against L_t. To a better level of approximation $[L]$ is estimated (14) by neglecting the influence of the 1:2 complex and calculating $[L] \approx L_t - [SL] \approx L_t - (S_t - s_0)$. Yet another approximation transforms the quadratic (8.30) into a linear function by means of a Taylor's series expansion; this is described in Section 2.3.

The exact solution of Eq. (8.30) is given in Eq. (8.32).

$$[L] = \frac{-(1 + K_{11}s_0) + [(1 + K_{11}s_0)^2 + 8K_{11}K_{12}s_0L_t]^{1/2}}{4K_{11}K_{12}s_0} \qquad (8.32)$$

Several workers (21) have developed iterative calculations based on Eq. (8.32). Initial estimates of K_{11} and K_{12} are obtained by means of one of the approximations described in the preceding paragraph. With Eq. (8.32), improved values of $[L]$ are now calculated, and better estimates of K_{11} and K_{12} are obtained via Eq. (8.31). Iterations are carried out until the parameter estimates converge.

Figure 8.5 shows the solubility diagram for the biphenyl (S): α-cyclodextrin (L) system (22). The nonlinear isotherm with concave-upward curvature is characteristic of systems with $n > 1$. Table 8.1 gives the experimental data for this system and lists the free ligand concentrations calculated with Eq. (8.32) for the final iteration. The linear plot according to Eq. (8.31) is shown in Figure 8.6. This analysis yielded the binding constants $K_{11} = 50 \ M^{-1}$, $K_{12} = 63 \ M^{-1}$.

Iga et al. (23) have shown that Eqs. (8.29) and (8.30) can be combined to give the linear plotting form Eq. (8.33), which involves no approximations.

$$\frac{S_t - s_0}{L_t - 2(S_t - s_0)} = \alpha + \beta[L_t - 2(S_t - s_0)] \qquad (8.33)$$

$$\alpha = \frac{K_{11}s_0}{1 - K_{11}s_0} \qquad (8.34)$$

$$\beta = \frac{K_{11}K_{12}s_0}{(1 - K_{11}s_0)^2} \qquad (8.35)$$

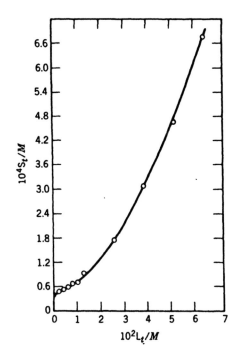

Figure 8.5. Solubility diagram for the system biphenyl (S):α-cyclodextrin (L) at 25°C. The data are from Table 8.1.

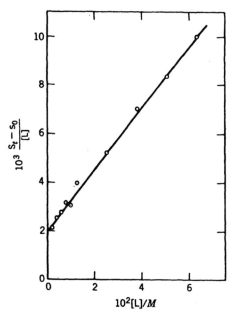

Figure 8.6. Plot of the data in Table 8.1 according to Eq. (8.31).

Table 8.2 gives solubility data for 1,4-dimethylterephthalate (S): α-cyclodextrin (L); these data were used to plot the isotherm in Figure 8.2. The table also lists the plotting variables needed according to Eq. (8.33), and the linear plot of these variables is shown in Figure 8.7. The constants $K_{11} = 464 \ M^{-1}$, $K_{12} = 102 \ M^{-1}$ are obtained.

For some systems these methods of data treatment fail. Thus if $K_{11}s_0$ is very close to unity, the plot of Eq. (8.33) cannot be made, as shown by Eqs. (8.34) and (8.35). Another problem arises if s_0 is so small that the product $K_{11}s_0$ is also small; then the intercepts of plots according to Eqs. (8.31) and (8.33) may not be significantly different from zero, preventing calculation of the stability constants even though the plots can be made. Another problem sometimes encountered is that S_t can be measured but s_0 cannot, for very slightly soluble substrates. Since s_0 is very small, $(S_t - s_0) \approx S_t$, and the plots according to Eqs. (8.31) and (8.33) can be made, but without an estimate of s_0 the stability constant K_{11} cannot be calculated (although K_{12} is accessible as the ratio slope/intercept of Eq. 8.31). In such instances a plot of $S_t^{1/2}$ against L_t may give a useful extrapolated estimate of s_0.

Equation (8.29) suggests that the characteristic shape of the solubility diagram for the $1:1 + 1:2$ system is a concave-upward curve, as seen in Figures 8.2 and 8.5. However, the solubility diagram is a plot of s_t against L_t rather than [L], and the distinction may lead to curves of other shapes.

Table 8.1. Concentration Variables in the Solubility Study of the Biphenyl: α-Cyclodextrin System[a]

$10^2 \ L_t/M$	$10^4 \ S_t/M$	$10^2 [L]/M^b$
0.000	0.40	0.000
0.200	0.44	0.200
0.400	0.50	0.399
0.600	0.56	0.598
0.800	0.65	0.796
1.00	0.70	0.995
1.29	0.91	1.28
2.58	1.74	2.56
3.87	3.07	3.83
5.16	4.65	5.09
6.45	6.76	6.34

[a] At 25°C in 0.10 M NaCl.
[b] Final calculated values.

Table 8.2. Solubility Data for 1,4-Dimethylterephthalate (S): α-Cyclodextrin (L)

$10^2\,L_t/M$	$10^3\,S_t/M$	$10^2[L_t - 2(S_t - s_0)]$	$\dfrac{S_t - s_0}{L_t - 2(S_t - s_0)}$
0.000	0.169	–	–
0.663	0.867	0.523	0.1336
1.007	1.367	0.767	0.1563
1.326	1.898	0.980	0.1766
1.989	3.175	1.387	0.2167
2.517	4.292	1.692	0.2437
3.315	6.144	2.120	0.2819

For a system having a nonzero value of K_{12}, elimination of [L] between Eqs. (8.29) and (8.30) yields (24)

$$S_t = s_0 + \frac{L_t}{2} + \frac{ab}{c} - \frac{b}{c}\left(a^2 + 8K_{11}K_{12}s_0L_t\right)^{1/2} \tag{8.36}$$

where

$$a = 1 + K_{11}s_0 \tag{8.37}$$

$$b = \frac{a}{2K_{11}s_0} - 1 \tag{8.38}$$

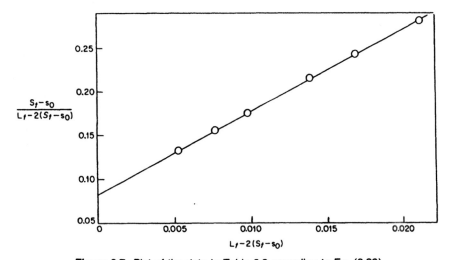

Figure 8.7. Plot of the data in Table 8.2 according to Eq. (8.33).

$$c = 4K_{12} \tag{8.39}$$

The slope of the solubility diagram is therefore given by

$$\frac{dS_t}{dL_t} = \frac{1}{2}\left[1 - \frac{8bK_{11}K_{12}s_0}{c(a^2 + 8K_{11}K_{12}s_0L_t)^{1/2}}\right] \tag{8.40}$$

Since a and c must be positive, the curvature in the slope depends on the sign of b. If $b = 0$, the slope is constant with the value 0.5; this implies that $K_{11} = 1/s_0$. If b is negative, the slope decreases as L_t increases; if b is positive, the slope increases as L_t increases. Whether these cases can be distinguished experimentally will be determined by the magnitudes of the parameters and the range of L_t over which the variable S_t can be observed. The value of the slope at $L_t = 0$ is, from Eq. (8.40),

$$\left(\frac{dS_t}{dL_t}\right)_{L_t=0} = \frac{K_{11}s_0}{1 + K_{11}s_0} \tag{8.41}$$

Equation (8.41) can be helpful in those instances, cited above, in which the product $K_{11}s_0$ is so small that Eqs. (8.31) or (8.33) do not allow the binding constants to be calculated; estimation of the slope, at $L_t = 0$, of the plot of S_t against L_t may permit $K_{11}s_0$ to be obtained.

Figure 8.8 is a solubility diagram for the *trans*-cinnamic acid (S): α-cyclodextrin (L) system (24). There is strong evidence that a 1:1 complex stoichiometry is not adequate to describe this system, despite the linear solubility diagram; note, for example, the two-level plateau region, and the spectral data for this system shown in Figure 4.11. Moreover, chemical analysis of the solid phase reveals that, under the second plateau, the mole fraction of L in the solid is greater than 0.5. Hence the 1:1 + 1:2 model is reasonable. Since the rising portion of the diagram appears to be linear, the analysis of Eq. (8.40) suggests that its slope should be 0.5; actually the slope is 0.8. Equations (8.31) and (8.33) are inapplicable to this system, and the solubility data were treated in conjunction with spectral and potentiometric data on the same system (24).

Let us analyze the plateau regions of Figure 8.8. We consider the discontinuity giving rise to the first plateau. As L_t increases, S_t rises because of the formation of soluble SL and SL_2 complexes. Ultimately, the solubility limit of one of these complexes is reached (as long as the solid S phase is not exhausted first and the solubility limit of L is not

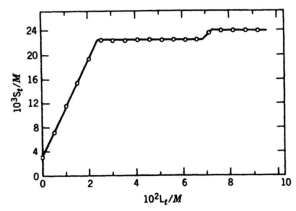

Figure 8.8. Solubility diagram of the system *trans*-cinnamic acid (S):α-cyclodextrin (L) at 25°C.

reached). Suppose that the solubility of SL is reached before that of SL_2. Then $[S] = s_0$ and $[SL] = s_{11}$, and from the mass balance on substrate we get

$$S_t = s_0 + s_{11} + \frac{K_{12}(s_{11})^2}{K_{11}s_0} \qquad (8.42)$$

That is, S_t becomes invariant as soon as *one* of the complexes reaches its solubility limit, with the complexation equilibria ensuring that the second complex concentration then also is fixed. This conclusion is general for any number of complexes and no matter which complex reaches its solubility first as long as solid S is present. Thus the first plateau is accounted for. Of course, the interpretation of S_t in this region depends on which complex has reached its solubility limit. Because of the solid-phase data on this system, the argument leading to Eq. (8.42) appears to apply to Figure 8.8.

As L_t continues to increase, the solid S eventually is depleted by its conversion into complexes and its precipitation as solid SL. At the point where solid S disappears, the constraint on S_t is lost (although the condition $[SL] = s_{11}$ still applies). The S_t value rises until the solubility limit of the second complex (SL_2) is reached, when S_t again becomes invariant; its value, given by Eq. (8.43), is reached by an argument similar to that leading to Eq. (8.42).

$$S_t = \frac{K_{12}(s_{11})^2}{K_{11}s_{12}} + s_{11} + s_{12} \qquad (8.43)$$

Thus a "second plateau" is observed. (If L_t could be made sufficiently large, eventually all solid SL would disappear, and S_t would decrease to a final value approximately equal to s_{12}.)

If $n = 3$ for a system of the type being considered, the isotherm is, from Eq. (8.27),

$$S_t = s_0 + \beta_{11}s_0[L] + \beta_{12}s_0[L]^2 + \beta_{13}s_0[L]^3 \qquad (8.44)$$

This is rearranged to Eq. (8.45) to give a plotting form with which β_{11} can be estimated as the extrapolated intercept, although the plot of $(S_t - s_0)/[L]$ against $[L]$ is not linear.

$$\frac{S_t - s_0}{[L]} = \beta_{11}s_0 + \beta_{12}s_0[L] + \beta_{13}s_0[L]^2 \qquad (8.45)$$

Alternatively, β_{11} $(= K_{11})$ can be estimated from the initial slope of the solubility diagram according to Eq. (8.41). Next Eq. (8.45) is rearranged to Eq. (8.46), a linear plotting form that allows K_{12} and K_{13} to be estimated (25).

$$\frac{(S_t - s_0) - K_{11}s_0[L]}{[L]^2} = K_{11}K_{12}s_0 + K_{11}K_{12}K_{13}s_0[L] \qquad (8.46)$$

These graphical techniques can provide good initial estimates of the parameters for a nonlinear regression analysis of Eq. (8.44) together with the corresponding equation, based on Eq. (8.28), for ligand concentration. In treating systems of this type Miyahara and Takahashi (21c) concluded that it is sounder to use the experimental value of s_0 than to treat s_0 as an adjustable parameter.

REFERENCES

1. J. E. Gordon, *The Organic Chemistry of Electrolyte Solutions*, Wiley, New York, 1975, pp. 10–30.
2. H. Schneider, *Solute–Solvent Interactions*, Vol. 2, J. F. Coetzee and C. D. Ritchie, Eds., Dekker, New York, 1976, pp. 205–218.

3. S. Glasstone, *Thermodynamics for Chemists*, Van Nostrand, Princeton, 1947, p. 398.

4. C. W. Davies, *Ion Association*, Butterworth, Washington, 1962, pp. 48–53.

5. R. D. Srivastava and P. D. Gupta, *J. Indian Chem. Soc.*, **44**, 960 (1967).

6. G. R. Samayajula and S. R. Palit, *J. Phys. Chem.*, **58**, 417 (1954).

7. K. E. Schulte, P. Rohdewald, and M. Baumeister, *Pharmazie*, **25**, 762 (1970).

8. P. Mukerjee and J. R. Cardinal, *J. Pharm. Sci.*, **65**, 882 (1976).

9. (a) H. A. Krebs and J. C. Speakman, *J. Chem. Soc.*, 593 (1945); (b) I. Zimmermann, *Intern. J. Pharmaceut.*, **13**, 57 (1983).

10. W. H. Streng, S. K. Hsi, P. E. Helms, and H. G. H. Tan, *J. Pharm. Sci.*, **73**, 1679 (1984).

11. (a) I. Leden, Proceedings of the Symposium on Co-ordination Chemistry, Copenhagen, 1953, Danish Chemical Society, Copenhagen (pub. 1954), pp. 77–85; (b) S. Ahrland and I. Grenthe, *Acta Chem. Scand.*, **11**, 1111 (1957); (c) G. P. Haight, *Acta Chem. Scand.*, **16**, 209 (1962).

12. (a) I. Leden, *Acta Chem. Scand.*, **10**, 812 (1956); (b) K. H. Lieser, *Z. Anorg. Allgem. Chem.*, **292**, 97 (1957).

13. L. Johansson, *Acta Chem. Scand.*, **24**, 1572 (1970).

14. L. J. Andrews and R. M. Keefer, *J. Am. Chem. Soc.*, **71**, 3644 (1949).

15. T. Higuchi and K. A. Connors, *Advan. Anal. Chem. Instrum.*, **4**, 117 (1965).

16. J. L. Cohen, Ph.D. Dissertation, University of Wisconsin-Madison, 1969, p. 101.

17. D. D. Pendergast, Ph.D. Dissertation, University of Wisconsin-Madison, 1983, p. 92.

18. T. Higuchi and D. A. Zuck, *J. Am. Pharm. Assoc., Sci. Ed.*, **42**, 138 (1953).

19. E. H. Gans and T. Higuchi, *J. Am. Pharm. Assoc., Sci. Ed.*, **46**, 458 (1957).

20. K. A. Connors and J. A. Mollica, *J. Pharm. Sci.*, **55**, 772 (1966).

21. (a) K. Kakemi, H. Sezaki, T. Mitsunaga, and M. Nakano, *J. Pharm. Sci.*, **59**, 1597 (1970); (b) T. Higuchi and H. Kristiansen, *J. Pharm. Sci.*, **59**, 1601 (1970); (c) M. Miyahara and T. Takahashi, *Chem. Pharm. Bull.*, **30**, 288 (1982).

22. A. Paulson, M.S. Thesis, University of Wisconsin-Madison, 1986.

23. K. Iga, A. Hussain, and T. Kashihara, *J. Pharm. Sci.*, **70**, 108 (1981).

24. K. A. Connors and T. W. Rosanske, *J. Pharm. Sci.*, **69**, 173 (1980).

25. T. Higuchi and S. Bolton, *J. Am. Pharm. Assoc., Sci, Ed.*, **48**, 557 (1959).

9

LIQUID–LIQUID PARTITIONING

9.1. INTERPRETATION OF PARTITION ISOTHERMS

Principle of the Method

The equilibrium distribution of a single species between two immiscible liquid phases can be written

$$\text{Species } i \text{ in phase } 1 \rightleftharpoons \text{species } i \text{ in phase } 2$$

The equilibrium constant for this process is

$$\mathbf{P} = \frac{(a_i)_2}{(a_i)_1} \tag{9.1}$$

Equation (9.1) is the Nernst distribution law, and the equilibrium constant is called the partition coefficient. Usually the ratio of activities is approximated by a ratio of concentrations,

$$P_0 = \frac{(c_i)_2}{(c_i)_1} \tag{9.2}$$

A plot of concentration in phase 2 against concentration in phase 1 is called a partition isotherm or distribution isotherm. Evidently, if P_0 is a constant, independent of concentration, the partition isotherm will be linear with slope P_0. (The isotherm may also be rendered as a plot of P_0 against c_1 or of $\log P_0$ against $\log c_1$.) Although it does not necessarily follow that a linear isotherm means that P_0 is constant and that a single species is present, this is often a reasonable conclusion.

Experimentally, we measure total concentrations of the solute, no matter what its molecular disposition may be, and we call the resulting ratio P the apparent partition coefficient or the partition ratio. If the partition isotherm is nonlinear, either the ratio of activity coefficients must be varying or more than one species must be present. The dependence of P on concentration then offers a means for the study of the deviation from ideal behavior. Thus the introduction of the second liquid phase provides a way to study complex formation in the first liquid phase by, in effect, creating the property P, whose value is a function of system concentrations, and therefore of the extent of binding.

Alternatively, the method may be described as follows. We measure P_0 for a substrate S in the absence of ligand L whose interaction with S we wish to study in phase 1. Then L is incorporated into the solution of S in phase 1, and after equilibration the concentration of S in phase 2 is measured. With Eq. (9.2) the free (unbound) concentration of S in phase 1 is calculated, and this together with the material balance expressions gives the extent of binding. This calculation assumes that bound S cannot partition into phase 2.

The introduction of phase 2 as a "probe phase" enabling us to study equilibria in phase 1 carries with it both advantages and disadvantages. As we have noted, an advantage is that it provides a new property (moreover, this is a thermodynamic property) for the study of complex formation. Another advantage is that P_0 is an amplification factor allowing extremely small free concentrations in phase 1 to be determined indirectly by analysis of phase 2, if P_0 is greater than unity. There are, however, some important disadvantages. One of these is that addition of the second, probe, phase inevitably alters the nature of the first phase (the medium of interest) because of the mutual solubility of the two phases; it is even possible that the dissolved probe phase may function as a competitively binding substrate, a complicating factor that must be taken into account. A second disadvantage is that in general all species present may partition into the second phase, and therefore equilibria among these species in phase 2 may have to be accounted for in order to describe the equilibria of interest in phase 1.

Usually one of the phases is aqueous and the other is organic, and it is common practice to describe partitioning as an equilibrium between aqueous and organic solvents. In some applications, for example, phase-transfer catalysis, purification by extraction, and absorption by biological tissues, emphasis is placed on the extraction efficiency, usually for transfer from the aqueous to the organic phase. Our concern is with partitioning as a tool for the study of binding equilibria. We may be interested in these equilibria in either the aqueous or the nonaqueous phase, so we shall adopt the noncommittal designations phase 1 and phase 2, defining the partition coefficient as in Eq. (9.2). Phase 1 is the phase in which we wish to study the equilibria, and phase 2 is the added probe phase. We adopt the definition of partition coefficient as a ratio of the concentrations in phase 2 to phase 1 to maintain consistency·throughout this chapter; in a particular research situation, however, it may be convenient to invert the definition if this simplifies the appearance of equations.

1:1 Binding

We begin with the simplest possible case, in which substrate S and ligand L form 1:1 complex SL in phase 1. Free S, but neither L nor SL, can partition into phase 2. In the absence of L we measure $P_0 = [S]_2/[S]_1$. Now consider the system when ligand has been added to phase 1. The apparent partition coefficient is defined by Eq. (9.3), where S_t represents total molar concentration in the indicated phase.

$$P = \frac{(S_t)_2}{(S_t)_1} \tag{9.3}$$

Because of the assumptions defining this system, we have $(S_t)_2 = [S]_2$ and $(S_t)_1 = [S]_1 + [SL]_1$. Combining these expressions with the definitions of P_0, P, and K_{11} gives

$$P = \frac{P_0}{1 + K_{11}[L]_1} \tag{9.4}$$

Thus the experimentally determined apparent partition coefficient P is a function of free ligand concentration in phase 1. From the mass balance on ligand, $(L_t)_1 = [L]_1 + [SL]_1$, we obtain

$$[L]_1 = \frac{(L_t)_1 P_0}{P_0 + K_{11}[S]_2} \tag{9.5}$$

Equations (9.4) and (9.5) define the system. $(L_t)_1$ is a known quantity, and $[S]_2$ is measured experimentally. If $P_0 \gg K_{11}[S]_2$, then $[L]_1$ is essentially equal to $(L_t)_1$; this convenient condition may be achieved by working with a very low substrate concentration if the analytical method is sufficiently sensitive. The stability constant is found from the dependence of P on $(L_t)_1$ by methods that have been given in detail in earlier chapters; for example, the linear plot of $1/P$ against $[L]_1$ yields both K_{11} and P_0.

The applicability of Eq. (9.4) depends on the validity of the assumptions underlying it, and these are to some extent under the control of the experimentalist, who can select the probe phase (phase 2) to suit the solubility characteristics of the solutes. In a perfectly controlled situation the value of K_{11} should be independent of the nature of phase 2, but obviously this is nearly an impossible condition, owing to the solubility of phase 2 in phase 1. Serious discrepancies observed with changes in the probe phase indicate that the simple Eq. (9.4) is not applicable to all of the phases studied.

In general it is possible for both the free and the bound forms of substrate to partition into the probe phase. (In this circumstance the free substrate concentration $[S]_1$ cannot be calculated directly via P_0.) The material balance equations are now $(S_t)_1 = [S]_1 + [SL]_1$ and $(S_t)_2 = [S]_2 + [SL]_2$. Carrying out the development as before yields

$$\frac{P}{P_0} = \frac{1 + (K_{11})_2[L]_2}{1 + (K_{11})_1[L]_1} \qquad (9.6)$$

Since both S and SL are present in phase 2, they are related by an equilibrium, having the stability constant $(K_{11})_2$ in this phase. Evidently $[L]_1$ and $[L]_2$ are related by a partition coefficient. Equation (9.6) is identical in form with Eq. (7.24), and Eq. (9.4) is a special case of Eq. (9.6). An alternative formulation of this system is developed later (see Eq. 9.29).

Let us consider the evaluation of P according to Eq. (9.3). Often the system is designed so that the concentration $(S_t)_2$ in the probe phase is measured and the concentration $(S_t)_1$ is obtained by difference. It is then necessary to take into account the ratio of phase volumes. Suppose the substrate is initially present, at concentration $(S_i^\circ)_1$, only in phase 1. At equilibrium the concentration in phase 2 is $(S_t)_2$. The total amount of substrate is therefore $(S_i^\circ)_1 V_1$, and the amount in phase 2 is $(S_t)_2 V_2$, where

V_1, V_2 are the volumes of phases 1 and 2. We therefore obtain

$$P = \frac{(S_t)_2}{(S_t^\circ)_1 - (S_t)_2 R} \tag{9.7}$$

where $R = V_2/V_1$. If P is very large, this method is inaccurate because the difference in the denominator is subject to large relative error, and a direct measurement of $(S_t)_1$ is then required.

The partition method has found occasional use in the study of simple molecular complex systems (1–5). The treatment may have to be modified to suit the system under study. For example, Higuchi and Zuck (3) studied the benzoic acid (S):caffeine (L) system in water (phase 1), using Skellysolve-C® as the probe phase. A constant P_0 was obtained when it was defined $P_0 = (S_t)_2/(S_t)_1^2$. This result implies that the benzoic acid exists in the organic phase mainly as the dimer. Moreover, in calculating P_0 the analytical value for $(S_t)_1$ was corrected for the dissociation of benzoic acid in water. In another type of modification Nakajima et al. (5) took into account the competitive complexing of the ligand (cyclodextrin) with probe solvent (octanol) dissolved in the aqueous phase.

pK_a Determination

Acid–base strength in water can be measured by the partitioning method (6–8). Consider the neutral weak acid HA having acid dissociation constant K_a in water (phase 1). It is assumed that the conjugate base A^- and the hydronium ion cannot partition into phase 2. In a development similar to that of the preceding treatment we define the apparent partition coefficient P:

$$P = \frac{[HA]_2}{[HA]_1 + [A^-]_1} \tag{9.8}$$

Combining this with the expressions $P_0 = [HA]_2/[HA]_1$ and $K_a = a_{H^+}[A^-]_1/[HA]_1$, where K_a is the apparent or "mixed" dissociation constant, we get

$$P = \frac{P_0 a_{H^+}}{a_{H^+} + K_a} \tag{9.9}$$

which can be succinctly written $P = P_0 f_{HA}$, where f_{HA} is the fraction of solute in the conjugate acid form. Equation (9.9) can be written

$$\frac{1}{P} = \frac{1}{P_0} + \frac{K_a}{P_0 a_{H^+}}$$ (9.10)

showing that K_a can be obtained from a plot of $1/P$ against $1/a_{H^+}$ (8a) or from one of the other linearized forms of the rectangular hyperbola. In the region where $a_{H^+} \ll K_a$, Eq. (9.9) becomes, approximately,

$$\log P \approx \log P_0 + pK_a - pH$$

so a plot of $\log P$ against pH has a slope of -1 in this region. More generally, Eq. (9.9) is rearranged to

$$\log \left(\frac{P}{P_0 - P} \right) = -pH + pK_a$$ (9.11)

The intrinsic partition coefficient P_0 can be obtained from the data set of P as a function of pH, as with Eq. (9.10), or by direct measurement in solutions several pH units more acidic than pK_a. Analogous equations are easily derived for neutral bases.

For a neutral dibasic acid H_2A we have $P = P_0 f_{H_2A}$, which becomes

$$P = \frac{P_0 a_{H^+}^2}{a_{H^+}^2 + K_1 a_{H^+} + K_1 K_2}$$ (9.12)

The double-reciprocal form is

$$\frac{1}{P} = \frac{1}{P_0} + \frac{K_1}{P_0 a_{H^+}} + \frac{K_1 K_2}{P_0 a_{H^+}^2}$$ (9.13)

which may be compared with Eq. (9.10). Equation (9.14) is the logarithmic form.

$$\log \left(\frac{P}{P_0 - P} \right) = -2pH + pK_1 - \log (K_2 + a_{H^+})$$ (9.14)

Equation (9.11) is a special case of Eq. (9.14). The pH dependencies of these functions can be useful in demonstrating the number of ionizable groups on a molecule.

The applicability of the partition method depends on the validity of the definition of the apparent partition coefficient. It also requires that P_0 be accurately known, and this may be a limiting factor if P_0 is extremely large or small. The K_a value obtained by this method obviously refers to an aqueous phase that is saturated with respect to the nonaqueous probe phase.

The treatment can be expressed in terms of the solute concentration in the probe phase, $(S_t)_2$, rather than P, and this is often experimentally convenient (9). Define f_2 as the fraction of solute in phase 2 at equilibrium. Letting A_t be the total amount of solute in the system, we have $f_2 = (S_t)_2 V_2 / A_t$, or

$$f_2 = \frac{(S_t)_2 V_2}{(S_t)_1 V_1 + (S_t)_2 V_2}$$

Combining this with $P = (S_t)_2 / (S_t)_1$ gives

$$f_2 = \frac{PR}{1 + PR} \tag{9.15}$$

where $R = V_2 / V_1$. For the neutral weak acid HA, Eqs. (9.9) and (9.15) are combined, giving

$$(S_t)_2 = \frac{k_1 a_{H^+}}{k_2 + a_{H^+}} \tag{9.16}$$

where

$$k_1 = \frac{A_t P_0}{V_1 (1 + P_0 R)} \tag{9.17}$$

$$k_2 = \frac{K_a}{1 + P_0 R} \tag{9.18}$$

Equation (9.16) is of the usual hyperbolic form. From the dependence of $(S_t)_2$ on a_{H^+}, in a system designed to have fixed R and A_t / V_1, the dissociation constant is obtained from $K_a = k_2 P_0 A_t / k_1 V_1$.

Ion-Pair Extraction

Thus far in the treatment we have assumed that ions are restricted to the aqueous phase. Many systems are known, however, in which ions—or rather, ion pairs—can partition into organic phases. This phenomenon is important synthetically in phase-transfer catalysis, and analytically in ion-pairing reverse-phase liquid chromatography and in the acid-dye spectrophotometric method for amines. This transfer of ions to a nonpolar medium is accomplished by complexing them with a hydrophobic substance. In one approach an ion in the aqueous phase is transferred to the organic phase by formation of an ion pair with a large hydrophobic counterion. Another method is to form an inclusion complex of the ion with a macrocyclic "host" compound, thus presenting a hydrophobic exterior to the organic medium; of course, the aqueous counterion is also transferred and can be considered part of the complex.

Several groups have dealt with these kinds of systems (10–13). A general treatment is difficult because of the many possible equilibria, so data treatment is usually adapted to the requirements of a specific case. The ion-pair extraction is commonly discussed in terms of the following composite equilibrium:

$$_1M^+ + {_1}A^- + {_2}L \xrightleftharpoons{E} {_2}MLA \qquad\qquad Scheme\ I$$

The subscripts 1 and 2 represent the two phases, phase 1 being the polar (usually aqueous) phase. In this equation M^+ and A^- are the cation and anion, L is a neutral ligand, and MLA is an ion-pair:ligand complex. The equilibrium constant E is neither a partition coefficient nor a stability constant, but is called the extraction constant.

Let us consider the simplest case. We assume that the substrate ions are completely dissociated in phase 1, that the ligand does not partition into phase 1, that no higher complexes form, and that the ion-pair complex does not dissociate in phase 2. Thus the preceding scheme is a complete description of the system. The extraction constant is defined as

$$E = \frac{[MLA]_2}{[M^+]_1[A^-]_1[L]_2} \qquad\qquad (9.19)$$

By electroneutrality, $[M^+]_1 = [A^-]_1$. The usual analytical procedure is to measure either total M or total A in phase 2; this result together with the mass balance relationships on substrate and ligand gives the information necessary to calculate E. Thus let A_0 be the total amount (molar basis) of anion initially taken. Then for this system

$$A_0 = [A^-]_1 V_1 + [MLA]_2 V_2 \qquad\qquad (9.20)$$

This gives $[A^-]_1$, since $[MLA]_2$ is determined analytically. Similarly for the ligand,

$$L_0 = ([L]_2 + [MLA]_2)V_2 \tag{9.21}$$

Cram and his group (12) have modified this by measuring the ratio of substrate to ligand in phase 2 by nuclear magnetic resonance. Let this ratio be r. Then

$$r = \frac{[MLA]_2}{[L]_2 + [MLA]_2} = \frac{[MLA]_2}{L_t} \tag{9.22}$$

From this we find

$$[L]_2 = L_t(1 - r) \tag{9.23}$$

Combining Eqs. (9.20) and (9.22) gives $[A^-]_1$.

The situation can be much more complicated if the ligand can partition into both phases and if other reasonable equilibria are included in the description. The following equations show these equilibria:

Equilibria in phase 1

$$_1M^+ + {}_1L \rightleftharpoons {}_1ML^+$$

$$_1ML^+ + {}_1A^- \rightleftharpoons {}_1MLA$$

Scheme II

Equilibria in phase 2

$$_2M^+ + {}_2L \rightleftharpoons {}_2ML^+$$

$$_2ML^+ + {}_2A^- \rightleftharpoons {}_2MLA$$

$$_2M^+ + {}_2A^- \rightleftharpoons {}_2MA$$

Scheme III

Partitioning equilibria

$$_1L \rightleftharpoons {}_2L$$

$$_1MLA \rightleftharpoons {}_2MLA$$

$$_1M^+ + {}_1A^- \rightleftharpoons {}_2MA$$

Scheme IV

The species $_2MA$ is a solvent-stabilized ion pair in phase 2. The preceding schemes allow for the formation of ion-pair complex MLA in both phases, and for its dissociation into ions in each phase. It is easily shown that the extraction scheme I is composed of four of the preceding equilibria. Frensdorf (11) has given a treatment of a system incorporating most of these equilibria. The approach is more complicated but is similar to that used for the simple case treated earlier: the equilibrium constants, electroneutrality, and mass balance relationships are combined, and the resulting expression is subjected to nonlinear regression analysis.

Multiple Equilibria

The binding of small molecules to macromolecules nearly always involves multiple binding and stoichiometric ratios greater than 1:1. In 1951 Karush (14) applied the liquid–liquid partitioning method to protein binding, proposing it as an alternative to equilibrium dialysis. The partition method is faster than dialysis and it eliminates the problem of adsorption to the membrane; however, denaturation of the protein at the interface is a possibility. The method has found occasional use (15–18). Let the small molecule be the substrate and the macromolecule the ligand. The intrinsic partition coefficient $P_0 = [S]_2/[S]_1$ is determined in the absence of ligand. Then the macromolecule is incorporated into phase 1. If the macromolecule cannot partition into phase 2, the probe phase, then measurement of substrate concentration in phase 2 gives $[S]_2$, and $[S]_1$ is then calculated via P_0. The total substrate concentration in phase 1 is determined experimentally or by difference, and is written:

$$(S_t)_1 = [S]_1 + h[S_h L]_1 \qquad (9.24)$$

Thus the difference $(S_t)_1 - [S]_1$ gives the concentration of substrate bound to ligand. The quantity $\bar{h}$ (see Section 2.4), interpreted as the average number of substrate molecules bound per ligand molecule, is calculated with

$$\bar{h} = \frac{(S_t)_1 - [S]_1}{(L_t)_1} \qquad (9.25)$$

Therefore the experiment yields $\bar{h}$ as a function of $[S]_1$. Subsequent data treatment proceeds as outlined in Section 2.6, "Multiple Equilibria."

This method has considerable flexibility in its application to macromolecular binding because the experimentalist may choose the organic solvent (phase 2), which may be a mixed solvent, to suit the properties of the particular system. Waring et al. (17a) and Albertsson (17b) have discussed the selection of solvent. Krugh et al. (18) further controlled the partitioning properties of the system by incorporating an ion-pairing reagent in phase 2. Karush had actually used this strategem in his work (14), the benzyltrimethylammonium cation increasing the partition coefficient of methyl orange anion.

Let us consider the more general case in which complexes may partition between both phases. We will treat the system in which complexes SL, SL_2, SL_3, . . . may form. There are equivalent ways to express the equilibria. We define overall stability constants β_{1i} for the formation of SL_i in phase 1 (see Eq. 2.4) and partition coefficients P_{1i} for the partitioning of SL_i between phases 1 and 2. The apparent partition coefficient is

$$P = \frac{(S_t)_2}{(S_t)_1} = \frac{\sum_{i=0}^{n} [SL_i]_2}{\sum_{i=0}^{n} [SL_i]_1} \tag{9.26}$$

where the species having $i = 0$ is unbound substrate. Combining (9.26) with the equilibrium expressions gives Eq. (9.27), where the summations run from 0 to n.

$$P = \frac{\sum P_{1i}\beta_{1i}[L]_1^i}{\sum \beta_{1i}[L]_1^i} \tag{9.27}$$

The mass balance on ligand in phase 1 is similarly developed to yield

$$(L_t)_1 = [L]_1 + [S]_1 \sum \beta_{1i}[L]_1^i \tag{9.28}$$

Equations (9.27) and (9.28) describe the system. The number of parameters to be estimated may often be reduced by means of experimental observations showing that some species do not partition into phase 2, and data treatment is simplified if the substrate concentration is so low that the assumption $[L]_1 = [L_t]_1$ is justified.

Workers in the field of metal ion complexes have developed this approach extensively (19). To show one of the forms of data treatment we apply Eq. (9.27) to the case $n = 2$, obtaining

$$P = \frac{P_{10} + P_{11}\beta_{11}[L]_1 + P_{12}\beta_{12}[L]_1^2}{1 + \beta_{11}[L] + \beta_{12}[L]_1^2} \tag{9.29}$$

A plot of $\log P$ against $\log [L]_1$ can be helpful in establishing the identities of complexes and in estimating the parameters. We take the derivative $d \log P / d \log [L]_1 = [L]_1 \, dP/Pd[L]_1$.

$$\frac{d \log P}{d \log [L]_1}$$

$$= \frac{(P_{11} - P_{10})\beta_{11}[L]_1 + 2(P_{12} - P_{10})\beta_{12}[L]_1^2 + (P_{12} - 2P_{11})\beta_{11}\beta_{12}[L]_1^3}{(P_{10} + P_{11}\beta_{11}[L]_1 + P_{12}\beta_{12}[L]_1^2)(1 + \beta_{11}[L]_1 + \beta_{12}[L]_1^2)}$$

Now suppose that only species SL_2 can partition into phase 2, as might happen if S is a dication and L is a monoanion. This derivative simplifies to

$$\frac{d \log P}{d \log [L]_1} = \frac{2 + \beta_{11}[L]_1}{1 + \beta_{11}[L]_1 + \beta_{12}[L]_1^2} \tag{9.30}$$

For this system the average number of L molecules bound per S molecule, is, from Eq. (2.47),

$$\bar{i} = \frac{\beta_{11}[L]_1 + 2\beta_{12}[L]_1^2}{1 + \beta_{11}[L]_1 + \beta_{12}[L]_1^2} \tag{9.31}$$

Comparison of Eqs. (9.30) and (9.31) gives

$$\bar{i} = 2 - \left(\frac{d \log P}{d \log [L]_1} \right) \tag{9.32}$$

Note that $n = 2$ for this system. If, in a system described by Eq. (9.27), the highest-order complex $(i = n)$ is the only one that partitions into phase 2, then

$$\bar{i} = n - \left(\frac{d \log P}{d \log [L]_1} \right) \tag{9.33}$$

This equation can be used to obtain $\bar{i}$ as a function of ligand concentration if n is known (20).

Sometimes it is possible to simplify the description still further when it is reasonable to postulate the existence of the complex only in phase 2. For example, the diphenol hexestrol is extracted from a polar phase (ethylene glycol) by a nonpolar phase containing H-bond acceptor species (21). Letting S be hexestrol, L the H-bond acceptor, and SL_n the putative complex in the nonpolar phase, we define

$$P = \frac{[SL_n]_2}{[S]_1}$$

Combining this with the equilibrium expressions gives

$$P = P_0(\beta_{1n})_2[L]_2^n$$

or

$$\log P = \log P_0(\beta_{1n})_2 + n \log [L]_2 \qquad (9.34)$$

so that a log–log plot gives an estimate of n.

Equation (9.33) was obtained for the case in which only the complex of highest stoichiometry (SL_n) partitions into phase 2. In general, however, more than one complex may enter the probe phase. Irving et al. (22) have given a treatment whose principle is demonstrated in the following. We consider the system described by Eqs. (9.26) and (9.27). Let us suppose that P is written in terms of the partitioning of a single complex of average composition, which will generally be different in each phase. Thus we write

$$P = \frac{[SL_{\mathbf{i}}]_2}{[SL_i]_1} \qquad (9.35)$$

where this P is identical with that in Eq. (9.26). The italic subscript i is the average stoichiometry (with respect to L) in phase 1, and the boldface $\mathbf{i}$ refers to the average stoichiometry in phase 2. We define the following average quantities:

$$\beta_{1i} = \frac{[SL]_i]_1}{[S]_1[L]_1^i} \qquad (9.36a)$$

$$\beta_{1\mathbf{i}} = \frac{[SL_{\mathbf{i}}]_1}{[S]_1[L]_1^{\mathbf{i}}} \qquad (9.36b)$$

$$P_{1\bar{i}} = \frac{[SL_{\bar{i}}]_2}{[SL_{\bar{i}}]_1} \tag{9.36c}$$

Combining Eqs. (9.35) and (9.36) yields

$$P = \frac{P_{1\bar{i}}\beta_{1\bar{i}}[L]_1^{\bar{i}}}{\beta_{1\bar{i}}[L]_1^{\bar{i}}} \tag{9.37}$$

From Eq. (9.37) we obtain

$$\frac{d\log P}{d\log [L]_1} = \bar{i} - \bar{\bar{i}} \tag{9.38}$$

If complex SL_n is the only species that is extractable into phase 2, then $\bar{i} = n$ and Eq. (9.38) becomes Eq. (9.33). In general, however, any or all of the species may partition, and $\bar{i}$ then depends on $[L]_i$, as does $\bar{\bar{i}}$.

The appearance of the plot of $\log P$ against $\log [L]_1$, interpreted with the aid of Eq. (9.38), can be diagnostically helpful. Figure 9.1 shows the nature of this plot for some hypothetical systems as calculated with Eq. (9.29). These curves are drawn supposing that only a single substrate

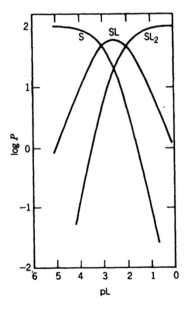

Figure 9.1. Log–log plot of apparent partition coefficient P against $pL = -\log[L]$, for the hypothetical system described by Eq. (9.29). Curve S: $P_{10} = 100$, $P_{11} = 0$, $P_{12} = 0$; curve SL: $P_{10} = 0$, $P_{11} = 100$, $P_{12} = 0$; curve SL_2: $P_{10} = 0$, $P_{11} = 0$, $P_{12} = 100$. For all curves $\beta_{11} = 10^3$, $\beta_{12} = 10^5$.

species, S, SL, or SL_2 in turn, can partition into phase 2. Consider the curve for partitioning of species S. At very low ligand concentration $\bar{i} = 0$ in phase 1; moreover only S partitions, so $\bar{i} = 0$, and the slope is zero, in accord with Eq. (9.38). As $[L]_1$ increases, $\bar{i}$ increases, but $\bar{i} = 0$, so the slope of this curve is negative, approaching the limiting value -2. If more than one species is extractable, the graph will be a composite of the special cases seen in Figure 9.1. There is a practical difficulty in that wide variations in concentration will very likely produce nonconstancy of composition of the phases (23).

Much use has been made of these $\log P - pL$ plots to study metal ion complexes (20, 22–25). The principle outlined in the preceding treatment can be extended to describe stoichiometric variations in more than one component of the complex. For example, complexes of the general structure $H_a M_b X_c$, where M is a metal ion and X is an anion, may be studied by holding two of the three concentrations (pH, pM, pX) constant, and varying the third; then the derivative in Eq. (9.38) is a partial derivative. Equation (9.38) can be generalized to include variations in all of the concentrations (22).

9.2. SELF-ASSOCIATION

Liquid–liquid distribution has long been applied to the study of self-association (26). To demonstrate the basis of the method, let us suppose that a substrate S associates strongly to form aggregate S_m in phase 1. Phase 2, the probe phase, is chosen so that only the monomer partitions into it, with partition coefficient $P_0 = [S]_2/[S]_1$. The apparent partition coefficient is $P = (S_t)_2/(S_t)_1$, or

$$P = \frac{[S]_2}{[S]_1 + m[S_m]_1} \tag{9.39}$$

The stability constant is $\beta_m = [S_m]_1/[S]_1^m$. With the assumption $[S]_1 \gg m[S_m]_1$ we get

$$P = \frac{P_0^m}{m\beta_m[S]_2^{m-1}} \tag{9.40}$$

on combining these relationships. From Eq. (9.40) in logarithmic form,

$$\log P = m \log \frac{P_0}{m\beta_m} - (m - 1) \log [S]_2 \qquad (9.41)$$

we see that the slope of a log–log plot yields m:

$$\frac{d \log P}{d \log [S]_2} = 1 - m \qquad (9.42)$$

With m evaluated, Eq. (9.40) can be solved for β_m.

In general, there may be more than one associated species. A model of self-association is described in Section 2.4, "Examples of Models." With this model, and the assumption that only the monomer partitions into phase 2, the apparent partition coefficient is defined

$$P = \frac{[S]_2}{[S]_1 + 2[S_2]_1 + 3[S_3]_1 + \cdots} \qquad (9.43)$$

This can be expressed in terms of the stability constants, giving

$$P = \frac{P_0}{\Sigma \, h\beta_h [S]_1^{h-1}} \qquad (9.44)$$

where the summation is from 1 to m. The intrinsic partition coefficient is determined by extrapolating P to zero concentration. The concentration of monomer in phase 1 can then be calculated from $[S]_1 = [S]_2 / P_0$. The experimental data consist of P as a function of $[S]_2$.

In principle it is only necessary to fit the data to Eq. (9.44) by regression analysis. In practice the problem may be much simplified by making use of graphical techniques to obtain limits on the possible values of h and to estimate the stability constants (27, 28). For this purpose we test the data against special cases of Eq. (9.44), which can be written

$$\frac{P_0}{P} = 1 + 2\beta_2 [S]_1 + 3\beta_3 [S]_1^2 + 4\beta_4 [S]_1^3 + \cdots \qquad (9.45)$$

Dimerization is the simplest and most likely special case. For this case ($m = 2$), the appropriate form of Eq. (9.45) is

$$\frac{P_0}{P} = 1 + \frac{2\beta_2 [S]_2}{P_0} \qquad (9.46)$$

Figure 9.2 shows a plot according to Eq. (9.46) for the system caffeine (S):pH 9.2 borate buffer (phase 1):isooctane (phase 2) at 25 °C (29). For this system $P_0 = 0.00465$. It is evident that P_0/P is a reasonably linear function of $[S]_2$ at low concentrations. The slope is equal to 8100 M^{-1}, giving, from Eq. (9.46), $\beta_2 = 18.8\ M^{-1}$ as the dimerization constant.

Next we test for the presence of trimers. Letting $m = 3$ gives, from Eq. (9.45),

$$\frac{P_0 - P}{P[S]_2} = \frac{2\beta_2}{P_0} + \frac{3\beta_3[S]_2}{P_0^2} \tag{9.47}$$

Figure 9.3 shows the data from Figure 9.2 plotted according to Eq. (9.47). Within the precision of these results, there appears to be no evidence for trimer formation. The intercept in Figure 9.3 is equal to the slope in Figure 9.2.

Tetramer formation in this case can be examined by setting $m = 4$ and, from the preceding finding, $\beta_3 = 0$. These data do not require the addition of tetramers. Guttman and Higuchi, using an organic phase that provided a more precisely estimated partition coefficient, concluded that caffeine forms dimers and tetramers, but not trimers, in aqueous solution (27).

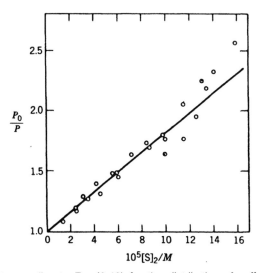

Figure 9.2. Plot according to Eq. (9.46) for the distribution of caffeine between pH 9 aqueous buffer (phase 1) and isooctane (phase 2).

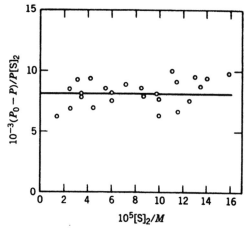

Figure 9.3. Plot according to Eq. (9.47) for the distribution of caffeine between pH 9 aqueous buffer (phase 1) and isooctane (phase 2).

It is a critical assumption of the method as based on Eq. (9.44) that only the monomeric species can partition into phase 2. Several useful checks of this assumption can be made. By changing the probe phase (phase 2), substantially the same stability constants in phase 1 should be obtained if the assumption is valid for both phase 2 solvents (27). Another requirement is that Beer's law must be rigorously followed by the substrate in phase 2 if only monomer is present. A further requirement is that the stability constant estimates should reproduce the experimental data within their limits of uncertainty. This is well illustrated by Goodman's study of the partitioning behavior of several long-chain fatty acids (30). These were distributed between water (buffered to a pH much higher than pK_a) and heptane. Since dimerization in heptane is the process being investigated, we denote heptane as phase 1 and water as phase 2. The development follows the lines of the earlier derivation, the apparent partition coefficient being defined

$$P = \frac{[HA]_2 + [A^-]_2}{[HA]_1 + 2[(HA)_2]_1} \qquad (9.48)$$

Since $[H^+] \ll K_a$, $[A^-]_2 \gg [HA]_2$, and Eq. (9.48) becomes

$$P = \frac{P_0'}{1 + 2\beta_2[HA]_1} \qquad (9.49)$$

where $P'_0 = P_0 K_a/[H^+]$. Equation (9.49) expresses P as a function of the concentration of monomer in the heptane layer. By using P'_0 this is converted to a function of the anion concentration in the aqueous phase.

$$\frac{1}{P} = \frac{1}{P'_0} + \frac{2\beta_2 [A^-]_2}{(P'_0)^2} \tag{9.50}$$

Equation (9.50) is similar to Eq. (9.46). To express P as a function of total solute concentration in the heptane phase we write Eq. (9.48) as $[A^-]_2 = P(S_t)_1$. This is substituted into (9.50) to give

$$\frac{1}{P} = \frac{(1/P'_0) + (1/P'_0)[1 + 8\beta_2 (S_t)_1]^{1/2}}{2} \tag{9.51}$$

For most of the acids studied by Goodman Eq. (9.51) gave a quantitative account of the data over most of the concentration range studied, suggesting the reasonableness of the assumptions. However, at high substrate concentrations significant deviations were observed, and for the longest-chain acids the data could not be accounted for by Eq. (9.51). This disagreement suggests that the assumptions underlying this treatment are not valid for these systems.

Mukerjee (31) later showed that all of Goodman's data could be quantitatively accounted for by assuming the formation of dimers by the anions in aqueous solution. The apparent partition coefficient now is defined as

$$Q = \frac{[HA]_2 + [A^-]_2 + 2[A_2^{-2}]_2}{[HA]_1 + 2[(HA)_2]_1} \tag{9.52}$$

As before, we neglect $[HA]_2$ in the numerator. Defining a dimerization constant for the aqueous phase as $\gamma_2 = [A_2^{-2}]_2/[A^-]_2^2$ and proceeding as before yields

$$\frac{1}{Q} = \frac{(1/P'_0) + 2\beta_2 [A^-]_2/(P'_0)^2}{1 + 2\gamma_2 [A^-]_2} \tag{9.53}$$

which should be compared with Eq. (9.50). From Eq. (9.48) it is possible to find $[A^-]_2$ as a function of γ_2, P, and $(S_t)_1$, and therefore through Eq. (9.53) to express Q as a function of $(S_t)_1$. The stability constant γ_2 can be estimated as follows (31). Let P be the quantity defined $P = [A^-]_2/(S_t)_1$

and calculated by means of Eq. (9.51) from β_2 values evaluated at low concentrations, and let $Q = ([A^-]_2 + 2[A_2^{-2}]_2)/(S_t)_1$. Therefore Q is the correct quantity by hypothesis, and P is the corresponding quantity lacking consideration of the aqueous dimerization. Combining these expressions gives

$$Q = P + 2\gamma_2 P^2 (S_t)_1 \qquad (9.54)$$

With the experimental data (Q) and calculated P values γ_2 is estimated.

The mutual solubility of the two liquid phases introduces a possible complication. At the simplest level this results in the medium being not pure phase 1, but rather phase 1 saturated with respect to phase 2. If the two solvents are very insoluble in each other, this may not be a serious factor. At a molecular level, however, the dissolved phase 2 may undergo specific interaction with the solute, introducing competing equilibria or changing the molecular nature of the solute species. For example, Goldman and Duer (32) studied the dimerization of nitrobenzene in benzene by partitioning it in a benzene–water system. The dimerization constant evaluated by the partition method was significantly different from that found by an independent method. The discrepancy was accounted for by supposing that the water dissolved in the benzene phase interacts with monomeric nitrobenzene.

Mukerjee and Ghosh (33) have devised the "isoextraction" method for the study of the self-association of methylene blue cation in aqueous solution (phase 1). Let D^+ represent the dye cation and A^- the anion that is extracted with it into the organic phase (phase 2). The procedure is to vary the total dye concentration $(D_t)_1$ and the total anion concentration $(A_t)_1$ in the aqueous phase (this latter by adding a salt M^+A^- that is not itself extracted); the quantities $(D_t)_1$ and $(A_t)_1$ are mutually adjusted in a number of systems so as to achieve the same total dye concentration $(D_t)_2$ in phase 2 in each of the systems, this being the isoextraction condition.

Although it is not known how the dye may be associated in phase 2, the distribution of species depends only on the total dye concentration, which is constant. It follows that the concentration of dye monomer is constant in phase 2. Since electroneutrality must be satisfied, a more complete statement is that the mean ionic molarity $[D^+]_2[A^-]_2$ is constant.

Since phase 1 is in equilibrium with phase 2, the product $[D^+]_1[A^-]_1$

must also be a constant in the series of systems. With the assumption that the anion does not associate in the aqueous phase we write $[A^-]_1 = (A_t)_1$, and define an ion product as

$$K = [D^+]_1 (A_t)_1 \qquad (9.55)$$

We also define an apparent ion product in terms of the total dye concentration in phase 1:

$$K' = (D_t)_1 (A_t)_1 \qquad (9.56)$$

The quantity K' is calculated according to Eq. (9.56), and K' is plotted against $(D_t)_1$. This plot is extrapolated to $(D_t)_1 = 0$ to yield K, since at vanishingly small concentration $(D_t)_1$ becomes equal to $[D^+]_1$. Then, from Eqs. (9.55) and (9.56), the monomer concentration is calculated:

$$[D^+]_1 = \frac{K}{K'} (D_t)_1 \qquad (9.57)$$

With $[D^+]_1$ known as a function of $(D_t)_1$ it is possible to obtain the average number of monomers per associated species (28, 33), for processing as in Section 2.4, "Examples of Models," or to carry out the analysis by treating the mass balance on dye,

$$(D_t)_1 = [D^+]_1 + 2\beta_2 [D^+]_1^2 + 3\beta_3 [D^+]_1^3 + \cdots \qquad (9.58)$$

REFERENCES

1. H. Morse, Z. Physik. Chem., **41**, 709 (1902).
2. T. S. Moore, F. Shepherd, and E. Goodall, J. Chem. Soc., 1447 (1931).
3. T. Higuchi and D. A. Zuck, J. Am. Pharm. Assoc., Sci. Ed., **41**, 10 (1952).
4. H. P. A. Buist, E. Tomlinson, and J. F. M. Kinkel, Int. J. Pharmaceut., **15**, 347 (1983).
5. T. Nakajima, M. Sunagawa, and T. Hirohashi, Chem. Pharm. Bull., **32**, 401 (1984).
6. (a) R. C. Farmer, J. Chem. Soc., **79**, 863 (1901); (b) R. C. Farmer and F. J. Warth, J. Chem. Soc., **85**, 1713 (1904).
7. C. Golumbic, M. Orchin, and S. Weller, J. Am. Chem. Soc., **71**, 2624 (1949).
8. (a) H. Irving, S. J. H. Cooke, S. C. Woodger, and R. J. P. Williams, J.

Chem. Soc., 1847 (1949); (b) H. M. Irving and C. F. Bell, *J. Chem. Soc.*, 1216 (1952).

9. K. Opong-Mensah, T. W. Woller, A. O. Obaseki, and W. R. Porter, *J. Pharm. Biomed. Anal.*, **2**, 381 (1984).

10. C. J. Pedersen, *J. Am. Chem. Soc.*, **92**, 391 (1970).

11. H. K. Frensdorf, *J. Am. Chem. Soc.*, **93**, 4684 (1971).

12. E. P. Kyba, R. C. Helgeson, K. Madan, G. W. Gokel, T. L. Tarnowski, S. S. Moore, and D. J. Cram, *J. Am. Chem. Soc.*, **99**, 2564 (1977).

13. Y. Marcus and L. E. Asher, *J. Phys. Chem.*, **82**, 1246 (1978).

14. F. Karush, *J. Am. Chem. Soc.*, **73**, 1246 (1951).

15. L. F. Cavaliere, A. Angelos, and M. E. Balis, *J. Am. Chem. Soc.*, **73**, 4902 (1951).

16. H. G. Heilweil and Q. van Winkle, *J. Phys. Chem.*, **59**, 939 (1955).

17. (a) M. J. Waring, L. P. G. Wakelin, and J. S. Lee, *Biochim. Biophys. Acta*, **407**, 200 (1975); (b) P.-A. Albertsson, in *Methods of Biochemical Analysis*, Vol. 29, D. Glick, Ed., Wiley-Interscience, 1983, pp. 1–24.

18. T. R. Krugh, S. A. Winkle, and D. E. Graves, *Biochem. Biophys. Res. Commun.*, **98**, 317 (1981).

19. Y. Marcus, *Chem. Rev.*, **63**, 139 (1963).

20. J. Rydberg, *Acta Chem. Scand.*, **4**, 1503 (1950).

21. K. G. Wahlund, *Acta Pharm. Suec.*, **12**, 425 (1975).

22. (a) H. Irving, F. J. C. Rossotti, and R. J. P. Williams, *J. Chem. Soc.*, 1906 (1955); (b) F. J. C. Rossotti, *Rec. Trav. Chim.*, **75**, 743 (1956).

23. H. Irving and F. J. C. Rossotti, *J. Chem. Soc.*, 1946 (1955).

24. (a) J. Rydberg, *Arkiv. Kem.*, **8**, 101, 113 (1955); (b) J. Rydberg, *Rec. Trav. Chim.*, **75**, 737 (1956).

25. (a) D. Dryssen and D. Johansson, *Acta Chem. Scand.*, **9**, 763 (1955); (b) M. Dryssen, *Rec. Trav. Chim.*, **75**, 748 (1956).

26. W. S. Hendixson, *Z. Anorg. Allgem. Chem.*, **13**, 73 (1897).

27. D. Guttman and T. Higuchi, *J. Am. Pharm. Assoc., Sci. Ed.*, **46**, 4 (1957).

28. F. J. C. Rossotti and H. Rossotti, *J. Phys. Chem.*, **65**, 926, 930 (1961).

29. J. A. Mollica, Ph.D. Dissertation, University of Wisconsin-Madison, 1966, p. 71.

30. D. S. Goodman, *J. Am. Chem. Soc.*, **80**, 3887 (1958).

31. P. Mukerjee, *J. Phys. Chem.*, **69**, 2821 (1965).

32. S. Goldman and W. C. Duer, *Can J. Chem.*, **52**, 3918 (1974).

33. P. Mukerjee and A. K. Ghosh, *J. Am. Chem. Soc.*, **92**, 6403 (1970).

10

DIALYSIS

Dialysis and the related techniques discussed in this chapter are based on differences in molecular size. Their primary uses are for isolation, purification, and characterization, but our concern is with their application to the study of binding. In nearly all such applications the substrate and ligand species differ greatly in size, and the methods have been widely applied to the binding of small molecules and ions to macromolecules.

10.1. EQUILIBRIUM DIALYSIS

Principle of the Method

In dialysis, two compartments are separated by a semipermeable membrane, the membrane being (ideally) perfectly permeable to one interactant, say the ligand L, and perfectly impermeable to the substrate S. After equilibrium is achieved, the two compartments are analyzed for their total concentrations of ligand. Since the activity of free (unbound) ligand is identical in the two compartments at equilibrium, the analytical data yield the concentration of free ligand and, by difference, the concentration of bound ligand in the compartment containing the substrate.

Let us formalize this treatment while continuing to view the system as behaving ideally. As in Chapter 9, we designate as compartment 1 that in which the binding equilibria of interest take place. Therefore compartment 1 contains the nondiffusible substrate. Ligand can freely distribute between compartments 1 and 2. From Eq. (2.46) we define $\bar{i}$ as the average number of ligand molecules bound per substrate molecule:

$$\bar{i} = \frac{(L_t)_1 - [L]_1}{(S_t)_1} \qquad (10.1)$$

where the subscripts denote the compartment. The basic premise of equilibrium dialysis is that $[L]_1 = [L]_2$, giving

$$\bar{i} = \frac{(L_t)_1 - [L]_2}{(S_t)_1} \qquad (10.2)$$

Since $(L_t)_1$ and $[L]_2$ are the analytical ligand concentrations in compartments 1 and 2 and $(S_t)_1$ is a known quantity by experimental design, $\bar{i}$ can be calculated. The experiment is carried out at several different ligand concentrations in order to evaluate $\bar{i}$ as a function of free ligand concentration. The data are then interpreted according to Eq. (2.47) by methods that have been discussed in Chapters 2 and 3.

Although the dialysis experiment may be carried out as described, it is more commonly executed by making use of a control study in which the substrate is omitted. The conditions for the test and for the control are otherwise identical. After equilibrium is reached, the ligand concentration is determined in compartment 2 for both the test and the control. From this information we can obtain the quantity $(L_t)_1$, which is needed in Eq. (10.2). Since the test and control experiments contain the same total amounts of ligand, we write

$$(L_t)_1 V_1 + (L_t)_2 V_2 = (L_t)_1^c V_1 + (L_t)_2^c V_2 \qquad (10.3)$$

where V_1 and V_2 are the compartment volumes and the superscript c denotes the control experiment. Combining Eq. (10.3) with the identities $(L_t)_2 = [L]_2$, $(L_t)_1^c = [L]_1^c = [L]_2^c$, and $(L_t)_2^c = [L]_2^c$ yields

$$(L_t)_1 = [L]_2^c(1 + R) - [L]_2 R \qquad (10.4)$$

where $R = V_2/V_1$.

We can demonstrate these calculations with data of Klotz et al. (1) on the binding of methyl orange anion (L) to bovine serum albumin (S). In this study $V_1 = 10.00 \, \text{mL}$, $V_2 = 20.00 \, \text{mL}$, and $1.394 \times 10^{-7} \, \text{mol}$ of protein was placed in compartment 1. At equilibrium, spectrophotometric analysis yielded $[L]_2^c = 9.06 \times 10^{-5} \, M$ and $[L]_2 = 7.46 \times 10^{-5} \, M$. With Eq. (10.4) we find $(L_t)_1 = 12.26 \times 10^{-5} \, M$, and Eq. (10.2) gives $\bar{i} = 3.44$.

Equipment and membranes for dialysis are commercially available. Cellulose membranes with molecular weight cutoffs from 1000 to 50,000 can be obtained. Reviews of dialysis as a separation technique provide much information on equipment design (2). Steinhardt and colleagues (3) have reviewed dialysis as a method for studying binding. Most of our information on the binding of drugs to proteins has come from equilibrium dialysis studies (4).

Complications

Although equilibrium dialysis is simple in concept, and is often simple in practice, it is not without problems and limitations. One class of problem is a general type that may be subjected to theoretical analysis. Thus at equilibrium it is valid to state that the chemical potential of unbound ligand is identical in the two compartments, and therefore that the activities of unbound ligand are the same, since we reasonably choose the same standard state for ligand in each compartment. But it does not necessarily follow that the concentrations of ligand are identical, because it is conceivable that the ligand activity coefficients are different in the two compartments, whose contents obviously have different compositions.

Another general problem arises if the solutes are charged, for then the equilibrium distribution of the diffusible species is subject to the Donnan effect (5). Suppose the nondiffusible substrate is cationic with a charge of $+z$ units, and the ligand is a 1:1 electrolyte M^+L^-; hydroxide ion is the counterion to the substrate. At equilibrium the ligand activities are equal; for a 1:1 electrolyte this is written

$$(a_{M^+})_1(a_{L^-})_1 = (a_{M^+})_2(a_{L^-})_2$$

Supposing that the mean molar ionic activity coefficients are equal in the two compartments, these activities can be replaced by molar concentrations.

$$[M^+]_1[L^-]_1 = [M^+]_2[L^-]_2 \tag{10.5}$$

But in addition to this constraint, electroneutrality must be satisfied in both compartments, giving these relationships, where the hydronium and hydroxide ion concentrations are neglected:

$$[M^+]_2 = [L^-]_2 \tag{10.6}$$

$$[M^+]_1 + z[S^{+z}]_1 = [L^-]_1 \tag{10.7}$$

Equations (10.5) to (10.7) are combined to give

$$[L^-]_2^2 = [L^-]_1([L^-]_1 - z[S^{+z}]_1) \tag{10.8}$$

Equation (10.8) shows that the ligand anion concentrations are not equal on the two sides of the membrane for this system. If, however, the product $z[S^{+z}]$ is much smaller than the ligand concentration, Eq. (10.8) approaches the equality $[L^-]_2 = [L^-]_1$. This conclusion can be generalized (5) to the statement that the difference between $[L^-]_2$ and $[L^-]_1$ can be minimized by adding high concentrations of any electrolyte to the system.

The preceding conclusion is the basis for the practice of adding an "indifferent" electrolyte to the dialysis system in order to swamp out the Donnan inequality effect. But this practice itself may introduce a specific type of problem in that the added electrolyte may not be without effect on the binding equilibrium. Thus is it possible that the anion of the added electrolyte may itself bind to the positively charged macromolecule in the preceding hypothetical example. This competitive binding effect would vitiate the interpretation of the substrate–ligand binding if it were unrecognized and not accounted for. Klotz and Urquhart (6) found that buffer anions can compete with methyl orange for binding sites on proteins.

It was pointed out earlier that control experiments can be carried out in which the macromolecule is omitted. Such paired controls provide information useful in statistical analysis (7). They also assist in the detection of ligand adsorption to the membrane, though the lack of such an effect in the absence of the macromolecule is not conclusive, for the presence of the macromolecule might itself alter the nature of the membrane, and thus alter the interaction between the membrane and the ligand (3b).

More specialized applications of equilibrium dialysis have some unusual features that must be taken into account. A study of the interaction of sodium dodecylsulfate with polyvinylpyrrolidone required consideration of micelle formation by the ligand (8). Investigation of the binding of ligands to biological membrane-bound receptors can lead to errors of interpretation if the volume of the biological membrane (which is not accessible to the ligand) is not taken into account (9).

Yet further problems may arise. Since the membrane is impermeable to the macromolecule, there is an osmotic pressure gradient between the two compartments, and a net solvent flow into compartment 1 (containing the macromolecule) is to be expected. This will change the volumes of the solutions in the two compartments. If the macromolecule concentration is very low, the volume change will be unimportant. However, in studies of the binding of drugs to plasma proteins, carried out to yield the fraction of unbound drug in plasma (a quantity needed to interpret the drug pharmacokinetics), it is found that the volume change is surprisingly large, with average changes ranging from $+10$ to $+31\%$ in the volume of the plasma compartment (10). The magnitude of the effect appears to depend on the equilibration time, and it has been suggested that osmotic equilibrium may not be achieved when dialysis equilibrium is reached; however, the two processes are not independent. The volume effect does not change the concentration of unbound drug, but it reduces the concentration of macromolecule and of bound drug. Unless the volume change is taken into account, erroneous estimates of total drug recovery (10a) and of fraction unbound (10b, 10c) can result. One way to measure the volume change is to measure the protein concentration before and after dialysis, if the protein cannot cross the membrane. This leads to another problem, namely, the possibility that the protein may pass through the membrane into compartment 2, where it can bind the drug. Correction for this effect can be made (11).

Another set of problems is associated with the rate of equilibration across the membrane. Typical dialysis equilibration times are in the range 12–48 hours; these long times are a consequence of the presence of the membrane, the binding equilibria themselves being relatively fast reactions. In some instances longer times are required. These kinetic effects have been most closely studied with protein binding systems. Among the important factors are the volumes of solution in the compartments, extent of agitation, presence or absence of added electrolytes and buffers, protein concentration, and, of considerable importance, initial compartmental placement of the ligand species (3b, 12, 13). Thin-film dialysis is helpful in reducing equilibration times (2e, 14).

10.2. RELATED TECHNIQUES

Ultrafiltration

Ultrafiltration is similar to equilibrium dialysis. The binding equilibrium between macromolecule and ligand is established, and then pressure is applied to force the solution through the ultrafiltration membrane. Since the macromolecule cannot penetrate the membrane, the ligand concentration in the filtrate is equal to the free (unbound) ligand concentration in the presence of the macromolecule. The total ligand concentration in the presence of macromolecule is also measured, and the molar ratio of bound ligand to substrate is calculated with Eq. (10.1). The experiment is repeated at different ligand concentrations.

The technique is also valuable as a method of separation and purification, and there has been much development of equipment and membranes (2d, 15). Ultrafiltration has the advantage over equilibrium dialysis that the elapsed time per experiment is much reduced as a consequence of the pressure filtration.

There has been some concern that ultrafiltration is not an exact equilibrium method because one of the components is selectively removed from the equilibrium system. Sophianopoulos et al. (16) have shown that ultrafiltration is an equilibrium method, provided the complexes SL_i between macromolecule S and ligand L contain only one S molecule. The binding constant definition gives

$$[L]^i = \frac{[SL_i]}{[S]\beta_{1i}} \qquad (10.9)$$

Neither S nor SL_i can pass through the membrane, so on ultrafiltration the ratio of their concentrations remains constant, although the volume changes. Since the right-hand side of Eq. (10.9) is constant, the free ligand concentration is constant on ultrafiltration.

Ultrafiltration can be described as a filtration process without replacement. If the filtration is carried out with replacement of the ligand, the technique is sometimes called diafiltration. Blatt et al. (17) describe a method by which the macromolecule is placed in the sample solution, a reservoir of ligand solution is fed into this, and the mixture is filtered until the filtrate stream has the same concentration as the reservoir solution; the experimental design holds the sample solution volume constant.

Adsorption of solutes to the ultrafiltration membrane can be a serious problem. It has been reported that numerous drugs are retained by the membrane when filtered at therapeutic concentration levels (18).

Dynamic Dialysis

The rate at which a solute diffuses through a membrane depends on the size of the solute molecule, its physicochemical properties, the nature of the membrane, and environmental factors. Rates of dialysis therefore provide information about molecular size, shape, and purity (19). During the 1960s several laboratories developed methods based on the kinetics of dialysis for the study of binding equilibria.

The basis of the method is that the rate of "escape" of a diffusible species across the membrane is directly proportional to the concentration of the species. Thus measurement of the rate of escape of ligand yields the concentration of the free (unbound) ligand. Stein (20) and Colowick and Womack (21) developed flow methods. In the technique of Colowick and Womack the macromolecular substrate and radiolabeled diffusible ligand are placed in a sample chamber, which is separated from a flow chamber by a semipermeable membrane. When a steady-state concentration of ligand is reached in the flow chamber, its radioactivity is measured. Since the rate measurement involves the loss of a negligible fraction of total ligand, the same sample can be used for additional data points by adding more unlabeled ligand to the sample chamber and repeating the measurement. Suppose N_f is the radioactivity (cpm) at steady state in the flow chamber when the ligand is totally unbound, N_b the count when it is totally bound, and N_i the count when it is partially bound. Then if f_i is the fraction of ligand that is unbound, we can write

$$N_i = f_i N_f + (1 - f_i) N_b$$

or

$$f_i = \frac{N_i - N_b}{N_f - N_b} \tag{10.10}$$

as shown by Mermier (22). If the bound form of the ligand cannot cross the membrane, this becomes $f_i = N_i / N_f$. The product $f_i L_t$ then gives the free concentration of ligand.

Other methods have been described that are carried out in a manner similar to conventional chemical kinetics experiments. The nondiffusible substrate and diffusible ligand are placed in compartment 1, and the appearance of ligand in compartment 2 is measured as a function of time. The system is simplified by maintaining "sink" conditions of negligible ligand concentration buildup in compartment 2, so as to eliminate significant back diffusion. Many authors have demonstrated that the ligand escape kinetics are first order in the absence of substrate. The following argument has been used by some authors (23, 24). If the observed kinetics are first order in total ligand concentration in the presence of substrate, we write

$$-\frac{d(L_t)_1}{dt} = k_{obs}(L_t)_1 \tag{10.11}$$

The theoretical rate equation for the same system is written

$$-\frac{d(L_t)_1}{dt} = k[L]_1 \tag{10.12}$$

where the subscripts denote the compartment. Combining these gives

$$[L]_1 = \frac{k_{obs}}{k}(L_t)_1 \tag{10.13}$$

The intrinsic rate constant k is determined in a separate experiment in which the substrate is omitted, and it is a key assumption of the calculation that this value applies also when substrate is present. The significance of k_{obs} can be developed as follows, where 1:1 binding is assumed for the purpose. The mass balance on ligand in compartment 1 is $(L_t)_1 = [L]_1 + [SL]_1$, giving $(L_t)_1 = [L]_1(1 + K_{11}[S]_1)$. This is combined with Eqs. (10.12) and (10.13):

$$k_{obs} = \frac{k}{1 + K_{11}[S]_1} \tag{10.14}$$

If $[S]_1$ is substantially constant with time, then k_{obs} will be constant and apparent first-order kinetics will be observed in the presence of substrate (13). In general, however, $[S]_1$ will change as $[L]_1$ changes. This general expression is obtained from the mass balance on ligand for this system:

$$\frac{d(L_t)_1}{dt} = (1 + K_{11}[S]_1)\frac{d[L]_1}{dt} + K_{11}[L]_1\frac{d[S]_1}{dt} \tag{10.15}$$

Thus first-order kinetics in the presence of substrate cannot in general be expected.

Meyer and Guttman (25) developed a method in which the rate $d(L_t)_1/dt$ is measured from the concentration–time profile. The kinetics were not first order, and the data were fitted to an empirical triexponential function for the purpose of extracting the instantaneous rates. The intrinsic rate constant k was measured from the first-order kinetics in the absence of substrate, and then the free ligand concentration was calculated from Eq. (10.12). A single kinetic run thus provided data for the calculation of $\bar{i}$ as a function of [L] over a wide range. Several alternative empirical equations have been examined, but the triexponential form gave the best results (26).

There exists an ambiguity in these methods in that Eq. (10.12) may not be the appropriate theoretical expression for the escape rate; rather Eq. (10.16) may be the rate equation in the presence of substrate.

$$-\frac{d[L]_1}{dt} = k[L]_1 \qquad (10.16)$$

We measure $d(L_t)_1/dt$ but not $d[L]_1/dt$. As Eq. (10.15) shows for the simplest binding system, these are not equal and they may not even be proportional to each other.

Dynamic dialysis appears to have some potential for the study of self-association, for the self-associated species is obviously larger than the monomer. Simulated escape curves have been calculated for such systems (27).

Reuning and Levy (28) have described a method in which compartment 1 contains fixed initial concentrations of S and L (in a series of experiments), and a range of concentrations of L (the diffusible species) is placed initially in compartment 2. The concentration of L in compartment 2 is measured as a function of time for each such system. In that system for which $[L]_2$ does not change with time, evidently $[L]_1 = [L]_2$; that is, the system is already at equilibrium. An interesting feature of their studies was that the rates of diffusion of S (caffeine) and L (salicylamide) across a nylon membrane were greatly different, although these molecules do not differ markedly in size. The membrane took up appreciable amounts of salicylamide, and it was necessary to preequilibrate the membrane.

Ultracentrifugation

Ultracentrifugation applied to binding is also called sedimentation equilibrium (29) or sedimentation dialysis (30). A solution of a macro-molecular substrate and a low-molecular-weight ligand is placed in a centrifugal field until an equilibrium distribution is achieved. Because the ligand is a small molecule its concentration remains essentially constant throughout the centrifuge tube, whereas the substrate responds to the field with a concentration profile having a boundary between a substrate-free region and a region of limiting substrate concentration. This sedimentation boundary is therefore analogous to the membrane in equilibrium dialysis.

A single sedimentation experiment is equivalent to a series of dialysis experiments, each position in the centrifuge tube corresponding to a different substrate concentration but the same ligand concentration (29). At a given position (x) in the tube the ratio of moles of ligand bound per substrate molecule is

$$\bar{i} = \frac{(L_t)_x - [L]}{(S_t)_x}$$

since $[L]$ is the same at all points. Thus we write

$$(L_t)_x = [L] + \bar{i}(S_t)_x \tag{10.17}$$

A plot according to Eq. (10.17) yields $[L]$ as the intercept and $\bar{i}$ as the slope. Alternatively, the study may be carried out at different ligand concentrations. The analysis is usually by absorption spectroscopy.

It is not strictly exact to suppose that the small ligand molecule does not respond to the centrifugal field. It has been reported (31) that drug–protein binding studies by ultracentrifugation are limited to drugs having molecular weights < 400. Ultracentrifugation has even been applied to the study of binding between two small molecules, hexamethyl-benzene (S) and tetracyanoethylene (L) (32). Evidence was obtained, in the form of average molecular weights, for the presence of both SL and S_2L complexes.

Gel Filtration

Four different experimental techniques have been developed for the study of binding with the aid of size exclusion gels. One of these is a batch process, being closely analogous to equilibrium dialysis. Let S be a substrate that is excluded from the gel interior and L the ligand that is freely diffusible into the interior of the gel. The equilibrium is established, and the external phase is analyzed for its total ligand concentration, which includes both free and bound ligand. From the mass balances on the known totals of substrate and ligand, and the assumption that the free ligand concentration in the external phase is equal to the interior ligand concentration, the moles of ligand bound per mole of substrate can be calculated. It is not unusual for the ligand concentration in the gel to be higher than that in the external phase as a consequence of adsorption to the gel, and this effect can be assessed in a separate study by omitting the substrate. Applications include studies of the binding of methyl paraben to polysorbate 80 (33), steroid–protein binding (34), and antigen–antibody binding (35). In this last study the gel phase and solution phase were quantitatively separated and the radioactivity in each was measured.

A second method is a conventional chromatographic separation of free ligand from bound ligand (36). This method requires that the rate of binding equilibration be much slower than the rate of chromatographic elution, so it is in general not applicable.

A chromatographic procedure utilizing frontal analysis has been described by Cooper and Wood (37). This method is thermodynamically sound in the sense that the binding equilibrium is maintained in the large volume of the reactants that is passed through the gel column.

The most widely used gel filtration technique is an ingenious method introduced by Hummel and Dreyer (38). The substrate S, which for the present we take to be completely excluded from the gel, is dissolved in a solution of ligand L, the ligand concentration being $[L]_0$. A portion of this solution is placed on the gel column, which has already been preequilibrated with the same ligand solution of concentration $[L]_0$. Elution is carried out with the ligand solution of concentration $[L]_0$.

Since S is excluded from the gel, so too is the complexed form (or forms) of S. These therefore are eluted at the column dead volume. The column effluent is analyzed for its total ligand concentration, relative to a background level equal to $[L]_0$, the concentration of the eluting solvent. Thus a peak appears above this background corresponding to the complex, and the area of this peak is proportional to the amount of ligand bound to substrate.

The elution is continued until, relative to the background level $[L]_0$, a *negative* peak appears. The area of this peak represents a deficiency of ligand that is equal to the amount of ligand bound to substrate. Thus the information required to calculate $\bar{i}$, the number of moles of ligand bound per mole of substrate, can be obtained from either the positive or the negative peak, depending on analytical convenience and the chromatographic characteristics of the species. The success of the method is based upon the establishment of equilibrium of complexed substrate with the background ligand concentration $[L]_0$ by the time the complex is eluted. Thus the system is at equilibrium with respect to binding, and the free ligand concentration is known by preparation.

The technique has been used to study the binding of small molecules to proteins (38–41), of methylene blue and sodium dodecyl sulfate (42) (in this case the complex is retained by the column, so the negative peak appears first), of inclusion complexes of cyclodextrins (43), and of metal ion complexes (44). It is not necessary that the complex or substrate be completely excluded from the gel interior.

To understand the physical basis of this method it is helpful to use a simplifying analogy. Let us replace the gel filtration experiment with a countercurrent distribution system; this is simpler because it is a discontinuous process for which distribution equilibrium can be achieved at each stage. As before, S is the large excluded interactant and L is the small ligand species. Let the tubes be numbered $0, 1, 2, \ldots, r$ and the transfers $0, 1, 2, \ldots, n$. Partition coefficients are defined $P = $ (concentration in upper phase)/(concentration in lower phase). Then the fraction of a solute in the upper phase, p, is given by

$$p = \frac{PR}{PR + 1} \qquad (10.18)$$

where $R = V_u/V_l$, the phase volume ratio, and $p + q = 1$, where q is the fraction in the lower phase. Elementary analysis of the countercurrent distribution experiment (45) then gives Eq. (10.19), a binomial distribution, in which T_{nr} is interpreted as the fraction of total solute contained in tube r after n transfers.

$$T_{nr} = \frac{n!}{r!(n-r)!} \, p^r q^{(n-r)} \qquad (10.19)$$

To carry out the hypothetical experiment, we load the tubes with ligand solution of concentration $[L]_0$. (Upper and lower phases are separated by a hypothetical phase boundary.) A sample solution of S in ligand solution of concentration $[L]_0$ is prepared, and the upper phase of tube 0 is charged with this solution. At this stage the free ligand concentration of the upper phase of tube 0 is less than $[L]_0$, and the complex is in equilibrium with this reduced ligand concentration.

The countercurrent distribution is now carried out by shaking the tubes to reach distribution equilibrium, transferring the upper phase of each tube to the next tube, and replenishing the upper phase of tube 0 with the same ligand solution of concentration $[L]_0$. Equation (10.19) then gives the distribution of each solute throughout the train of tubes after any number n of such steps.

The easiest way to cope with the propagation of the initial deficiency of ligand in the upper phase of tube 0 is to treat the deficiency as a solute, whose partition coefficient is equal to unity. As the countercurrent distribution proceeds, the deficiency travels down the train of tubes, assuming a binomial distribution according to Eq. (10.19). At the same time the complex species, which has a very large partition coefficient, moves independently of, and faster than, the deficiency, and, if n is large enough, eventually draws sufficiently far ahead that the complex species is in the presence of ligand at the background concentration $[L]_0$. Elution at this point would yield the desired result of a positive complex peak, a return to the baseline level $[L]_0$, followed by a negative deficiency peak.

The number of transfers necessary to achieve this separation is readily calculated. We define the resolution as $R_s = \Delta d / w$, where Δd is the distance between peak maxima and w is average peak width. The mean of the binomial distribution is np, so for this system $\Delta d = n(p_+ - p_-)$, where p_+ and p_- denote the fractions for the positive and negative peaks. The standard deviation of the binomial distribution is $\sigma = \sqrt{npq}$, and w may be taken as 4σ. Combining these expressions gives a relationship for R_s as a function of n, p_+, and p_-. We may then choose a value for R_s as a criterion for resolution of the complex and the deficiency peaks and solve for n, the number of transfers needed to achieve this separation. Suppose, for example, we set $R_s = 2$ as the criterion. For an ideal gel filtration system we have $P = 1$ for the deficiency and $P = \infty$ for the fully excluded substrate species. If $R = 1$, we calculate that 16 transfers are necessary to achieve the desired separation.

In the chromatographic setting the retention time of a solute on a gel filtration column is given by

$$V_R = V_0 + K_D V_i \qquad (10.20)$$

where V_R is retention volume, V_0 is the column dead volume, V_i is volume of the gel interior, and K_D is the distribution coefficient, defined as the ratio of concentrations of the internal to external phases. For an ideal gel filtration system the ligand has $K_D = 1$ and the large substrate species has $K_D = 0$. As noted earlier, it is not essential that the substrate be completely excluded from the gel in order to apply this method.

A variant of this gel filtration method adds substrate and extra ligand as the sample (46); then the negative peak is converted to a positive peak.

REFERENCES

1. I. M. Klotz, F. M. Walker, and R. B. Pivan, *J. Am. Chem. Soc.*, **68**, 1486 (1946).
2. (a) R. E. Stauffer, in *Technique of Organic Chemistry*, Vol. III, Part, 1, 2nd ed., Interscience, New York, 1956, pp. 65–119; (b) L. C. Craig and T. P. King, *Methods of Biochemical Analysis*, Vol. 10, D. Glick, Ed., Wiley-Interscience, New York, 1962, pp. 175–199; (c) L. C. Graig, H.-C. Chen, and E. J. Harfenist, *Progress in Separation and Purification*, Vol. 2, T. Gerritsen, Ed., Wiley-Interscience, New York, 1969, pp. 219–237; (d) S.-T. Hwang and K. Kammermeyer, *Membranes in Separations*, Vol. VII, *Techniques of Chemistry*, Wiley-Interscience, New York, 1975; (e) K. K. Stewart, in *Advances in Protein Chemistry*, Vol. 31, C. B. Anfinsen, J. T. Edsall, and F. M. Richards, Eds., Academic, New York, 1977, pp. 135–187.
3. (a) J. Steinhardt and S. Beychok, in *The Proteins*, 2nd ed., H. Neurath, Ed., Academic, New York, 1964, Ch. 8; (b) J. Steinhardt and J. A. Reynolds, *Multiple Equilibria in Proteins*, Academic, New York, 1969, pp. 45–50.
4. (a) A. Goldstein, *Pharmacol. Rev.*, **1**, 102 (1949); (b) M. C. Meyer and D. E. Guttman, *J. Pharm. Sci.*, **57**, 895 (1968); (c) M. Rowland, *Therap. Drug Monitoring*, **2**, 29 (1980).
5. C. Tanford, *Physical Chemistry of Macromolecules*, Wiley, New York, 1961, pp. 225–227.
6. I. M. Klotz and J. M. Urquhart, *J. Phys. Chem.*, **53**, 100 (1949).
7. C. J. Briggs, J. W. Hubbard, C. Savage, and D. Smith, *J. Pharm. Sci.*, **72**, 918 (1983).
8. M. L. Fishman and F. R. Eirich, *J. Phys. Chem.*, **75**, 3135 (1971).
9. D. B. Donner, J.-J. L. Fu, and G. P. Hess, *Anal. Biochem.*, **75**, 454 (1976).
10. (a) G. F. Lockwood and J. G. Wagner, *J. Pharm. Pharmacol.*, **35**, 387 (1983); (b) J. Huang, *J. Pharm. Sci.*, **72**, 1368 (1983): (c) T. N. Tozer, J. G. Gambertoglio, D. E. Furst, D. S. Avery, and N. H. G. Holford, *J. Pharm. Sci.*, **72**, 1442 (1983).

11. S. P. Khor, H. J. Wu, and H. Boxenbaum, *Int. J. Pharmaceut.*, **23**, 109 (1985).

12. J. Cassel, J. Gallagher, J. Reynolds, and J. Steinhardt, *Biochemistry*, **8**, 1706 (1969).

13. T. J. Silhavy, S. Szmelcman, W. Boss, and M. Schwartz, *Proc. Nat. Acad. Sci. USA* **72**, 2120 (1975).

14. R. H. McNenamy, *Anal. Biochem.*, **23**, 122 (1968).

15. (a) A. S. Michaels, *Prog. Sep. Sci. Purific.*, Vol. 1, E. S. Perry, Ed., Wiley-Interscience, New York, 1968, pp. 297–334; (b) C. J. van Oss, *Prog. Sep. Sci. Purific.*, Vol. 3, E. S. Perry and C. J. van Oss, Eds., Wiley-Interscience, New York, 1970, pp. 97–132; (c) S. Jacobs, *Methods Biochem. Anal.*, Vol. 22, D. Glick, Ed., Wiley-Interscience, New York, 1974, p. 307.

16. J. A. Sophianopoulos, S. J. Durham, A. J. Sophianopoulos, H. L. Ragsdale, and W. P. Cropper, *Arch. Biochem. Biophys.*, **187**, 132 (1978).

17. W. F. Blatt, S. M. Robinson, and H. J. Bixler, *Anal. Biochem.*, **26**, 151 (1968).

18. Y. A. Zhirkov and V. K. Piotrovskii, *J. Pharm. Pharmacol.*, **36**, 844 (1984).

19. (a) L. C. Craig, *Adv. Anal. Chem. Instrum.*, **4**, 35 (1965); (b) L. C. Craig, *Methods Enzymol.*, **11**, 870 (1967); (c) A. Agren and E. R. Garrett, *Acta Pharm. Suec.*, **4**, 1 (1967); (d) S. Siggia, J. G. Hanna, and N. M. Serencha, *Anal. Chem.*, **36**, 638 (1964).

20. H. H. Stein, *Anal. Biochem.*, **13**, 305 (1965).

21. (a) S. P. Colowick and F. C. Womack, *J. Biol. Chem.*, **244**, 774 (1969); (b) F. C. Womack and S. P. Colowick, *Methods Enzymol.*, **27**, 464 (1973).

22. P. Mermier, *FEBS Lett.*, **55**, 75 (1975).

23. A. Agren and R. Elofsson, *Acta Pharm. Suec.*, **4**, 281 (1967).

24. C. F. Beyer, L. C. Craig, and W. A. Gibbons, *Biochemistry*, **11**, 4920 (1972).

25. (a) M. C. Meyer and D. E. Guttman, *J. Pharm. Sci.*, **57**, 1627 (1986); (b) M. C. Meyer and D. E. Guttman, *J. Pharm. Sci.*, **59**, 33, 39 (1970).

26. I. Kanfer and D. R. Cooper, *J. Pharm. Pharmacol.*, **28**, 58 (1976).

27. K. K. Stewart, L. C. Craig, and R. C. Williams, *Anal. Chem.*, **42**, 1252 (1970).

28. R. H. Reuning and G. Levy, *J. Pharm. Sci.*, **57**, 1556 (1968).

29. I. Z. Steinberg, and H. K. Schachman, *Biochemistry*, **5**, 3728 (1966).

30. P. H. Lloyd, R. N. Prutton, and A. R. Peacocke, *Biochem. J.*, **107**, 353 (1968).

31. Y. Matsushita and I. Moriguchi, *Chem. Pharm. Bull.*, **33**, 2948 (1985).

32. P. J. Trotter and D. A. Yphantis, *J. Phys. Chem.*, **74**, 1399 (1970).

33. R. W. Ashworth and D. D. Heard, *J. Pharm. Pharmacol.*, **18**, 98S (1966).

34. W. H. Pearlman and O. Crepy, *J. Biol. Chem.*, **242**, 182 (1967).

35. C. Souleil and A. Nisonoff, *Nature (London)*, **217**, 144 (1968).

36. T. L. Hardy and K. R. L. Mansford, *Biochem. J.*, **83**, 34P (1962).

37. P. F. Cooper and G. C. Wood, *J. Pharm. Pharmacol.*, **20**, 150S (1968).

38. J. P. Hummel and W. J. Dreyer, *Biochim. Biophys. Acta*, **63**, 530 (1962).

39. G. F. Fairclough and J. S. Fruton, *Biochemistry*, **5**, 673 (1966).

40. A. N. Glazer, *J. Biol. Chem.*, **242**, 4528 (1967).

41. S. F. Sun, S. W. Kuo, and R. A. Nash, *J. Chromatogr.*, **288**, 377 (1984).

42. P. K. Nandi, *J. Chromatogr.*, **116**, 93 (1976).

43. (a) T. K. Korpela and J. P. Himanen, *J. Chromatogr.*, **290**, 351 (1984); (b) I. Sanemasa, T. Mizoguchi, and T. Deguchi, *Bull. Chem. Soc. Jap.*, **57**, 1358 (1984).

44. N. Yoza, *J. Chem. Educ.*, **54**, 284 (1977).

45. K. A. Connors, *A Textbook of Pharmaceutical Analysis*, 3rd ed., Wiley-Interscience, New York, 1982, Ch. 16.

46. (a) N. Yoza, K. Kouchiyama, T. Miyajima, and S. Ohashi, *Anal. Lett.*, **8**, 641 (1975); (b) T. Miyajima, N. Yoza, and S. Ohashi, *Anal. Lett.*, **10**, 709 (1977).

11

CHROMATOGRAPHY

11.1. PRINCIPLE OF THE CHROMATOGRAPHIC METHOD

The Partition Coefficient

The basic retention volume equation of elution chromatography is Eq. (11.1), where V_R is the retention volume (of the peak maximum), V_0 is the column dead volume, V_{st} is the stationary-phase volume of the column, and P is the partition coefficient of the eluted substance, defined as in Eq. (11.2).

$$V_R = V_0 + PV_{st} \tag{11.1}$$

$$P = \frac{C_{st}}{C_{mo}} \tag{11.2}$$

The definition of the partition coefficient as the ratio of concentrations in the stationary (st) to mobile (mo) phases is standard in chromatography. The quantity $V_R - V_0$ is the adjusted retention volume; V_0 can be determined experimentally as the retention volume of a nonretained solute, for which $P = 0$. From the preparation of the column, V_{st} is known, hence the partition coeffficient can be determined chromatographically by means of Eq. (11.1). With P accessible in this manner, binding between species in either the stationary or the mobile phase (depending on the chemical and chromatographic system) may be studied according to experimental designs and calculational methods described in Chapter 9 for conventional liquid–liquid partitioning.

Partition coefficients determined chromatographically are often in good agreement with partition coefficients determined by the static method. Nevertheless, chromatography is a dynamic method, and in binding studies there are three classes of rate processes that may affect the validity of a thermodynamic measurement: these are the binding (association–dissociation) process, the distribution (mass-transfer) process, and the elution process. Distribution and elution rates are dealt with in kinetic theories of chromatography. It is usually reasonable to assume that binding rates are much faster than the rates of the chromatographic process.

Much of the interest in studying complex formation by chromatography has been stimulated by the possibility, and the practice, of modifying chromatographic resolution by means of specific intermolecular interactions. For example, a chromatographic separation of antipyrine and caffeine is based on the retention of antipyrine by complex formation on a cation-exchange column in the ferric ion form (1). Gas–liquid chromatographic columns having silver nitrate incorporated into the stationary phase selectively retard olefins (2). β-Cyclodextrin added to the mobile phase improved the ion-exchange resolution of some prostaglandins (3).

Classification of Systems

We may use chromatography to study binding in either the stationary phase or the mobile phase, as appropriate. Although it is in principle necessary that a species present in one phase must distribute, however minutely, into the other phase, it is reasonable to specify that we may have partition coefficients of zero or infinity. It is similarly reasonable to take zero as a possible value of stability constants. Clearly, the only systems of interest are those in which substrate, ligand, and complex(es) coexist in either the stationary or the mobile phase. With these stipulations, the possible types of systems are listed in Table 11.1.

Since the identification of an interactant as substrate or ligand is arbitrary, there is some redundancy in the table. It is not arbitrary, of course, to identify the chromatographic phases. The partition coefficient is defined for either the substrate or the ligand as appropriate. The design of the chromatographic experiment depends on the case to which the

Table 11.1. Classification of Chromatographic Systems in Which Complex Formation Might be Studied

Case	Species Present[a]	
	Stationary Phase	Mobile Phase
Ia	S, L, C	L
Ib	S, L, C	S
Ic	S, L, C	C
II	S, L, C	S, L
IIIa	S, L, C	L, C
IIIb	S, L, C	S, C
IVa	L	S, L, C
IVb	S	S, L, C
IVc	C	S, L, C
V	S, L	S, L, C
VIa	L, C	S, L, C
VIb	S, C	S, L, C
VII	S, L, C	S, L, C

[a] S, substrate; L, ligand; C, complex(es).

system belongs. For example, gas chromatographic binding studies are usually carried out on Case Ia (or Ib) systems, only one of the species, S or L, being volatile. It is then required that the binding kinetics be much faster than the elution rate so that the binding system is at equilibrium.

On the other hand, gel filtration, which is discussed in Section 10.2, is a Case IVa system, with only the ligand having access to the stationary phase. If chromatographic elution is carried out in the usual manner, by injecting a small sample of solution containing equilibrated S, L, and C followed by elution with a solvent, then if S and C are to be separated as species in the same amounts as in the original sample, it is necessary that the binding kinetics be much slower than the elution rate. Otherwise, dilution of the sample by the eluting solvent will result in dissociation of the complex.

When binding is studied in the chromatographic stationary phase, it must be considered whether this thin adsorbed solvent layer has the same properties as does bulk solvent. The agreement of chromatographic and static partition coefficients suggests that the stationary-phase solvent is equivalent to bulk solvent, but some reservations have been expressed concerning gas–liquid chromatography (4).

11.2. GAS CHROMATOGRAPHY

Principles

The gas chromatographic method makes use of gas–liquid chromatography (GLC) applied to Case Ia or Ib of Table 11.1. Thus the binding equilibrium is studied in the stationary phase. We shall adopt the Case Ib description. Although some studies have been carried out in which the chromatographic liquid phase itself serves as the ligand, usually the ligand is a nonvolatile solute dissolved in the nonvolatile liquid phase, which is the solvent for the binding process. The complex is nonvolatile, but the substrate is volatile and is the only system component that is detected in the chromatographic eluent. A good example is the study of silver ion–olefin complexes, the silver ion being the ligand and the olefin the substrate.

To develop the method, we suppose that a 1:1 complex SL is formed. The apparent partition coefficient of the substrate is defined by

$$P = \frac{(S_t)_{st}}{(S_t)_{mo}} \tag{11.3}$$

The mass balances give $(S_t)_{mo} = [S]_{mo}$ and $(S_t)_{st} = [S]_{st} + [SL]_{st}$. Combining these with Eq. (11.3) and the stability constant definition gives Eq. (11.4), where $P_0 = [S]_{st}/[S]_{mo}$ is the intrinsic partition coefficient.

$$P = P_0 + P_0 K_{11}[L]_{st} \tag{11.4}$$

P is obtained from the retention volume, corrected for column pressure drop, by Eq. (11.1). The free ligand concentration may be taken equal to the total ligand concentration because the substrate concentration can be held very low (owing to the high sensitivity of GC detectors). Thus a plot of P against $[L]_{st}$ is expected to yield a straight line, from which K_{11} is obtained. The extrapolated value of P_0 should agree with the value obtained in the absence of ligand. A separate column packing is required for each value of $[L]_{st}$.

Notice the essential equivalence of Eqs. (11.4) and (9.4). Most of the reported stability constants determined by GLC have been obtained by means of Eq. (11.4).

The GLC method has been reviewed by several authors (5–8). Purnell (9) has described the various stoichiometries that might be observed in the stationary phase and has derived equations for many of these cases. For example, suppose complexes SL, S_2L, and S_3L are formed; then the preceding treatment yields

$$P = P_0 + \beta_{11}P_0[L]_{st} + 2\beta_{21}P_0[L]_{st}[S]_{st} + 3\beta_{31}P_0[L]_{st}[S]_{st}^2 \quad (11.5)$$

If, on the other hand, complexes SL, SL_2, and SL_3 are formed, we get

$$P = P_0 + \beta_{11}P_0[L]_{st} + \beta_{12}P_0[L]_{st}^2 + \beta_{13}P_0[L]_{st}^3 \quad (11.6)$$

Figure 11.1 shows a plot according to Eq. (11.4) for the study of complexing between several alcohols and an amine (10). When such a wide range of ligand concentrations is studied, it is not surprising that some nonlinearity may be observed in these plots, and the 1:1 stability constant is then most reasonably based on the data at the low concentrations. From the lines in Figure 11.1 we obtain, by applying Eq. (11.4), $K_{11} = 7.6$ (methanol), 5.2 (ethanol), 3.5 (n-propanol), these quantities having the (mole fraction)$^{-1}$ dimension.

GLC measurements of this type have been applied to several types of complexing systems. Gil-Av and Herling (11) introduced this calculational method for the study of silver ion–olefin complexes, and it has since

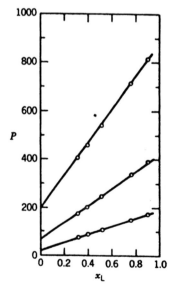

Figure 11.1. Plot according to Eq. (11.4) for the GLC study of complex formation between quinoline (ligand) and the substrates (top to bottom) n-propanol, ethanol, methanol at 45°C in squalene. Data from reference (10). The ligand concentration scale is in mole fraction units.

been used by several other workers to study silver ion complexation (argentation). Pronounced negative curvature of the plots according to Eq. (11.4) has been ascribed to salting out of the volatile ligand by the silver salt (12), but Schnecko (13) obtained linear plots when the ionic strength of the stationary phase was held constant.

Organic complexes of several types have been investigated by GLC; Figure 11.1 shows examples of hydrogen-bonded complexes (10). Further examples of H-bonded systems are didecyl sebacate–alcohol complexes (14) and N,N-diethylacetamide–alkyne complexes (15). Polar molecules of several types (16–18), and presumed charge-transfer complexes have been studied by GLC (19–21). Nearly all the reported K_{11} values are extremely small. Metal ion–amine interactions have been studied by GC (22).

Mathiasson (23) has devised a competitive gas chromatographic method that permits the study of binding between two nonvolatile solutes in the stationary phase, the volatile substrate serving as an indicator of this process.

Complications

The GLC method is not as straightforward as is suggested by the preceding development. One of the complications is a consequence of the physical nature of the stationary phase, which is a thin adsorbed layer. Thus Cadogan and Purnell (14) had to apply corrections for surface adsorption effects. But a large part of the literature on the GLC method is concerned with solution nonidealities. Let us redevelop the earlier treatment replacing concentrations by activities. Again we assume $1:1$ complexing of a volatile substrate S and a nonvolatile ligand L in the nonvolatile solvent (stationary) phase. The thermodynamic stability constant is defined

$$K_{11}^{th} = \frac{[SL]_{st}}{[S]_{st}[L]_{st}} \cdot \frac{(\gamma_{SL})_{st}}{(\gamma_S)_{st}(\gamma_L)_{st}} \tag{11.7}$$

Because of the nature of GLC, the observed partition coefficient is considered by some authors to be a ratio of activities, so Eq. (11.3) is replaced by

$$P = \frac{[S]_{st}(\gamma_S)_{st} + [SL]_{st}(\gamma_{SL})_{st}}{[S]_{mo}(\gamma_S)_{mo}} \tag{11.8}$$

Combining Eqs. (11.7) and (11.8),

$$P = P_0 + P_0 K_{11}^{th}[L]_{st}(\gamma_L)_{st} \tag{11.9}$$

which can be compared with Eq. (11.4). The close similarity of these equations is partly illusory, because P and P_0 in Eq. (11.4) are defined as ratios of concentrations, whereas in Eq. (11.9) they are interpreted to be ratios of activities; the experimental numbers remain the same.

From Eq. (11.7) we can write

$$K_{11}^{th} = K_{11} \frac{(\gamma_{SL})_{st}}{(\gamma_S)_{st}(\gamma_L)_{st}} \tag{11.10}$$

If a plot of P against $[L]_{st}$ is linear, the quotient slope/intercept can be interpreted either as K_{11} (Eq. 11.4) or as $K_{11}^{th}(\gamma_L)_{st}$ (Eq. 11.9). These are quite different interpretations, as Eq. (11.10) shows. Gil-Av and Herling (11) interpreted their results as K_{11} and measured $(\gamma_S)_{st}$ gas chromatographically (as shown later). The ratio $K_{11}/(\gamma_S)_{st}$ was considered to be proportional to K_{11}^{th} if the activity coefficient ratio $(\gamma_{SL})_{st}/(\gamma_L)_{st}$ was substantially constant.

To see how $(\gamma_S)_{st}$ can be estimated we use Purnell's treatment (24). For the volatile solute S in solution in the nonvolatile solvent M (no ligand L present), the vapor pressure p_S is given by Eq. (11.11), which is Raoult's law with an activity coefficient added to account for nonideal behavior. In Eq. (11.11) p_S^o is the vapor pressure of pure S and x_S is the mole fraction of S in the solution.

$$p_S = (\gamma_S)_{st} x_S p_S^o \tag{11.11}$$

Assuming that the vapor phase behaves ideally, the ideal gas law gives an expression for p_S,

$$p_S = \frac{(n_S)_{mo}}{V_{mo}} RT = [S]_{mo} RT \tag{11.12}$$

where $(n_S)_{mo}$ is the number of moles of S in V_{mo} liters of vapor phase (mobile phase in chromatography), and $[S]_{mo}$ is the molar concentration. The mole fraction is approximated by the mole ratio.

$$x_S = \frac{(n_S)_{st}}{(n_S)_{st} + n_M} \approx \frac{(n_S)_{st}}{n_M} \tag{11.13}$$

The volume of stationary phase solvent, V_{st}, is equal to the product of its partial molar volume $\bar{V}_M$ and the number of moles n_M, so $n_M = V_{st}/\bar{V}_M$, giving, from Eq. (11.13),

$$x_S = [S]_{st}\bar{V}_M \qquad (11.14)$$

Equations (11.12) and (11.14) are combined with Eq. (11.11), where $P_0 = [S]_{st}/[S]_{mo}$:

$$P_0 = \frac{RT}{(\gamma_S)_{st}\bar{V}_M P_S^\circ} \qquad (11.15)$$

With Eq. (11.15), or equivalent versions, $(\gamma_S)_{st}$ can be estimated by measuring the intrinsic partition coefficient P_0 of substrate S in solvent M.

According to Eq. (11.11), $(\gamma_S)_{st}$ is a Raoult's law activity coefficient; that is, $(\gamma_S)_{st} = 1$ when $x_S = 1$. This definition can lead to quite remarkable numerical values. Thus $(\gamma_S)_{st}$ values for olefins in ethylene glycol (11) were estimated to be in the range 10^2–10^3; these large values mean that departures from Raoult's law are extreme, and their use in estimating a thermodynamic equilibrium constant is uncertain.

It is evident that in a well-behaved system the P_0 value obtained by extrapolation of the P versus $[L]_{st}$ plot should agree with that measured directly (in the absence of L). Disagreement of these estimates may mean that the presence of L in the solvent is producing a medium effect (25). Further limitations on Eq. (11.15) as a means for estimating the activity coefficient of the volatile component are the use of partial pressure rather than fugacity, assumption of ideality of the vapor phase, approximation of mole fraction by mole ratio, and neglect of the carrier gas solubility in the stationary phase. These problems have been addressed (4, 6).

We have seen that interpretation of the gas chromatographically measured partition coefficient as a ratio of activities gives Eq. (11.9); if instead we consider it to be a ratio of concentrations, an analogous treatment yields

$$P = P_0 + \frac{P_0 K_{11}^{th}[L]_{st}(\gamma_L)_{st}(\gamma_S)_{st}}{(\gamma_{SL})_{st}} \qquad (11.16)$$

Since P_0 is a function of the activity coefficient $(\gamma_S)_{st}$, as shown by Eq. (11.15), evidently P_0 cannot be considered a constant independent of $[L]_{st}$

in a rigorous interpretation. Thus Eq. (11.16) contains activity coefficients that cannot be evaluated chromatographically. Several authors have devised means to overcome this inability. Martire and Riedl (26), de Ligny (8), and Reznikov et al. (27) make use of "control" systems in which a complexing solute is replaced by a solute that is of similar size but that presumably does not complex (e.g., replacement of an alkene by an alkane). Ratios of partition coefficients are then manipulated to achieve cancellation of activity coefficients. Eon et al. (28), Castells (29), and Mathiasson and Jonsson (17, 30) have applied theoretical estimates of activity coefficients. Since S is volatile whereas L and M are nonvolatile, it is usual that the molar volume of S is much smaller than those of L and M, and this difference is in part responsible for the nonideal behavior. Flory–Huggins theory was used to account for this effect.

More attention has been given to nonideal behavior in the GLC method than in any other method for studying binding constants, probably for two reasons: in these systems deviations from Raoult's law are common, and most of the systems examined yield very small binding constants, necessitating the experimental study of a wide range of ligand concentrations.

GLC-determined binding constants were compared with those measured by nuclear magnetic resonance and optical spectroscopy by Purnell and Srivastava (19), and this comparison has given rise to controversy and differences of interpretation. Disagreement among the GLC, NMR, and spectroscopic K_{11} values was observed, and it was concluded that the GLC results, expressed on the molar scale, were the correct ones. Martire (31) argued that the NMR and spectroscopic values only include "specific chemical" interactions, whereas the GLC values include both "chemical" and "physical" contributions; however, as noted in Section 2.6, "Random Association," this was a consequence of assigning randomly associated species to the class S (substrate) for NMR and spectroscopy, and to class S̄ (not-substrate) for GLC. Crowne et al. (21) used Martire's treatment in discussing their results on complexes of picric acid with naphthalenes. Mathiasson (30) claimed to separate the chemical from the physical contributions to the GLC result. It was pointed out in Section 2.6 that it is not possible to conclude whether or not the randomly associated pairs (contact pairs) make a contribution to K_{11} on the basis of derivations proceeding from mass balance relationships.

The stability constants reported in the foregoing accounts are very small quantities, usually $<1 \, M^{-1}$. These are of the magnitude to be expected from the purely random association of species, as shown in Section 2.6. Purnell (18, 32) observed that plots according to Eq. (11.4) were linear over the entire concentration range for many species and was led to the concept that these binary solutions are essentially mixtures of two immiscible phases ("microscopic partition theory"); it was concluded that complexation may not occur at all in these systems. Martire questioned the sensitivity of Purnell's data analysis (33).

Many chromatographers have concluded that GLC is one of the best experimental methods for measuring binding constants, but relatively few results have been obtained in this way. Possibly the technique is more valuable for the study of the structures of solutions than for binding phenomena.

11.3. LIQUID CHROMATOGRAPHY

Column Chromatography

Binding equilibria are studied in the bulk mobile phase with most forms of liquid chromatography, so this technique seems on this account alone to be of wider applicability than is gas chromatography. Gel filtration chromatography is treated in Section 10.2.

The conventional measure of retention in liquid column chromatography is the capacity factor k', defined by Eq. (11.17), where P is the partition coefficient (Eq. 11.2) and $R = V_{st}/V_0$ is the phase volume ratio of the column, V_0 being the dead volume (column mobile-phase volume).

$$k' = PR \qquad (11.17)$$

Combination of Eqs. (11.1) and (11.17) gives Eq. (11.18), with which k' may be evaluated.

$$k' = \frac{V_R - V_0}{V_0} \qquad (11.18)$$

If the mobile-phase flow velocity is constant, k' may also be expressed in terms of retention times:

$$k' = \frac{t_R - t_0}{t_0} \qquad (11.19)$$

Since k' is directly proportional to P, the quantitative description of retention in complexing systems can be carried out just as for liquid–liquid partitioning (Chapter 9) and gas–liquid chromatography (Section 11.2). It is assumed that the binding processes are much faster than the chromatographic rates so that the system is always at equilibrium with respect to complex formation. [If this condition is not met, the elution peaks will be distorted, and it is possible to obtain information about the rate of equilibration (34).]

To show the basis of the method we consider the determination of the pK_a of a weak acid HA on an ion exchange or reverse-phase column with an aqueous mobile phase. We call the solute the substrate and the hydrogen ion the ligand. Then the apparent partition coefficient is defined by

$$P = \frac{[HA]_{st} + [A^-]_{st}}{[HA]_{mo} + [A^-]_{mo}} \qquad (11.20)$$

Combination of Eq. (11.20) with the dissociation constant expression and with intrinsic partition coefficients for the conjugate acid and base species gives Eq. (11.21), where k'_{HA} and k'_A are capacity factors for these species, related to the intrinsic partition coefficients by equations of the form of Eq. (11.17).

$$k' = \frac{k'_{HA}[H^+] + k'_A K_a}{[H^+] + K_a} \qquad (11.21)$$

Eq. (11.21) has been used by several workers to measure pK_a values of acids and bases (35–37). One of the ways to analyze the data is to rearrange Eq. (11.21) to

$$k' = k'_{HA} - K_a \frac{(k' - k'_A)}{[H^+]} \qquad (11.22)$$

The capacity factor k'_A is measured in experiments at $pH \gg pK_a$; then k' is plotted against $(k' - k'_A)/[H^+]$.

Equation (11.21) can also be written as Eq. (11.23), where f_{HA} and f_A are the fractions of solute present as the conjugate acid and base, respectively.

$$k' = k'_{HA} f_{HA} + k'_A f_A \qquad (11.23)$$

Thus k' is a weighted average of k'_{HA} and k'_A; this is a consequence of the assumption that the system is at binding equilibrium. Analogous equations have been written for polyprotic acids.

Let us next consider a liquid chromatographic system in which substrate S and ligand L form a 1:1 complex SL in the mobile phase and both S and SL can partition into the stationary phase. The apparent partition coefficient is

$$P = \frac{[S]_{st} + [SL]_{st}}{[S]_{mo} + [SL]_{mo}} \tag{11.24}$$

Proceeding as earlier leads to Eq. (11.25), where K_{11} is the stability constant for the formation of SL in the mobile phase, and k'_S and k'_{11} are capacity factors for uncomplexed and complexed substrate, respectively.

$$k' = \frac{k'_S + k'_{11}K_{11}[L]_{mo}}{1 + K_{11}[L]_{mo}} \tag{11.25}$$

In this equation $[L]_{mo}$ is the free ligand concentration in the mobile phase; since S_t is usually very small, this can be set equal to L_t. Note that the method involves addition of ligand, at this known concentration, to the mobile phase. The concentration of complex is in equilibrium with this set concentration of ligand, which is varied to generate the full data set.

Again we see that k' is a weighted average, for Eq. (11.25) can be written $k' = k'_S f_{10} + k'_{11} f_{11}$, where these fractions are given by Eq. (2.51). Equation (11.25) is a hyperbola, and K_{11} can be estimated by methods described in earlier chapters. For example, k'_S can be measured in the absence of ligand, and a linear plot made according to

$$\frac{k' - k'_S}{[L]_{mo}} = -K_{11}k' + k'_{11}K_{11} \tag{11.26}$$

The method has been used to study a variety of complexing systems (38–40). Horváth and co-workers have extensively applied reverse-phase liquid chromatography to the study of complex equilibria (41) and have commented on some of its limitations, including the difficulty of detecting multiple equilibria and the effects of sample overloading (39).

Another complication arises when the stationary phase does not behave as a bulk phase into which the solute distributes by a partitioning mechanism. It is possible for the stationary phase to possess discrete binding sites. Let such a site be denoted B, and define stability constants K_{SB} and K_{SLB} for the formation, in the stationary phase, of the species SB and SLB. Then the apparent partition coefficient is

$$P = \frac{[S]_{st} + [SL]_{st} + [SB]_{st} + [SLB]_{st}}{[S]_{mo} + [SL]_{mo}}$$

Proceeding as before gives

$$k' = \frac{k'_S + k'_{11}K_{11}[L]_{mo} + [B](k'_S K_{SB} + k'_{11}K_{SLB}K_{11}[L]_{mo})}{1 + K_{11}[L]_{mo}}$$

$$(11.27)$$

Evidently, Eq. (11.25) is a special case of (11.27), describing the limit of pure partitioning behavior. The other limit, that of pure adsorption or binding to the B sites, is obtained by neglecting the first two terms in the numerator of Eq. (11.27). Then on writing the mass balance for binding sites as $B_t = [B] + [LB]$, where LB represents the binding of a ligand species to site B, the concentration of free B sites in the stationary phase is given by

$$[B] = \frac{B_t}{1 + K_{LB}[L]_{mo}} \qquad (11.28)$$

where K_{LB} describes the distribution of L between the phases according to $K_{LB} = [LB]/[L]_{mo}[B]$. Combining Eq. (11.27) for this case with Eq. (11.28):

$$k' = \frac{B_t(k'_S K_{SB} + k'_{11}K_{SLB}K_{11}[L]_{mo})}{(1 + K_{11}[L]_{mo})(1 + K_{LB}[L]_{mo})} \qquad (11.29)$$

Equation (11.29) describes k' as a function of $[L]_{mo}$. This function displays a maximum because of the form of the denominator. Equations of the same form were encountered as Eqs. (6.44) and (6.54); this functional form arises because a ternary complex is present. Horváth et al. (42) have observed chromatographic behavior consistent with Eq. (11.29). Such behavior is undesirable if one wishes to measure K_{11} in the mobile phase; however, it is of considerable interest for the study of chromatographic retention mechanisms.

Affinity chromatography is a form of column chromatography that by design exhibits specific binding in the stationary phase, so the equilibrium hypothesis is reasonably applied. Binding constants for binding of the retained species to the stationary-phase sites can be obtained by a treatment similar to the preceding (43, 44).

Related Methods

Countercurrent distribution possesses features of both liquid–liquid partitioning (two bulk phases brought to true distribution equilibrium) and liquid chromatography (stationary and mobile phases contacted in a cascade process). This procedure may therefore be applied to the study of complex formation. For example, Mold et al. (45) found that the countercurrent distribution patterns of polycyclic aromatic compounds depended on the concentration of the complexing agent tetramethyluric acid in the lower (aqueous) phase. Since the lower phase is the stationary phase, this is a Case Ib system in the Table 11.1 classification, tetramethyluric acid being the ligand. Equation (11.4), developed for gas chromatography, is therefore applicable. For the system in which pyrene is the substrate, with distribution between cyclohexane (upper phase) and 90% aqueous methanol containing the ligand, a plot according to Eq. (11.4) reveals upward curvature, consistent with the presence of complexes SL and SL_2.

Paper electrophoresis and related methods have provided some complex stability constants. In all these methods the assumption is made that the observed mobility of the substrate is an average of the intrinsic mobilities of the migrating species weighted by the fractional concentrations of these species. This definition, namely, $\mu_{obs} = \Sigma \, \mu_i f_i$, is combined with the appropriate stability constant definitions to give μ_{obs} as a function of stability constant and ligand concentration. Acid dissociation constants have been studied in this way (46), but most applications have been to the complexes of metal ions with anionic ligands (47–50). The methods of data treatment are identical with those described for similar functions, as in Sections 2.5 and 2.6.

Complexing agents have been incorporated into the stationary phase to influence resolution in thin-layer chromatography (51), and relative values of complex stability can be obtained by TLC (52), but it is not a powerful method. Armstrong and Stine (53) carried out thin-layer chromatography with a mobile phase in which micelles of a surfactant constituted an additive, and evaluated the partition coefficient of a substrate between water and the micelle. Most of the reasoning necessary to give a quantitative description of this system has been presented in Section 6.3. Armstrong and Stine treat the mobile phase as a two-phase system, namely, the aqueous phase (phase A) and the micellar phase (phase M). The substrate can partition between the TLC stationary phase and phases A and M, these processes being described by partition coefficients $P_A = [S]_{st}/[S]_A$, $P_M = [S]_{st}/[S]_M$, $P_{AM} = [S]_M/[S]_A$; it follows that $P_A = P_M P_{AM}$. The apparent partition coefficient is defined $P = [S]_{st}/[S]_{mo}$. An expression for $[S]_{mo}$ is available as Eq. (6.36), which we write

$$[S]_{mo} = [S]_M CV + [S]_A (1 - CV) \qquad (11.30)$$

where C is the molar concentration of surfactant present as micelles and V is the molar volume of surfactant. Equation (11.30) is combined with Eq. (11.20) and the definition of P to yield

$$\frac{V_{st}}{V_R - V_0} = \frac{1}{P_A} + \frac{(P_{AM} - 1)CV}{P_A} \qquad (11.31)$$

In thin-layer chromatography the measured retention parameter is the R_f value, which is related to chromatographic quantities by $R_f = V_0/V_R$. This is combined with Eq. (11.31):

$$\frac{R_f}{1 - R_f} = \frac{V_0}{V_{st}P_A} + \frac{V_0(P_{AM} - 1)CV}{V_{st}P_A} \qquad (11.32)$$

From a plot according to Eq. (11.32) the partition coefficient P_{AM} can be determined. This is not a binding constant, but it can be related to an apparent binding constant by Eq. (6.38).

This treatment was applied to surfactant systems and to mobile phases consisting of α-cyclodextrin in water. The two-phase description may be appropriate for a micellar system but does not seem applicable to an aqueous solution of a cyclodextrin.

An ion-exchange procedure is carried out as a batch process, so it is not chromatographic, but it is conveniently considered here. Suppose that complexing of a metal ion with an anionic ligand is to be studied. This can be done with an anion-exchange resin. The basic premise is that the free ligand can be taken up by the resin but that the complexed ligand cannot.

Let complexes SL and S_2L be formed, where S is the cation. For this system we observe the distribution of ligand between the resin and the bulk solvent, so we define an apparent partition coefficient by $P = (L_t)_{resin}/(L_t)_{bulk}$. The total concentration of ligand in the system is known, and $(L_t)_{bulk}$ is determined by measurement after equilibration has been achieved. Thus $(L_t)_{resin}$ is obtained by difference, so P can be measured.

Combining the stability constant expressions with the definition of P yields

$$P = \frac{P_0}{1 + K_{11}[S]_{mo} + K_{11}K_{21}[S]_{mo}^2} \qquad (11.33)$$

Taking the reciprocal shows that $1/P$ is a polynomial in the substrate concentration. A linear plotting form can be obtained, or least-squares regression analysis may be used (54). This method has been used to study the complexes of metal ions (55, 56) and polyamines (57) with nucleotides.

REFERENCES

1. E. Sjöström and L. Nykanen, *J. Am. Pharm. Assoc., Sci. Ed.*, **47**, 248 (1958).

2. B. W. Bradford, D. Harvey, and D. E. Chalkley, *J. Inst. Petrol.*, **41**, 80 (1955).

3. K. Uekama, F. Hirayama, K. Ikeda, and K. Inaba, *J. Pharm. Sci.*, **66**, 706 (1977).

4. D. C. Locke, *Adv. Chromatogr.*, **14**, 95 (1976).

5. (a) J. R. Conder, *Adv. Anal. Chem. Instrum.*, **6**, 209 (1968); (b) J. R. Conder and C. L. Young, *Physicochemical Measurement by Gas Chromatography*, Wiley, New York, 1979, Ch. 6.

6. C. A. Wellington, *Adv. Anal. Chem. Instrum.*, **11**, 237 (1973).

7. D. C. Locke, *Adv. Chromatogr.*, **14**, 161 (1976).

8. C. L. deLigny, *Adv. Chromatogr.*, **14**, 265 (1976).

9. J. H. Purnell, *Gas Chromatography 1966*, A. B. Littlewood, Ed., The Institute of Petroleum, London, 1967, p. 3.

10. Z. Kopacka-Kozak and W. Waclawek, *J. Chromatogr.*, **298**, 319 (1984).

11. E. Gil-Av and J. Herling, *J. Phys. Chem.*, **66**, 1208 (1962).

12. (a) M. A. Muhs and F. T. Weiss, *J. Am. Chem. Soc.*, **84**, 4697 (1962); (b) R.

J. Cvetanović, F. J. Duncan, W. E. Falconer, and R. S. Irwin, *J. Am. Chem. Soc.*, **87**, 1827 (1965); (c) S. P. Wasik and W. Tsang, *J. Phys. Chem.*, **74**, 2970 (1970).

13. H. Schnecko, *Anal. Chem.*, **40**, 1391 (1968).
14. D. F. Cadogan and J. H. Purnell, *J. Phys. Chem.*, **73**, 3849 (1969).
15. R. Queignec and B. Wojtkowiak, *Bull. Soc. Chim. Fr.*, 860 (1970).
16. (a) A. B. Littlewood and F. W. Willmott, *Anal. Chem.*, **38**, 1031 (1966).
    (b) D. F. Cadogan and J. H. Purnell, *J. Chem. Soc.*, *A*, 2133 (1968).
17. L. Mathiasson and R. Jönsson, *J. Chromatogr.*, **101**, 339 (1974).
18. J. H. Purnell and J. M. Vargas de Andrade, *J. Am. Chem. Soc.*, **97**, 3585 (1975).
19. J. H. Purnell and O. P. Srivastava, *Anal. Chem.*, **45**, 1111 (1973).
20. R. J. Laub and R. L. Pecsok, *J. Chromatogr.*, **113**, 47 (1975).
21. C. W. Crowne, M. F. Harper, and P. G. Farrell, *J. Chromatogr. Sci.*, **14**, 321 (1976).
22. R. C. Castells and J. A. Catoggio, *Anal. Chem.*, **42**, 1268 (1970).
23. L. Mathiasson, *J. Chromatogr.*, **114**, 47 (1975).
24. H. Purnell, *Gas Chromatography*, Wiley, New York, 1962, pp. 206–207.
25. S. V. Vitt, V. B. Bondarev, E. A. Paskonova, T. M. Shcherbina, and M. G. Bezrukov, *Izv. Akad. Nauk SSR, Ser. Khim.*, 441 (1972) (Engl. transl., p. 387).
26. (a) D. E. Martire and P. Riedl, *J. Phys. Chem.*, **72**, 3478 (1968); (b) H.-L. Liao, D. E. Martire, and J. P. Sheridan, *Anal. Chem.*, **45**, 2087 (1973).
27. S. A. Reznikov, Y. A. Batyrev, and R. I. Sidorov, *Zh. Fiz. Khim.*, **49**, 440 (1975) (Engl. transl., p. 255).
28. (a) C. Eon, C. Pommier, and G. Guiochon, *J. Phys. Chem.*, **75**, 2632 (1971); (b) C. Eon and B. L. Karger, *J. Chromatogr. Sci.*, **10**, 140 (1972).
29. R. C. Castells, *Chromatographia*, **6**, 57 (1973).
30. L. Mathiasson, *J. Chromatogr.*, **114**, 39 (1975).
31. D. E. Martire, *Anal. Chem.*, **46**, 1712 (1974).
32. R. J. Laub and J. H. Purnell, *J. Am. Chem. Soc.*, **98**, 35 (1976).
33. D. E. Martire, *Anal. Chem.*, **48**, 398 (1976).
34. (a) M. van Sway, *Advan. Chromatogr.*, **8**, 363 (1969); (b) A. T. Melenevskii, G. E. El'kin, and G. V. Samsonov, *J. Chromatogr.*, **148**, 299 (1978).
35. (a) M. D. Grieser and D. J. Pietrzyk, *Anal. Chem.*, **45**, 1348 (1973); (b) E. P. Kroeff and D. J. Pietrzyk, *Anal. Chem.*, **50**, 502 (1978).
36. C. Horváth, W. Melander, and I. Molnár, *Anal. Chem.*, **49**, 142 (1977).
37. K. Miyake, K. Okumura, and H. Terada, *Chem. Pharm. Bull.*, **33**, 769 (1985).
38. (a) K. Uekama, F. Hirayama, and T. Irie, *Chem. Lett.*, 661 (1978); (b) K. Uekama, F. Hirayama, S. Nasu, N. Matsuo, and T. Irie, *Chem. Pharm. Bull.*, **26**, 3477 (1978).

39. C. Horváth, W. Melander, and A. Nahum, *J. Chromatogr.*, **186**, 371 (1979).

40. M. Arunyanart and L. J. Love, *Anal. Chem.*, **56**, 1557 (1984).

41. W. R. Melander and C. Horváth, in *High-Performance Liquid Chromatography: Advances and Perspectives*, Vol. 2, C. Horváth, Ed., Academic, New York, 1980, pp. 114–319.

42. C. Horváth, W. Melander, I. Molnár, and P. Molnár, *Anal. Chem.*, **49**, 2295 (1977).

43. B. M. Dunn and I. M. Chaiken, *Proc. Nat. Acad. Sci. USA*, **71**, 2382 (1974).

44. J. Turková, *Affinity Chromatography*, Elsevier, Amsterdam, 1978, Ch. 3.

45. J. O. Mold, T. B. Walker, and L. G. Veasey, *Anal. Chem.*, **35**, 2071 (1963).

46. M. E. Tate, *Biochem. J.*, **195**, 419 (1981).

47. R. A. Alberty and E. L. King, *J. Am. Chem. Soc.*, **73**, 517 (1951).

48. (a) V. Jokl, *J. Chromatogr.*, **14**, 71 (1964); (b) V. Jokl, *J. Chromatogr.*, **71**, 523 (1972); (c) V. Jokl and I. Valaskova, *J. Chromatogr.*, **72**, 373 (1972).

49. M. Sakanoue and M. Nakatani, *Bull. Chem. Soc. Jpn.*, **45**, 3429 (1972).

50. Y. Kiso and T. Hirokawa, *Chem. Lett.*, 745 (1980).

51. J. G. Kirchner, *Thin-Layer Chromatography*, 2nd ed., *Techniques of Chemistry*, Vol. XIV, Wiley-Interscience, New York, 1978, pp. 47–54.

52. A. P. G. Kieboom, N. DeKruyf, and H. Van Bekkum, *J. Chromatogr.*, **95**, 175 (1974).

53. (a) D. W. Armstrong and G. Y. Stine, *J. Am. Chem. Soc.*, **105**, 2962 (1983); (b) D. W. Armstrong and F. Nome, *Anal. Chem.*, **53**, 1662 (1981).

54. L. G. Clary, W. H. Voige, and J. J. Leary, *Anal. Biochem.*, **129**, 228 (1983).

55. J. Schubert, *J. Phys. Chem.*, **56**, 113 (1952).

56. E. Walaas, *Acta Chem. Scand.*, **12**, 528 (1958).

57. W. H. Voige and R. I. Elliott, *J. Chem. Educ.*, **59**, 257 (1982).

12

ADDITIONAL
METHODS

The methods gathered in this chapter tend to be of less general applicability than those treated in Chapters 4 to 11, or they do not fit easily in those chapters. Some of these methods are of importance and have provided many results; others are of limited interest. The treatment here is less thorough than that in the earlier chapters, but the documentation leads to the literature of the subject.

12.1. OPTICAL METHODS

Fluorimetry

Several types of fluorescence measurements can be used in the study of binding. Fluorescence polarization is a measure of rotational relaxation time, so if a small molecule binds to a large molecule its fluorescence polarization will change to reflect the relaxation time of the large molecule; this effect is direct evidence of binding (1, 2). Excited-state lifetime is sensitive to environment, and therefore it changes on binding (3). Most applications of fluorimetry to binding have made use of fluorescence intensity measurements. Several types of experimental arrangements are used:

a. The system components are substrate S and ligand L. The quantum yield of fluorescence of either S or L increases on their interaction to form a complex; therefore the fluorescence intensity of the solution increases. Fluorescence probes of microenvironments, such as 2-toluidinylnaphthalene-6-sulfonate (TNS) or 8-anilino-1-naphthalenesulfonate (ANS), have very low quantum yields in aqueous solution, but have higher yields in nonpolar solvents. On binding to proteins their quantum yields increase markedly, indicating that the binding sites are relatively nonpolar (4).

b. The system components are S and L, one of which is fluorescent. On binding of these species, a decrease in fluorescence occurs. This phenomenon is called quenching, and it takes place when the emission band of the fluorescing species (say S) overlaps the absorption band of the quencher (L); at short intermolecular distances, as in the complex between S and L, a radiationless energy transfer from the excited state of S to the ground state of L takes place, thus reducing the fluorescence intensity. For example, the fluorescence of serum albumin is quenched by thyroxine via this mechanism (5).

c. The system components are fluorescent S, nonfluorescent L, and a quencher Q. S and Q both bind to L, so competitive equilibria are set up, and at constant total substrate concentration the fluorescence intensity decreases as the concentration of quencher is increased. The binding of either L or Q to S may be the phenomenon of interest. Binding to and partitioning into micelles have been studied in this way (6, 7).

At low concentrations the fluorescence intensity is directly proportional to fluorescent solute concentration according to $F = 2.3\,I_0 \phi \epsilon b c$, where F is fluorescence intensity, I_0 is the intensity of the excitation source, ϵ is molar absorptivity at the excitation wavelength, b is path length, c is molar concentration, and ϕ is fluorescence quantum yield. Because of this direct proportionality, the quantitative treatment of fluorescent binding data can be carried out in the same manner as that for absorption spectroscopy (Chapter 4). There are important quantitative differences, however, between absorption and emission spectroscopies, and one of these is that the direct proportionality between F and c is lost when the concentration becomes high enough that more than a few percentage of the incident radiation is absorbed. Correction of the observed fluorescence intensity for the absorption effect can be carried out (8, 9).

Quantitative treatment of fluorescent binding data starts with the premise that the intensity is a sum of contributions. Suppose we consider a system containing S, L, and complex SL. Then we write

$$F = k_S[S] + k_{11}[SL] + k_L[L] \qquad (12.1)$$

where the k_i represent proportionality constants connecting the intensities and concentrations of the species. In the absence of ligand but in the presence of the same total substrate concentration S_t, the intensity is

$$F_0 = k_S S_t \qquad (12.2)$$

The mass balance on substrate is $S_t = [S] + [SL]$, and the binding constant is $K_{11} = [SL]/[S][L]$. Combining these equations gives

$$\frac{F}{F_0} = \frac{1 + (k_{11}/k_S)K_{11}[L]}{1 + K_{11}[L]} + \frac{k_L}{k_S}[L] \qquad (12.3)$$

Equation (12.3) includes several types of behavior. It describes cases a and b earlier, depending on the relative sizes of k_{11} and k_S. When $k_L = 0$ and $k_{11} = 0$, Eq. (12.3) can be written as

$$\frac{F_0}{F} = 1 + K_{11}[L] \qquad (12.4)$$

From the linear plot of F_0/F against $[L]$ the binding constant K_{11} is obtained. McCormick (10) has reported many binding constants for flavine–purine complexes obtained by means of Eq. (12.4).

When $k_L = 0$, Eq. (12.3) becomes

$$\frac{F}{F_0} = \frac{1 + (k_{11}/k_S)K_{11}[L]}{1 + K_{11}[L]} \qquad (12.5)$$

which is equivalent to an equation given by Oster (11), who studied the binding of acriflavine to nucleic acids. Heilweil and van Winkle (12) observed deviations from Eq. (12.5) but made use of initial slope estimates. In these systems the acriflavine is the fluorescing substrate and the DNA is the quenching ligand.

Since Eq. (12.5) is a hyperbola of the form of Eq. (7.24), three linear transformations, corresponding to Eqs. (7.26) to (7.28), can be written. For example, Eq. (12.6) shows the Scatchard-type plotting form.

$$\frac{(F/F_0) - 1}{[L]} = \frac{k_{11}}{k_S} K_{11} - K_{11} \frac{F}{F_0} \qquad (12.6)$$

If $k_L \neq 0$, Eq. (12.3) is applicable. Since the intensive quantities k_L and k_S can be determined by measurements on ligand and substrate separately, correction for emission by ligand can be carried out by evaluating the dependent variable $F/F_0 - (k_L/k_S)[L]$ and treating this as described for Eq. (12.5). Takla et al. (13) describe alternative approaches for this type of system.

Fluorescence data are sometimes treated in the following manner. Again we take the system S, L, and SL for illustration. Equation (12.2) defines F_0, the intensity in the absence of ligand. For the same total substrate concentration S_t, the intensity F_∞ when the substrate is completely converted to the bound form is $F_\infty = k_{11}[SL] + k_L[L]$, or, since there is no free substrate in this condition,

$$F_\infty = k_{11}S_t + k_L[L] \tag{12.7}$$

Thus F_∞ is not a constant unless $k_L = 0$.

Equation (12.1) describes the intensity at intermediate states where both S and SL are present. Combination of Eqs. (12.1), (12.2), (12.7) and the mass balance on substrate gives Eq. (12.8), where $f_{11} = [SL]/S_t$ is the fraction of substrate in the bound form.

$$f_{11} = \frac{F - F_0 - k_L[L]}{F_\infty - F_0 - k_L[L]} \tag{12.8}$$

Several special cases are of practical interest. If $k_S = 0$ and $k_L = 0$, then $f_{11} = F/F_\infty$; this form was used by Weber and Young to calculate the fraction of ANS bound to serum albumin (1). When $k_S = 0$ and $k_L \neq 0$, Eq. (12.8) becomes

$$f_{11} = \frac{F - k_L[L]}{F_\infty - k_L[L]} \tag{12.9}$$

Equation (12.9) was used by Naik et al. (8) in their study of ANS–serum albumin binding. When $k_L = 0$ and $k_S \neq 0$,

$$f_{11} = \frac{F - F_0}{F_\infty - F_0} \tag{12.10}$$

Takla et al. (13) used Eq. (12.10) for the flurbiprofen–serum albumin system.

With f_{11} as a function of ligand concentration, the binding constant for this system is readily obtained by means of Eq. (2.57). Note that F_∞ is required. This quantity is accessible through direct measurement only if the binding constant is very large, for then a sufficiently high concentration of ligand will force the binding equilibrium to favor the complex species almost exclusively. This condition is commonly met in protein binding studies (8, 13, 14).

The derivation of Eq. (12.8) is for a $1:1$ complex, and the mass balance $S_t = [S] + [SL]$ was employed. If multiple complexes of the formula SL_i are formed, the derivation remains valid provided the intensive quantities $k_{11}, k_{12}, \ldots, k_{1i}$ are identical. If, however, complexes of the formula $S_h L$ are present, the mass balance expression must reflect the quantities h. If these are carried also into the definition of F, the same derivation follows. For example, consider the $SL + S_2L$ system, for which $S_t = [S] + [SL] + 2[S_2L]$. We define $F = k_S[S] + k_{11}[SL] + 2k_{21}[S_2L] + k_L[L]$. Then provided $k_{11} = k_{21}$, we obtain

$$F_{\bar{S}} = \frac{F - F_0 - k_L[L]}{F_\infty - F_0 - k_L[L]} \qquad (12.11)$$

where $F_{\bar{S}} = 1 - [S]/S_t$ is the sum of fractions of complexes; see Eq. (2.54). There is experimental support for the condition $k_{11} = k_{21} = k_{31}$ in the ANS–serum albumin system (8).

The stoichiometry of strongly binding systems can be investigated by fluorimetric titration. Drug–protein (8), antibody–hapten (15), and metal ion–oxine sulfonate (16) binding ratios have been estimated. A combined fluorimetry–solubility method yielded pyrene–protein binding ratios (17).

The microionization constants of tetracyclines were determined by fitting the fluorimetric titration data to an equation based on the ionization scheme (18).

Refractometry

There is no general theoretical basis for the form of the relationship between the refractive index n of a solution and the solute concentration (19), so empirical correlations must be sought. Linear correlations of n with molar concentration are often observed, but some authors plot n^2 against concentration (20). Other quantities that are used are the refraction per cubic centimeter, ϕ, defined by Eq. (12.12), the specific refraction $r = \phi/d$, where d is density, and the molar refraction $R = rM$, where M is molecular weight.

$$\phi = \frac{n^2 - 1}{n^2 + 2} \qquad (12.12)$$

The detection of complex formation by refractometry is based on the observation of nonadditivity of the contributions from the interactants (substrate and ligand), as in any other method. However, refractometry differs from many other techniques in that all solution components, including the solvent, make a contribution to the measured refractive index (21). Moreover, except in very dilute solutions, the solvent concentration changes as the solute concentrations are varied, so the amount of solvent is an important factor. Refractometry has been fairly widely used for the estimation of stoichiometric ratios in interacting systems by plotting n, n^2, or functions of ϕ against molar fractions in the method of continuous variations (see Section 2.1, "Determination of Stoichiometry") (20, 21–24). In this method the position of a discontinuity in the plotted function provides an estimate of the stoichiometric ratio. The extent of curvature, that is, the departure from a true discontinuity, is a measure of the stability constant, and refractive index studies of binding constants have made much use of this type of measurement and calculation (25, 26).

Grunwald and co-workers (27, 28) have used an approach that is similar to the experimental and calculational designs of most other methods. Let ϕ and ϕ_0 refer to the refractions of a solution and the solvent, respectively. Then for a solution containing substrate S, ligand L, and 1:1 complex SL, the differential refraction is written as a linear combination, Eq. (12.13), where the W_J are intensive quantities related to the apparent molar refraction and apparent molar volume of solute J.

$$10^3(\phi - \phi_0) = W_S[S] + W_L[L] + W_{11}[SL] \qquad (12.13)$$

Writing $10^3(\phi_L - \phi_0) = W_L[L]$ and defining $\Delta W = W_{11} - W_S - W_L$ allows Eq. (12.13) to be written as Eq. (12.14), where K_{11} is the stability constant.

$$10^3(\phi - \phi_L) = W_S S_t + \frac{\Delta W\, S_t K_{11}[L]}{1 + K_{11}[L]} \qquad (12.14)$$

In differential refractometry the difference in refractive indices of two solutions is measured. To relate $(\phi - \phi_L)$ to $(n - n_L)$, Grunwald and Haley (27) expand Eq. (12.12) in a Taylor's series and retain only the linear term, obtaining

$$10^3(\phi - \phi_L) = \frac{6000 n_L(n - n_L)}{(n_L^2 + 2)^2} \qquad (12.15)$$

Equation (12.14) can then be placed in the usual linear forms, or the stability constant can be obtained by nonlinear regression. The experiment consists of measuring $(n - n_L)$ for a pair of solutions containing the same total ligand concentration, one of them (n) containing substrate at total concentration S_t, the other (n_L) containing no substrate. This is repeated at other L_t values. Colter and Grunwald (28) were thus able to obtain K_{11} for a molecular complex, and from the experimental estimate of ΔW they calculated a value for the change in molar volume on binding.

Polarimetry and Related Methods

Polarimetry is the measurement of the angle through which the plane of linearly polarized light is rotated on passage through a sample; this rotation is a consequence of a difference in refractive indices of left and right circularly polarized light. The dependence of rotation on wavelength is called optical rotatory dispersion (ORD). If the absorption coefficients (molar absorptivities) of the sample differ for left and right circularly polarized light, the linearly polarized incident light will be transformed to elliptically polarized light; this phenomenon is called circular dichroism (CD). Rotation by a substance requires a center of chirality, but it is possible for a nonchiral molecule to exhibit induced ORD and CD on binding to a chiral molecule.

The most extensive applications of ORD and CD to binding have been in the field of binding to macromolecules, with the principal goal being the qualitative features of macromolecular structure and conformation, and complex structure. Chignell (29) and Perrin and Hart (30) have reviewed this subject. In the 1960s several laboratories described methods for the measurement of stability constants by polarimetry. Meier and Higuchi (31) studied camphor-phenol interactions in carbon tetrachloride. Martin (32) and Ramel and Paris (33) investigated complexes of metal ions with amino acids. All these systems required consideration of 1:1

and 1:2 stoichiometry. Examples of CD studies are the cyclodextrin-induced CD effects on achiral drugs studied by Han and Purdie (34).

All quantitative treatments of the ORD and CD effects of binding start with the premise that the observed effect is a sum of contributions of effects from the solution species and that each term is linear in concentration, with its characteristic intrinsic quantity. Thus the treatment is exactly analogous to the description of absorption spectroscopy, and it does not seem necessary to give a further description.

Linear dichroism (LD) is the difference in absorption of light that is polarized parallel to and perpendicular to a fixed axis, by an oriented sample. If the sample substance is oriented by binding to a macromolecule, for example, and if the macromolecule is itself oriented in space, an LD signal may be induced. This can provide information about the nature of binding and the strength of binding (35, 36).

12.2. ELECTRICAL METHODS

Conductometry

The measurement of solution conductance is the basis for a classical determination of dissociation constants of neutral weak acids and of stability constants for ion association reactions. The usual development proceeds as follows, where we take the formation of the 1:1 product ML from M^+ and L^- as an example. The fraction of solute present in the ionic form, α, is defined by Eq. (12.16), where c is the total (formal) concentration.

$$\alpha = \frac{[M^+]}{c} = \frac{[L^-]}{c} \tag{12.16}$$

α is sometimes called the degree of ionization. Combination of Eq. (12.16) with the thermodynamic definition of the stability constant gives (12.17), where $\gamma_\pm$ is the mean ionic activity coefficient, which is estimated, at low ionic strengths, with the Debye–Hückel equation.

$$K'_{11} = \frac{1 - \alpha}{\alpha^2 c \gamma_\pm^2} \tag{12.17}$$

The conductance experiment consists of measuring the equivalent conductance Λ of a dilute solution of total concentration c molar. This measurement is repeated at other c values, for example, by successive dilutions of the first solution. No additional electrolyte is present in these measurements, so the ionic strength is given by $I = \alpha c$.

A first approximation to the thermodynamic constant is achieved by calculating α with Eq. (12.18), where Λ_0 is the limiting conductance, namely, the conductance at infinite dilution.

$$\alpha = \frac{\Lambda}{\Lambda_0} \tag{12.18}$$

Λ_0 is obtained by extrapolation to infinite dilution or by calculation from the known conductances of the individual ions.

Very careful studies have shown that equilibrium constants obtained in this way are not truly thermodynamic constants, because the equivalent conductance itself is dependent on ionic strength, so that Eq. (12.18) does not yield an accurate measure of α. Several refinements have been proposed, and we show here the use of the limiting Onsager equation, applicable at very low ionic strengths. The limiting Onsager equation for ion conductance is

$$\Lambda = \Lambda_0 - SI^{1/2} \tag{12.19}$$

where $S = a\Lambda_0 + b$, and a, b are constants whose values are known from theory (37). For a 1:1 electrolyte, $I = \alpha c$; moreover, only the fraction α of c contributes to the conductance, so Eq. (12.19) becomes

$$\Lambda = \alpha[\Lambda_0 - S(\alpha c)^{1/2}] \tag{12.20}$$

Solving Eq. (12.20) for α gives

$$\alpha = \frac{\Lambda}{\Lambda_0 - S(\alpha c)^{1/2}} \tag{12.21}$$

which may be compared with Eq. (12.18). An iterative solution is used. With Eq. (12.18) a first estimate of α is made. This is inserted into the right-hand side of Eq. (12.21) together with the experimental Λ value, and an improved value of α is calculated. This calculation is repeated until a constant value of α is obtained, and this is used in Eq. (12.17).

Such refinements as this were developed during the first half of this century, in part because both experiment and theory were capable of coping with the subtle effects. As a consequence conductance measurements have provided some of the most accurately known equilibrium constants, both in acid–base chemistry (38, 39) and in ion-pair formation (37, 40). The method has some important limitations, however, one of these being the restriction that a series of measurements cannot be carried out at constant ionic strength. In pK_a determinations the presence of carbon dioxide from the atmosphere limits the method to acids whose strength is greater than that of carbonic acid (39). On the other hand, conductometry is applicable to very dilute solutions.

An alternative method of data analysis is to write Λ as a function of Λ_0, c, and K_{11} and to use nonlinear regression of the data to this function. Beronius (41) has pointed out an interesting complication. In his calculation an extension of the limiting Onsager equation, incorporating an interionic distance R, was used, and Λ_0, K_{11}, and R were treated as adjustable parameters. With many sets of conductance data for 1:1 ion pairs it was found that the least-squares parameter surface contained two minima. These minima gave significantly different parameter estimates, and there was no statistical basis for a choice between them.

The implication to this point is that conductance measurement is applicable when one state (say reactants) consists of ions and the other state is neutral, but the method is more versatile than this. All that is required is that the conductances of the reactant and product states be different. Conductometric titrations in electron donor–acceptor systems have yielded maxima corresponding to 1:1 stoichiometry of the interacting species (42, 43). The interpretation is that the donor and acceptor form a charge-transfer complex, which undergoes partial dissociation into ions. In another application an ion complexes with a neutral ligand to form an ionic complex whose conductivity is different from that of the reactant state. The quantitative description starts with the expression of the solution conductance as the sum of terms contributed by each ion in the solution, each term being the product of the ion concentration and its equivalent ionic conductance. This relationship is combined with mass balance expressions and stability constant definitions, and the experimental data are fit to this by nonlinear regression. In this way stability constants have been determined for complexes of crown ethers with alkali metal ions (44) and of α-cyclodextrin with 4-biphenylcarboxylate (45).

Polarography

The free and bound forms of an electroactive substance usually exhibit different half-wave potentials, the complexed form being more difficult to reduce at the dropping mercury electrode. This phenomenon provides a means for measuring complex stability constants. Most applications have made use of the quantitative treatment of DeFord and Hume (46), which we shall develop for a $1:1 + 1:2$ system, the extension to higher complexes being obvious.

It is a basic premise that the system is electrochemically reversible, and that all chemical equilibria are much faster than the diffusional mass transfer to the electrode surface. The symbol $[SL_i]$ signifies molar concentration in the bulk solution, and $[SL_i]_0$ is the concentration at the electrode surface. The stability constants are defined

$$\beta_{11} = \frac{[SL]}{[S][L]} = \frac{[SL]_0}{[S]_0[L]} \tag{12.22}$$

$$\beta_{12} = \frac{[SL_2]}{[S][L]^2} = \frac{[SL_2]_0}{[S]_0[L]^2} \tag{12.23}$$

where S is the electroactive substrate and L is the ligand. At sufficiently high ligand concentration it is reasonable to set $[L]_0 = [L]$, as in the preceding equations. We do not need to specify the nature of the electrochemical reaction of S, but simply let n be the number of electrons involved per molecule of S.

At any point on the polarographic wave the total current is equal to the sum of the currents produced by all electroactive species. In the usual formulation of polarography this condition gives Eq. (12.24) for the system being considered.

$$i = k_c\{([S] - [S]_0) + ([SL] - [SL]_0) + ([SL_2] - [SL_2]_0)\} \tag{12.24}$$

Each of the differences is a concentration gradient. The proportionality constant k_c is a function of electrode characteristics and of the diffusion coefficient of the electroactive substance, so implicit in Eq. (12.24) is the assumption that the diffusion coefficients of the free and bound forms are all equal. The limiting diffusion current is

$$i_d = k_c([S] + [SL] + [SL_2]) \tag{12.25}$$

Since the polarographic system is reversible, we also may write for the current

$$i = k_a[S(Hg)] \tag{12.26}$$

where S(Hg) signifies the reduced amalgam. Combination of the preceding equations gives

$$\frac{[S(Hg)]}{[S]_0} = \left(\frac{i}{i_d - i}\right)\left(\frac{k_c}{k_a}\right)(1 + \beta_{11}[L] + \beta_{12}[L]^2) \qquad (12.27)$$

Equation (12.28) is the Nernst equation for this system.

$$E_{dme} = E° - \frac{RT}{nF} \ln \frac{[S(Hg)]}{[S]_0} \qquad (12.28)$$

Combining Eqs. (12.27) and (12.28):

$$E_{dme} = E° - \frac{RT}{nF} \ln \left(\frac{i}{i_d - i}\right)\left(\frac{k_c}{k_a}\right)(1 + \beta_{11}[L] + \beta_{12}[L]^2) \qquad (12.29)$$

Equation (12.29) can be written in the conventional polarographic form as

$$E_{dme} = E^*_{1/2} - \frac{RT}{nF} \ln \left(\frac{i}{i_d - i}\right) \qquad (12.30)$$

where

$$E^*_{1/2} = E° - \frac{RT}{nF} \ln \frac{k_c}{k_a} - \frac{RT}{nF} \ln (1 + \beta_{11}[L] + \beta_{12}[L]^2) \qquad (12.31)$$

In the absence of ligand the half-wave potential is given by

$$E_{1/2} = E° - \frac{RT}{nF} \ln \frac{k_c}{k_a} \qquad (12.32)$$

[Some authors distinguish between the k_c values in Eqs. (12.31) and (12.32), but in view of the earlier assumption that all diffusion coefficients are equal there seems little point to this except in those cases where, in the presence of ligand, essentially all substrate is in the form of a single complex.] Defining $\Delta E_{1/2} = E_{1/2} - E^*_{1/2}$ gives

$$\Delta E_{1/2} = \frac{RT}{nF} \ln (1 + \beta_{11}[L] + \beta_{12}[L]^2) \qquad (12.33)$$

Experimentally $\Delta E_{1/2}$, the shift in half-wave potential, is measured as a function of ligand concentration.

A special case is of interest. Suppose a single complex SL_i is formed, and that the inequality $\beta_{1i}[L]^i \gg 1$ is satisfied. Then Eq. (12.33) becomes

$$\Delta E_{1/2} = \frac{RT}{nF} \ln \beta_{1i} + \frac{iRT}{nF} \ln [L] \qquad (12.34)$$

Lingane (47) developed this equation, which permits the stoichiometric coefficient i to be estimated from a plot of $\Delta E_{1/2}$ against $\log[L]$; the stability constant is obtained from the value of $\Delta E_{1/2}$ when $\log[L] = 0$. Lingane (47) shows examples of the use of Eq. (12.34) to study metal ion complexes, and Hojo and Imai use plots of this type to infer the stoichiometry of amine–carboxylic acid interaction products (48). Stackelberg and Freyhold (49) developed a similar method for systems in which both the oxidized and reduced forms undergo complex formation.

Returning to the general case exemplified by Eq. (12.33), we rearrange to

$$\exp\left(\frac{nF \, \Delta E_{1/2}}{RT}\right) = 1 + \beta_{11}[L] + \beta_{12}[L]^2 \qquad (12.35)$$

The quantity on the left is known experimentally as a function of ligand concentration, so the stability constants can be estimated by graphical or regression techniques as described in other chapters. For example, Senise and de Almeida Neves (50) identified five complexes of cadmium with azide ion by means of this method. Frost et al. (51) studied Cu(I) complexes of substituted thioureas. Electron donor–acceptor complexes have also been studied polarographically (52–54). Nonlinear least-squares regression analysis of the polarographic data has been developed by Momoki et al. (55) and by Klatt and Rouseff (56). Bilinski et al. (57) treated data from anodic stripping voltammetry and differential pulse polarography as for conventional polarographic data.

Dielectrometry

From measurements of the dielectric constant ϵ of solutions of interacting components it is possible to estimate the dipole moment of a complex, and this is valuable structural information. Dielectric constant measurements also can provide estimates of binding constants. Gur'yanova et al. (58) have reviewed this subject.

The basic premise of the method is that the dielectric constant of a solution may be written as a linear combination of contributions from each of the solute species plus the solvent, or

$$\epsilon = \epsilon_M + \sum k_J[J] \tag{12.36}$$

where M represents the solvent, and the k_J are intensive quantities. Several workers have made use of Eq. (12.36) to measure stability constants (59–61); here we show the method of Kopecni et al. (61), as developed for the 1:1 complex SL. Equation (12.36) becomes

$$\epsilon = \epsilon_M + k_S[S] + k_L[L] + k_{11}[SL] \tag{12.37}$$

From the mass balance on ligand Eq. (12.38) is obtained.

$$[SL] = L_t \left(\frac{K_{11}[S]}{1 + K_{11}[S]} \right) \tag{12.38}$$

Combining Eqs. (12.37) and (12.38) yields Eq. (12.39), where the mass balance on L has again been used.

$$\epsilon = \epsilon_M + k_S[S] + L_t \left\{ k_L + \frac{(k_{11} - k_L)K_{11}[S]}{1 + K_{11}[S]} \right\} \tag{12.39}$$

The dielectric constant is measured at constant S_t and varying L_t with the condition $S_t \gg L_t$, so [S] may be taken equal to S_t. A plot of ϵ against L_t gives a straight line, whose slope A is given by

$$A = k_L + \frac{(k_{11} - k_L)K_{11}[S]}{1 + K_{11}[S]} \tag{12.40}$$

This experiment is repeated at other S_t values. According to Eq. (12.40), A is a hyperbolic function of [S], and the usual linear plotting forms are applicable. For example, the Scatchard-type plot is made according to

$$\frac{A - k_L}{[S]} = k_{11}K_{11} - AK_{11} \tag{12.41}$$

The required parameter k_L is found by measuring the dielectric constant in dilute solutions of L, where $\epsilon = \epsilon_M + k_L L_t$.

The molar polarization P of a solution is a function of the dielectric constant of the solution, and is itself an additive property that can provide binding constants. The usual method of data treatment is that of Few and Smith (62). Again we treat the 1:1 case, for which we write

$$P = P_S[S] + P_L[L] + P_{11}[SL] + P_M[M] \qquad (12.42)$$

Few and Smith then define an *apparent* molar polarization of S, P_S^*, calculated on the assumption that no complex association has occurred; this gives

$$P = P_S^* S_t + P_L L_t + P_M(M) \qquad (12.43)$$

Using the mass balance on S and the stability constant definition, Eqs. (12.42) and (12.43) are combined to give

$$P_S^* - P_S = \frac{\Delta P \, K_{11}[L]}{1 + K_{11}[L]} \qquad (12.44)$$

where $\Delta P = P_{11} - P_S - P_L$. P_S is the molar polarization of S, an intensive quantity determined in the absence of ligand, whereas P_S^* is determined in the presence of ligand, assuming no complex formation. Equation (12.44) has the familiar hyperbolic form. It has been applied to hydrogen-bonding systems (63) and donor–acceptor complexes (61).

The molar polarization is linearly related to μ^2, the square of the dipole moment, and some authors have calculated the apparent dipole moment of a solute as a function of concentration, then written μ_{app}^2 as a weighted average of the dipole moments of solute species. The self-association of alcohols in nonpolar solvents has been studied in this way by expressing μ_{app}^2 in terms of association constants for various stoichio-metric models (64, 65).

12.3. THERMAL METHODS

In calorimetric studies of complex formation solutions of substrate and ligand are mixed and the resulting change in temperature is measured. By means of a separate determination of the effective heat capacity of the calorimeter, the temperature change is converted to a quantity of heat evolved or absorbed. This represents heat produced by the chemical reaction(s) plus extraneous heat (of dilution, for example). The extraneous contribution can be independently established, so by subtraction the heat generated by the reaction is obtained. We call this Q.

In complex-forming systems this heat is related to the concentrations of the products (complexes) formed by the sum

$$Q = V \sum \Delta H_{hi}^{\circ}[S_h L_i] \tag{12.45}$$

where V is the total volume of the solution in liters, ΔH_{hi}° is the molar heat of reaction (enthalpy change) for the formation of $S_h L_i$, and the summation is from $h = 1$ to m and $i = 1$ to n. It is evident, from Eq. (12.45), that Q, the experimental observable, is determined both by the stability constant (through the concentration of complex) and by the standard enthalpy change.

If a single complex is formed, and if it is formed practically quantitatively when the interactants are mixed (as would happen if the stability constant is very large), then the concentration of complex produced is known, and from measurement of Q the standard enthalpy change can be calculated. This is an application of direct calorimetry. But to estimate binding constants it is necessary that the reaction not go to completion, the extent of reaction then being a function of the observed heat relative to the heat produced if reaction had been quantitative. Let us take a 1:1 binding system in illustration; then Eq. (12.45) becomes $Q = V \Delta H_{11}^{\circ}[SL]$, or $[SL] = Q/V \Delta H_{11}^{\circ}$. This relationship forms the basis of methods for the calorimetric determination of stability constants.

In the usual procedure a solution of substate S is titrated with ligand L, and the temperature of the reaction mixture is recorded as a function of titrant volume. This method (which is also analytically useful) is called thermometric titrimetry, calorimetric titration, titration calorimetry, enthalpy titration, and (somewhat confusingly) entropy titration. We develop the quantitative description for a 1:1 system, for which $Q = V \Delta H_{11}^{\circ}[SL]$. This is combined with the mass balance on substrate and the stability constant definition to give the familiar form

$$Q = \frac{V S_t \Delta H_{11}^{\circ} K_{11}[L]}{1 + K_{11}[L]} \tag{12.46}$$

Equation (12.46) can be linearized in the usual ways. Note that Q is given as a function of free ligand concentration, which is not known. The mass balance on ligand is therefore combined with the stability constant expression to give total ligand concentration L_t as a function of $[L]$ [see Eq. (4.9)], and these two equations can be solved iteratively to yield estimates of both ΔH°_{11} and K_{11}. From this information the standard entropy change can be calculated. The generalization to multiple complexes is obvious.

Many authors have developed calorimetric methods for the simultaneous estimation of the stability constant and the heat of reaction (66–71). Christensen and Izatt and their co-workers (72, 73) have been very active in this field. Many of the published equations appear to be much more complicated than is Eq. (12.46); this is because they incorporate L_t rather than $[L]$.

There has been considerable discussion of the range of applicability of this method. The Christensen–Izatt group state that the stability constant should be between 10 and $10^4 \, M^{-1}$ for optimum results, and of course ΔH° must be large enough to provide a usable response (73). Cabani and Giani (74) carried out an analysis of errors treatment, reaching pessimistic conclusions, but Christensen et al. (75) responded more optimistically. Eatough (76) extended the range of accessible stability constants by setting up competitive equilibria and measuring calorimetrically the equilibrium constant for the competition; by knowing the stability constant for one of the two competitive reactions, the second could be calculated.

Many applications and variations have been reported. Izatt et al. (77) studied the complexing of Ag^+ and pyridine; this is a $1:1 + 1:2$ system. Marenchic and Sturtevant (78) investigated the association of purine bases, assuming a model of indefinite polymerization with all steps having the same binding constant. Anderson et al. (79) treated calorimetric data on the self-association of alcohols in isooctane with several stoichiometric models. Complexes of cyclodextrins have been studied with the assumption of $1:1$ stoichiometry (80, 81). Small molecule–macromolecule (82) and micelle formation equilibria (83) have been studied in this way. A nonlinear least-squares program for calorimetric data has been published (84).

12.4. MISCELLANEOUS METHODS

The colligative properties depend on the numbers of solute particles but not on their identities. The observed effect is a sum of effects contributed by all solute species, with each species possessing the same intrinsic proportionality constant. These properties therefore provide, in principle, methods of simplicity and generality. In practice they are not widely useful.

The measurement of freezing-point depression (cryoscopy) is limited to the temperature of the solvent freezing point. Rossotti and Rossotti (85) have discussed the data analysis for stepwise equilibria. Tobias (86) has made a thermodynamic analysis of the method, concluding that its results are not very reliable. The boiling-point elevation method (ebulliometry) has provided a few stability constants (87).

Vapor pressure osmometry has been applied to the study of self-association, for example, the aggregation of n-alkanols in n-octane (88) and of nicotinamide and isonicotinic acid hydrazide in water (89). Wachter and Simon (90) have described a graphical method of data treatment based on families of curves for different models of self-association. Ogston and Winzor (91) analyzed the assumption of cancellation of activity coefficients in determining equilibrium constants by osmometry, and found that the assumption is usually a good one.

The vapor pressure of a solute over a solution can be a useful property. To show the principle of the method we suppose that the free substrate S is volatile, that bound forms are not volatile, and that $1:1$ (SL) and $1:2$ (SL$_2$) complexes are formed. We also suppose that the ligand L is nonvolatile, and we neglect the vapor pressure of the solvent. Then the vapor pressure over the solution is given by Henry's law, $p = k[S]$, where k is the Henry's law constant. Combination with the mass balance on substrate and the definitions of the stability constants yields

$$p = \frac{kS_t}{1 + K_{11}[L] + K_{11}K_{12}[L]^2} \tag{12.47}$$

where [L] is the free ligand concentration and S_t is the total substrate concentration in the solution phase. In practice, the solvent contributes to the total vapor pressure, and the Henry's law constant may vary over the experimental concentration range. Tucker and Christian (92) studied the dimerization of benzene in aqueous solution by this method. They corrected the benzene vapor pressure to fugacity f, determining f as a function of mole fraction of benzene in the aqueous phase, x_B. Extrapolation of f/x_B against x_B to $x_B = 0$ gave the infinite dilution Henry's law constant, k_0. Since all of the benzene exists as the monomer at infinite dilution, the monomer concentration x_M can be calculated at any concentration with the expression $x_M = f/k_0$; thus the nonideality is ascribed to self-association. A monomer–dimer equilibrium accounted for the data. This gave, for the mass balance, $x_B = x_M + 2x_D = x_M + 2K_2 x_M^2$, where K_2 is the dimerization constant. The data set of f as a function of x_B was treated by least-squares analysis to yield K_2. A similar method gave stability constants for complexes of benzene with cyclodextrins in aqueous solution (93).

In some methods closely related to the vapor pressure method, the concentration of a volatile solute in the solution phase is established by measurement of its vapor-phase concentration by infrared spectroscopy (94), ultraviolet spectroscopy (95), or mass spectrometry (96). Consider the IR study by Christian and Stevens (94) of the association of trifluoroacetic acid in both the vapor and solution phases. Let $A_v = \epsilon_v b c_v$ be the absorbance due to monomer in the vapor phase, where b is path length and c_v is molar concentration in the vapor phase. Henry's law is assumed for the monomer, $p = k[S]$, where $[S]$ is the molar concentration of monomer in solution. These expressions are combined with the gas law $p = c_v RT$ to give

$$[S] = \frac{A_v RT}{\epsilon_v bk} \tag{12.48}$$

For the dimerization model of self-association in the solution phase the mass balance on solute is $S_t = [S] + 2[S_2]$ or $S_t = [S] + 2K_2[S]$, where K_2 is the dimerization constant. Combining this with Eq. (12.48) gives Eq. (12.49), showing that K_2 can be evaluated from a plot of S_t/A_v against A_v.

$$\frac{S_t}{A_v} = 2K_2 \left(\frac{RT}{\epsilon_v bk} \right)^2 A_v + \left(\frac{RT}{\epsilon_v bk} \right) \tag{12.49}$$

Note that S_t is the total concentration in the solution phase, and this is less than the total initial concentration, which must be corrected for the amount in the vapor phase at equilibrium.

When a solute associates, either through self-association or complex formation with a different species, the product is larger and of different shape, so that its diffusion coefficient is different. The measurement of self-diffusion can therefore provide information about the association. The premise is that the observed self-diffusion coefficient is a weighted average of the self-diffusion coefficients of the observed species, namely, $D_{obs} = \Sigma\, f_i D_i$, where f_i is the fraction in the ith state. This is combined with stability constant definitions based on the assumed stoichiometric model to generate a relationship between D_{obs} and solute concentration; the model is tested, and the constants evaluated, by fitting the data to this relationship. Both self-association and complex formation can be studied (97, 98).

The measurement of binding constants by means of NMR relaxation studies is considered in Section 5.2. Other relaxation techniques have also been used. The temperature-jump relaxation method can measure very fast reaction rates, and Pogonin and Chibisov (99) have reviewed its application to the study of the kinetics of metal ion–ligand binding. Such studies yield the rate constants of the association and dissociation steps, and from these the equilibrium constant can be calculated, though usually easier methods can be used. Malcolm determined the dissociation constant (not K_m) of an enzyme complex by relaxation kinetics (100). Ultrasonic relaxation provides another technique; this has been applied to self-association (101, 102). Gormally et al. (102) combined light scattering and ultrasonic relaxation measurements to study the aggregation of a planar drug molecule.

Other techniques of limited applicability are based on measurements of surface tension (103), viscosity (104), resonance Raman spectra (105), and positron annihilation (106).

REFERENCES

1. G. Weber and L. B. Young, *J. Biol. Chem.*, **239**, 1415 (1964).
2. R. J. C. Levine, D. N. Teller, and H. C. B. Denber, *Mol. Pharmacol.*, **4**, 435 (1968).
3. M. Baumann, P. Becker, and B. A. Bilal, *J. Solution Chem.*, **14**, 67 (1985).
4. G. M. Edelman and W. O. McClure, *Acc. Chem. Res.*, **1**, 65 (1968).
5. R. F. Steiner, J. Roth, and J. Robbins, *J. Biol. Chem.*, **241**, 560 (1966).
6. M. V. Encinas and E. A. Lissi, *Chem. Phys. Lett.*, **91**, 55 (1982).
7. E. Blatt, R. C. Chatelier, and W. H. Sawyer, *Chem. Phys. Lett.*, **108**, 397 (1984).
8. D. V. Naik, W. L. Paul, R. M. Threatte, and S. G. Schulman, *Anal. Chem.*, **47**, 267 (1975).
9. K. J. Wiechelman, *Am. Lab.*, **18**, No. 2, 49 (1986).

10. D. B. McCormick, in *Molecular Associations in Biology*, B. Pullman, Ed., Academic, New York, 1968, pp. 377–392.

11. G. Oster, *Trans. Faraday Soc.*, **47**, 660 (1951).

12. H. G. Heilweil and Q. van Winkle, *J. Phys. Chem.*, **59**, 939 (1955).

13. P. G. Takla, S. G. Schulman, and J. H. Perrin, *J. Pharm. Biomed. Anal.*, **3**, 41 (1985).

14. C. F. Chignell, *Mol. Pharmacol.*, **5**, 244 (1969).

15. S. F. Velick, C. W. Parker, and H. N. Eisen, *Proc. Nat. Acad. Sci. USA*, **46**, 1470 (1960).

16. (a) J. A. Bishop, *Anal. Chim. Acta*, **53**, 456 (1971); (b) J. A. Bishop, *Anal. Chim. Acta*, **63**, 305 (1973).

17. L. B. McGown and D. I. Ueda, *Anal. Chim. Acta*, **142**, 313 (1982).

18. (a) B. M. Ahmed and R. D. Jee, *Anal. Chim. Acta*, **156**, 263 (1984); (b) V. K. Bhatt and R. D. Jee, *Anal. Chim. Acta*, **167**, 233 (1985).

19. N. Bauer and K. Fajans, *Physical Methods of Organic Chemistry*, 2nd ed., A. Weissberger, Ed., Vol. I, Part II, Ch. XX, 1949.

20. F. M. Arshid, C. H. Giles, E. C. McLure, A. Oglivie, and T. J. Rose, *J. Chem. Soc.*, 67 (1955).

21. L. Barcza, *J. Phys. Chem.*, **80**, 821 (1976).

22. (a) F. M. Arshid, C. H. Giles, S. K. Jain, and A. S. A. Hassan, *J. Chem. Soc.*, 72 (1956); (b) F. M. Arshid, C. H. Giles, and S. K. Jain, *J. Chem. Soc.*, 559 (1956).

23. M. Donbrow and Z. A. Jan, *J. Chem. Soc.*, 3845 (1963).

24. (a) R. Sahai, M. Chauhan, and V. Singh, *Monatsh. Chem.*, **112**, 935 (1981); (b) V. Singh and R. Sahai, *Monatsh. Chem.*, **113**, 557 (1982).

25. Z. Yoshida and E. Osawa, *Bull. Chem. Soc. Jpn.*, **38**, 140 (1965).

26. (a) R. A. Singh and S. N. Bhat, *Indian J. Chem.*, **15A**, 1106 (1977); (b) R. Sahai, V. Singh, and R. Verma, *Bull. Chem. Soc. Jpn.*, **53**, 2995 (1980); (c) R. A. Singh and S. N. Bhat, *Bull. Chem. Soc. Jpn.*, **55**, 1624 (1982).

27. E. Grunwald and J. F. Haley, *J. Phys. Chem.*, **72**, 1944 (1968).

28. A. K. Colter and E. Grunwald, *J. Phys. Chem.*, **74**, 3637 (1970).

29. C. F. Chignell, *Concepts in Biochemical Pharmacology*, Part 1, B. B. Brodie and J. R. Gillette, Eds., Springer-Verlag, Berlin, 1971, pp. 197–203.

30. J. H. Perrin and P. A. Hart, *J. Pharm. Sci.*, **59**, 431 (1970).

31. J. Meier and T. Higuchi, *J. Pharm. Sci.*, **54**, 1183 (1965).

32. R. P. Martin, *Bull. Soc. Chim. Fr.*, 1354 (1967).

33. M. M. Ramel and M. R. Paris, *Bull. Soc. Chim. Fr.*, 1359 (1967).

34. (a) S. M. Han and N. Purdie, *Anal. Chem.*, **56**, 2822 (1984); (b) S. M. Han and N. Purdie, *Anal. Chem.*, **56**, 2825 (1984).

35. A. Yogev, L. Margulies, and Y. Mazur, *J. Am. Chem. Soc.*, **92**, 6059 (1970).

36. B. Norden and F. Tjerneld, *Biophys. Chem.*, **4**, 191 (1976).
37. C. W. Davies, *Ion Association*, Butterworths, London, 1962, Ch. 2.
38. G. Kortüm, W. Vogel, and K. Andrussow, *Dissociation Constants of Organic Acids in Aqueous Solution*, Butterworths, London, 1961, pp. 196–204.
39. A. Albert and E. P. Serjeant, *The Determination of Ionization Constants*, 3rd ed., Chapman and Hall, London, 1984, Ch. 6.
40. G. H. Nancollas, *Interactions in Electrolyte Solutions*, Elsevier, Amsterdam, 1966, pp. 26–32.
41. P. Beronius, *Acta Chem. Scand.*, **29A**, 289 (1975).
42. (a) F. Gutmann and H. Keyzer, *Electrochim. Acta*, **11**, 1163 (1966); (b) F. Gutmann and H. Keyzer, *Electrochim. Acta*, **12**, 1255 (1967).
43. M. V. Murti and A. Qayum, *Indian J. Chem.*, **12**, 1308 (1974).
44. N. Matsura, K. Umemoto, T. Takeda, and A. Sasaki, *Bull. Chem. Soc. Jpn.*, **49**, 1246 (1976).
45. R. I. Gelb, L. M. Schwartz, C. T. Murray, and D. A. Laufer, *J. Am. Chem. Soc.*, **100**, 3553 (1978).
46. D. D. DeFord and D. N. Hume, *J. Am. Chem. Soc.*, **73**, 5321 (1951).
47. (a) J. J. Lingane, *Chem. Rev.*, **29**, 1 (1941); (b) I. M. Kolthoff and J. J. Lingane, *Polarography*, 2nd ed., Interscience, New York, 1952, Ch. XII.
48. M. Hojo and Y. Imai, *Anal. Chem.*, **57**, 509 (1985).
49. M. von Stackelberg and H. von Freyhold, *Z. Elektrochem.*, **46**, 120 (1940).
50. P. Senise and E. F. de Almeida Neves, *J. Am. Chem. Soc.*, **83**, 4146 (1961).
51. J. G. Frost, M. B. Lawson, and W. G. McPherson, *Inorg. Chem.*, **15**, 940 (1976).
52. M. E. Peover, *Trans. Faraday Soc.*, **60**, 417 (1964).
53. R. D. Holm, W. R. Carper, and J. A. Blancher, *J. Phys. Chem.*, **71**, 3960 (1967).
54. L. Ramaley and S. Gaul, *Can. J. Chem.*, **56**, 2381 (1978).
55. K. Momoki, H. Sato, and H. Ogewa, *Anal. Chem.*, **39**, 1072 (1967).
56. L. N. Klatt and R. L. Rouseff, *Anal. Chem.*, **42**, 1234 (1970).
57. H. Bilinski, R. Huston, and W. Stumm, *Anal. Chim. Acta*, **84**, 157 (1976).
58. E. N. Gur'yanova, I. P. Gol'dshtein, and I. P. Romm, *Donor–Acceptor Bond*, Wiley (Halsted Press), New York, 1975, pp. 59–65.
59. A. A. Maryott, *J. Res. Nat. Bur. Stand. US*, **41**, 7 (1948).
60. I. P. Gol'dshtein, E. S. Shcherbakova, E. N. Gu'yanova, and L. A. Muzychenko, *Teor. Eksp. Khim.*, **6**, 634 (1970) (Engl. transl., p. 518).
61. M. M. Kopecni, *J. Phys. Chem.*, **86**, 1008 (1982).
62. A. V. Few and J. W. Smith, *J. Chem. Soc.*, 2781 (1949).
63. (a) B. Cleverdon, G. B. Collins, and J. W. Smith, *J. Chem. Soc.*, 4499 (1956); (b) A. H. Boud and J. W. Smith, *J. Chem. Soc.*, 4507 (1956).
64. P. Bordewijk, M. Kunst, and A. Rip, *J. Phys. Chem.*, **77**, 548 (1973).

65. C. Campbell, G. Brink, and L. Glasser, *J. Phys. Chem.*, **79**, 660 (1975).

66. F. J. Cioffi and S. T. Zenchelsky, *J. Phys. Chem.*, **67**, 357 (1963).

67. (a) T. F. Bolles and R. S. Drago, *J. Am. Chem. Soc.*, **87**, 5015 (1965); (b) T. F. Bolles and R. S. Drago, *J. Am. Chem. Soc.*, **88**, 3921 (1966).

68. A. Kolbe, *Z. Phys. Chem.*, **N.F.58**, 75 (1968).

69. D. Neerinck, A. Van-Audenhauge, L. Lamberts, and P. Huyskens, *Nature (London)*, **218**, 461 (1968).

70. I. G. Orlov, V. M. Ketskalo, A. S. Vavilkin, A. A. Vichutinskii, and M. I. Cherkashin, *Dokl. Acad. Nauk. S.S.S.R.*, **218**, 143 (1974) (Engl. transl., p. 849).

71. J. Skerjanc, A. Regent, and B. Plesnicar, *Chem. Commun.*, 1007 (1980).

72. (a) L. D. Hansen, J. J. Christensen, and R. M. Izatt, *Chem. Commun.*, 36 (1965); (b) J. J. Christensen, R. M. Izatt, L. D. Hansen, and J. A. Partridge, *J. Phys. Chem.*, **70**, 2003 (1966).

73. (a) J. J. Christensen, J. Ruckman, D. J. Eatough, and R. M. Izatt, *Thermochim. Acta*, **3**, 203 (1972); (b) D. J. Eatough, J. J. Christensen, and R. M. Izatt, *Thermochim. Acta*, **3**, 219 (1972); (c) D. J. Eatough, R. M. Izatt, and J. J. Christensen, *Thermochim. Acta*, **3**, 233 (1972).

74. S. Cabani and P. Giani, *J. Chem. Soc.*, A, 547 (1968).

75. J. J. Christensen, J. H. Rytting, and R. M. Izatt, *J. Chem. Soc.*, A, 861 (1969).

76. D. J. Eatough, *Anal. Chem.*, **42**, 635 (1970).

77. R. M. Izatt, D. Eatough, R. L. Snow, and J. J. Christensen, *J. Phys. Chem.*, **72**, 1208 (1968).

78. M. G. Marenchic and J. M. Sturtevant, *J. Phys. Chem.*, **77**, 544 (1973).

79. B. D. Anderson, J. H. Rytting, S. Lindenbaum, and T. Higuchi, *J. Phys. Chem.*, **79**, 2340 (1975).

80. E. A. Lewis and L. D. Hansen, *J. Chem. Soc. Perkin Trans.* 2, 2081 (1973).

81. (a) G. E. Hardee, M. Otagiri, and J. H. Perrin, *Acta Pharm. Suec.*, **15**, 188 (1978); (b) H. Ueda and J. H. Perrin, *J. Pharm. Biomed. Anal.*, **4**, 107 (1986).

82. D. W. Bolen, M. Flogel, and R. Biltonen, *Biochemistry*, **10**, 4136 (1971).

83. J. Jagur-Grodzinski, R. Frame, and R. M. Izatt, *J. Colloid Interface Sci.*, **105**, 73 (1985).

84. V. Cerda, J. M. Estela, R. Jara, and J. Lumbiarres, *Thermochim. Acta*, **87**, 13 (1985).

85. F. J. C. Rossotti and H. Rossotti, *J. Phys. Chem.*, **63**, 1041 (1959).

86. S. R. Tobias, *J. Inorg. Nucl. Chem.*, **19**, 348 (1961).

87. W. A. DeOliveira and T. S. M. Omoto, *Analyst*, **109**, 1617 (1984).

88. R. Aveyard, B. J. Briscoe, and J. Chapman, *J. Chem. Soc., Faraday Trans.*, I, **69**, 1772 (1973).

89. F. Kopecky, M. Vojtekova, and M. Bednarova-Hyttnerova, *Coll. Czech. Chem. Commun.*, **43**, 37 (1978).

90. A. H. Wachter and W. Simon, *Helv. Chim. Acta*, **52**, 371 (1969).

91. A. G. Ogston and D. J. Winzor, *J. Phys. Chem.*, **79**, 2496 (1975).

92. (a) E. E. Tucker and S. D. Christian, *J. Phys. Chem.*, **83**, 426 (1979); (b) E. E. Tucker, E. H. Lane, and S. D. Christian, *J. Solution Chem.*, **10**, 1 (1981).

93. E. E. Tucker and S. D. Christian, *J. Am. Chem. Soc.*, **106**, 1942 (1984).

94. S. D. Christian and T. L. Stevens, *J. Phys. Chem.*, **76**, 2039 (1972).

95. L.-N. Lin, S. D. Christian, and E. E. Tucker, *J. Phys. Chem.*, **82**, 1897 (1978).

96. A. Polaczek and E. Brandowska, *J. Inorg. Nucl. Chem.*, **27**, 1649 (1965).

97. B. Levay, *J. Phys. Chem.*, **77**, 2118 (1973).

98. R. Rymden and P. Stilbs, *Biophys. Chem.*, **21**, 145 (1985).

99. V. I. Pogonin and A. K. Chibisov, *Usp. Khim.*, **53**, 1601 (1984) (Engl. transl., p. 929).

100. A. D. Malcolm, *Anal. Biochem.*, **77**, 529 (1977).

101. P. Hemmes, A. A. Mayevski, V. A. Bucklin, and A. P. Sarvazyan, *J. Phys. Chem.*, **84**, 699 (1980).

102. J. Gormally, N. Natarajan, E. Wyn-Jones, D. Attwood, J. Gibson, and D. G. Hall, *J. Chem. Soc., Faraday Trans.*, *II*, **80**, 243 (1984).

103. R. A. Singh, S. P. Mishra, and S. N. Bhat, *Proc. Indian Acad. Sci.*, **89**, 139 (1980).

104. H. M. N. H. Irving and J. S. Smith, *J. Inorg. Nucl. Chem.*, **31**, 3163 (1969).

105. K. H. Michaelian, K. E. Rieckhoff, and E.-M. Voight, *J. Phys. Chem.*, **81**, 1489 (1977).

106. E. S. Hall and H. J. Ache, *J. Phys. Chem.*, **83**, 1805 (1979).

13

THE METHOD
OF CHOICE

13.1. COMPARISON OF METHODS

Kinds of Reproducibility

Since we can never know the "true" value of a physical quantity such as a binding constant, we attempt to eliminate all significant systematic errors in its determination and then express its reliability in terms of a measure of its reproducibility, usually its standard deviation or a related quantity such as a confidence interval. When comparing binding constants reported by different workers, or measured by different methods, it is not uncommon to find that they are (statistically) significantly different. In such a case there are two extreme possibilities: either the reported standard deviations are unrealistic, being too small, or unrecognized systematic errors exist. We must examine in a more detailed way the nature of such comparisons.

Table 13.1 shows a matrix of possible types of reproducibility, the variations being in the number of runs (individual data sets), the number of workers, and the number of methods. (Implicit in the table is a further variable, since the workers may be in the same or in different laboratories.) In general, we might expect standard deviations to increase as we proceed from left to right and from top to bottom in the tab e, for the number of variables increases in these directions. The possibilit es for systematic errors also increase. The most commonly quoted standard

Table 13.1. Types of Reproducibility

No. of Runs	Worker	Methods Used	
		Single	Multiple
One	Same	Within run, single method, same worker	Within run, among methods, same worker
	Different	Within run, single method, different workers	Within run, among methods, different workers
Several	Same	Among runs, single method, same worker	Among runs, among methods, same worker
	Different	Among runs, single method, different workers	Among runs, among methods, different workers

deviation is probably that described by the first entry in Table 13.1, namely, a standard deviation based on a single run. This is evaluated as shown in Section 3.2.

In Table 13.2 are listed experimental results for the system 1,4-dimethylterephthalate (the substrate) and α-cyclodextrin (the ligand), obtained by three workers using the same method in the same laboratory. This is a 1:1 + 1:2 system; it was studied by the solubility method as discussed in Chapter 8. These results are typical of such a comparison. The standard deviations in the table are within-run estimates. Although an analysis of variance can be helpful in identifying sources of variability, it is useful simply to note the good precision in K_{11} among workers (mean

Table 13.2. Binding Constants for the 1,4-Dimethylterephthalate: α-Cyclodextrin System[a,b]

Worker	K_{11}/M^{-1}	K_{12}/M^{-1}	$10^4 s_0/M$
A	480 (16)	86 (4)	1.74 (0.02)
A	471 (17)	97 (4)	
A	476 (25)	89 (6)	
B	464 (8)	109 (5)	1.69 (0.02)
C	490 (10)	114 (2)	1.56 (0.02)
C	484 (10)	117 (2)	
C	470 (11)	122 (3)	

[a]At 25.0°C in 0.10 M NaCl.
[b]Standard deviations in parentheses.

$476\ M^{-1}$, standard deviation $8\ M^{-1}$) and the considerable spread in K_{12} values (mean $105\ M^{-1}$, standard deviation $13\ M^{-1}$). As shown in Chapter 8, the solubility method actually generates the quantities $K_{11}s_0$ and $K_{11}K_{12}s_0$, where s_0 is the molar solubility; K_{11} and K_{12} are derived from these quantities, and variations in binding constants may arise in part from differences in solubility estimates. These particular data will not be analyzed further here; the point to be made is that in interpreting results and comparing quantities it is essential to be clear on how their uncertainties were obtained.

Comparison of Results

There are several reasons that an experimentalist might choose to use more than one method for the study of a binding system, and associated with these reasons are interpretations of why the results may fail to agree.

1. *New method development.* When a new experimental technique is developed or a modification is introduced, it is appropriate to compare its results with those produced by an independent established "reference" method. For example, MacInnes and Shedlovsky (1) made a conductance study of the dissociation of acetic acid, finding $pK_a = 4.756$, which was in excellent agreement with the potentiometric result $pK_a = 4.758$ reported by Harned and Owen (2). If the results should not agree, a systematic error may be present (presumably, though not certainly, in the new method). Extension of an established method to a previously unstudied class of equilibria should be validated in the same way. Thus Monk (3) has compared the dissociation constants of several cobalt dicarboxylates as determined by six experimental methods. Some agreements and some discrepancies were observed, and it was concluded that some of the disagreements arose from experimental errors, whereas others were the result of different theoretical treatments (such as estimates of activity coefficients). Many more examples are included in papers cited in Chapters 4 to 12.

2. *Confirmation of stoichiometric model.* If several experimental methods applied to the same system yield significantly different stability constants, it is possible that the wrong stoichiometric model was assumed. If the correct model is used in treating the data, all experimental methods should yield the same binding constants (subject to conditions treated elsewhere in this section). Thus the application of several independent

methods can be a powerful means for establishing the stoichiometric ratios. Of course, any additional information, of types discussed in earlier chapters, should be used in combination with the evidence provided by multiple methods.

The usual circumstance is that the system is first treated as a 1:1 binding system. Chapter 3 considers some ways to test this hypothesis. In the present context we evaluate K_{11} by several methods; if these differ significantly, it is probable that higher-order complexes are present. Not all methods are equally discriminating in this test (4). As examples of the approach, first consider the substrate methyl cinnamate complexed with the ligand 8-chlorotheophylline anion, for which was found $K_{11} = 26 \ M^{-1}$ (solubility), 25 M^{-1} (spectrophotometric), 22 M^{-1} (kinetic); these results support the assumption of simple 1:1 binding (5). In contrast, the corresponding K_{11} values with the ligand caffeine were 36 M^{-1} (solubility), 18 (spectrophotometric), 18 (kinetic). Even though the spectral and kinetic results agree, the binding cannot be purely 1:1. (It is pertinent that the spectrophotometric and kinetic methods are mathematically equivalent.) As is shown in Chapter 9, caffeine undergoes self-aggregation, and this effect must be taken into account. It was concluded that at low caffeine concentrations, the system is best described as including the complex species SL, SL_2, and L_2. Foster and co-workers (6, 7) have made effective use of comparative measurements of stability constants to deduce stoichiometric relationships.

3. *Environmental effects.* This is not so much a reason for making comparative measurements as a possible explanation of discrepancies in the results obtained by different methods. Insofar as is possible we choose to make comparative measurements by several methods under the same experimental conditions, for example, of solvent, temperature, ionic strength, pH, and solute concentrations. However, this desired constancy of the system composition is seldom attainable owing to the different requirements of the several experimental methods. These differences in environmental conditions may lead to measurably different binding constants. Such effects can be difficult to identify. The type of medium variation that may be encountered can be illustrated by the complexing of a weak acid substrate with a neutral ligand in aqueous solution. Suppose this system is studied by the solubility, potentiometric, and spectrophotometric methods. Of course, the ionic strength will be held constant. Since the binding of the neutral weak acid is of interest, in the solubility and spectrophotometric methods the pH must be held several

units below the pK_a of the acid, whereas in the potentiometric method the pH is near the pK_a. Furthermore, in the spectrophotometric method the substrate concentration is usually very low, in the potentiometric method it is moderate to high, and in the solubility method it is at the solubility limit. These variations may give rise to activity coefficient differences. However, in view of typical uncertainties in binding constant estimates, such environmental effects are not often the cause of major discrepancies between experimental methods. Specific effects may be described in terms of competitive binding rather than as medium effects.

4. *Sampling of the potential energy well.* There is a considerable literature dealing with the possibility that some experimental methods may be more "selective" than others, may distinguish between "chemical" and "physical" binding, or may (or may not) be able to detect random contact pairs. Some of these papers have been cited earlier (see Sections 2.6, "Random Association," and 11.2, "Complications.") In Section 2.6 we found that random association is expected to occur with an apparent association constant of the order $1\,M^{-1}$ or smaller, depending on the solvent and the details of solution structure. Such small association constants complicate experimental study and interpretation because of the likelihood of significant activity coefficient effects. In strongly complexed systems having large stability constants the activity coefficient variation is small, and the fractional contribution of the random association effect is also small. In these systems all experimental methods should therefore yield essentially the same results. Nevertheless it seems conceivable that different experimental probes may sample different populations of a complex "species," but the unambiguous demonstration of this effect may be difficult. Concerning this matter we note that arguments based on mass balance relationships cannot lead to a decision whether or not the random association process contributes to an observed binding constant; this conclusion is developed in Section 2.6, "Random Association."

It is instructive to examine binding constant estimates made by several methods in different laboratories, for such results give a sense of the variation that is typical in these data. Stability constants for the substrates p-nitrophenol (K_{11a}) and p-nitrophenolate (K_{11b}) with the ligand α-cyclodextrin in aqueous solution at 25°C are gathered in Table 13.3. Some workers have not reported uncertainty or precision estimates; others give within-run standard deviations, or among-run estimates, or ranges of unspecified meaning, so the table does not include these quantities. The ionic strengths range from zero to $0.5\,M$, and some of the solutions

Table 13.3. Reported Binding Constants for _p_-Nitrophenol (K_{11a}) and _p_-Nitrophenolate (K_{11b}) with _α_-Cyclodextrin at 25°C

Method	K_{11a}/M^{-1}	K_{11b}/M^{-1}	References
Spectrophotometry	250	2290	8
Titration calorimetry	126	—	9
Potentiometry	200	2200	10
Spectrophotometry	—	2500	11
Nuclear magnetic resonance	—	2700	11
Optical rotation	—	1590	11
Polarography	—	2439	12
Spectrophotometry	—	1890	13
Spectrophotometry	250	2720	14
Potentiometry	211	2143	15
Potentiometry	245	2408	16
Spectrophotometry	249	—	16
Competitive spectrophotometry	249	—	16
Gel filtration	210	2270	17
Gel filtration	177	3550	18

contained buffer substances. These differences in solution composition probably account for some of the variability.

All authors agree that the systems in Table 13.3 exhibit purely 1:1 stoichiometric binding. Some of the experimental methods are probably more suitable than others, but, with the exception of a few possible outliers, the results are fairly concordant. The means of all the estimates in Table 13.3 are $K_{11a} = 217\ M^{-1}$ (standard deviation 39, standard deviation of the mean 12) and $K_{11b} = 2392\ M^{-1}$ (standard deviation 463, standard deviation of the mean 134). Applying the _t_-distribution to these results, we find for the 95% confidence limits $K_{11a} = 217 \pm 27\ M^{-1}$ and $K_{11b} = 2392 \pm 295\ M^{-1}$. These may not be the best estimates of these parameters, because rejection of some of the outliers could be justified. However, this collection of results illustrates the kind of agreement that may be expected. From the point of view of some important chemical questions, no greater precision is needed. For example, the standard free energies of binding are reasonably well estimated, and the critical comparison between the stabilities of the complexes formed from the conjugate acid and base forms is conclusively demonstrated. Should a definitive estimate of binding constant be needed, further experimental work, making use of one or two of the most appropriate techniques for this system, presumably would have to be carried out, with careful attention given to ionic strength and buffer effects.

The comparative results can be much less satisfactory in complicated systems. Sun et al. (19) have compared binding parameters for the binding of warfarin to serum albumin as studied by equilibrium dialysis, gel filtration, and chromatography in several laboratories. This system gives a highly curved Scatchard plot, and it has been common practice to interpret this as a two-class binding site system, using Eq. (2.70) to estimate the numbers of sites n_I, n_{II} in each class and the binding constants k_I, k_{II} for each class. This model may not be appropriate. At any rate, considerable variation is seen in the reported parameters, with n_I ranging from 1 to 2 (not always integral) and n_{II} from 2 to 6. The binding constants showed order-of-magnitude discrepancies.

Method Selection

There is probably no general rule or flowchart for selecting the best experimental method, and many of the considerations are obvious, but it may be helpful to present some of the pertinent factors. There are two main features to attend to.

The first of these is the chemical and physical properties of the interactants. A checklist can be prepared simply from the methods described in Chapters 4 to 12. One should have information on solubilities, acid–base strength, absorption spectra in the UV, IR, and NMR regions, fluorescence spectra, partitioning behavior, optical activity, chemical reactivity, molecular size, and so on. Some limits or possibilities will be suggested by such information. For example, if the solubility is very low, methods of relatively low sensitivity, such as NMR, may be inapplicable, whereas a very high solubility will rule out use of the solubility technique. Lack of a significant UV spectrum indicates that spectroscopy will not be useful. If the substrate and ligand are similar in size, equilibrium dialysis will not be effective. On the other hand, a distinctive absorption spectrum may be promising. The next stage requires preliminary experiments to determine if the potentially useful properties undergo significant (in the sense of practically usable) changes on complex formation. The magnitude of a change that is usable depends upon the technique and upon instrumental capabilities; for example, UV and IR spectral changes that were too small for effective measurement two decades ago may be quite adequate for binding constant determinations now because of increased spectrophotometric sensitivity.

The second critical feature is the stoichiometry of the system. Some methods are quite successful when applied to binding systems of simple

1:1 stoichiometry, but are relatively ineffective in systems having multiple equilibria. In general, if there are more than two complex species present, it is desirable to make use of methods in which the observable property or variable depends only on the binding constants β_{hi} and the ligand concentration [L]. Potentiometry (Chapter 7) and dialysis (Chapter 10) are widely useful methods for the study of multiple equilibria. The difficulty in applying many techniques to systems having more than one or two complex species is that each species adds to the binding isotherm *two* unknowns, namely, a binding constant and an intensive factor such as a molar absorptivity, a chemical shift, or a chemical reactivity.

The complications produced by multiple equilibria may limit the concentration range that is practically accessible and in this way affect the choice of method. Suppose that in a study of binding between S and L, the substrate S is known to undergo self-association. It may be possible to work at a sufficiently low substrate concentration that the extent of self-association is negligible, but this may severely limit the methods that can be used. For example, methyl orange self-associates in acidic solutions at concentrations greater than 10^{-4} (20); consequently, a study of the complexing of methyl orange with a ligand is greatly simplified if it is carried out at concentrations much smaller than this. Spectrophotometry is evidently applicable, but the solubility method is not.

13.2. REFERENCE BINDING SYSTEMS

Having selected a technique for application to a problem of interest, an experimentalist unfamiliar with the technique may wish to gain experience with it before using it for research. It would therefore be desirable to identify binding systems that can serve as "training systems" for the many experimental methods of determining binding constants. In the fields of acid–base equilibria and metal ion coordination complexing there are many textbook examples of suitable systems and it is not necessary to add more. For other types of systems, such as interactions between small organic molecules and small molecule–macromolecule binding, there are probably no systems presently well enough understood to be presented as reference binding systems. Since, however, there seems to be a need for such systems, we will pursue the notion briefly.

First let us list those properties of a suitable reference system that are necessary or desirable:

1. The stoichiometry should be known. That is, all significant values of h and i in the formula S_hL_i should have been established.

2. The binding constant(s), together with reliability estimates, should be known under specified, easily controlled, conditions.

3. All stepwise binding constants should be greater than $1 \, M^{-1}$.

4. The analytical response (e.g., absorbance change, alteration in partition coefficient) should be substantial.

5. The required reagents should be commercially available or easily prepared by standard techniques.

Comparative studies in several laboratories will be required to satisfy requirements 1 and 2. For reference systems to be widely useful, systems must be available that exhibit complicated behaviors in addition to those that are simple, so that users may experience the collection and treatment of data in systems that closely resemble a research situation of interest.

Despite the current lack of systems meeting all the listed requirements, there are many binding systems that satisfy some of them, or that satisfy all of them at least approximately. These systems may be developed into reference systems, and they at least provide material for the development of experience. Many of the tables and figures in Chapters 4 to 12 describe such systems. Tables 13.2 and 13.3 give additional examples, and the accompanying discussion provides guidance to the type of variation that may be anticipated. Further suggestions may be made. Since the study by Klotz et al. (21) of the binding of methyl orange to bovine serum albumin, which was made by equilibrium dialysis, numerous workers have employed this system as a reference system when developing other methods for the study of binding to proteins. This is a complicated system, but it has been useful. An alternative approach in studying small molecule–macromolecule binding is to employ a reference system exhibiting simple 1:1 binding; Glazer (22) observed the 1:1 binding of the dye Biebrich Scarlet to α-chymotrypsin, obtaining good agreement between binding constants measured with spectrophotometry and gel filtration. The benzoic acid : α-cyclodextrin system is a satisfactory one for study by the potentiometric method of weak acid substrates (23). None of these binding systems has been studied carefully enough to constitute a reliable reference system in the sense of this section, but they serve as examples of systems with promise for further investigation.

REFERENCES

1. D. A. MacInnes and T. Shedlovsky, *J. Am. Chem. Soc.*, **54**, 1429 (1932).
2. H. S. Harned and B. B. Owen, *J. Am. Chem. Soc.*, **52**, 5079 (1930).
3. C. B. Monk, in *Chemical Physics of Ionic Solutions*, B. E. Conway and R. G. Barradas, Eds., Wiley, New York, 1966, pp. 175–195.
4. K. A. Connors and J. A. Mollica, *J. Pharm. Sci.*, **55**, 772 (1966).
5. J. A. Mollica and K. A. Connors, *J. Am. Chem. Soc.*, **89**, 308 (1967).
6. B. Dodson, R. Foster, A. A. S. Bright, M. I. Foreman, and J. Gorton, *J. Chem. Soc.*, B, 1283, 1971.
7. J. A. Chudek, R. Foster, and F. M. Jarrett, *J. Chem. Soc., Faraday Trans. I*, **79**, 2729 (1983).
8. F. Cramer, W. Saenger, and H.-Ch. Spatz, *J. Am. Chem. Soc.*, **89**, 14 (1967).
9. E. A. Lewis and L. D. Hansen, *J. Chem. Soc., Perkin Trans.*, 2, 2081 (1973).
10. K. A. Connors and J. M. Lipari, *J. Pharm. Sci.*, **65**, 379 (1976).
11. R. J. Bergeron, M. A. Channing, G. B. Gibeily, and D. M. Pillor, *J. Am. Chem. Soc.*, **99**, 5146 (1977).
12. T. Osa, T. Matsue, and M. Fujihara, *Heterocycles*, **6**, 1833 (1977).
13. A. Cooper and D. D. MacNicol, *J. Chem. Soc., Perkin Trans.*, 2, 760, (1978).
14. T. W. Rosanske, Ph.D. Thesis, University of Wisconsin-Madison, 1979, p. 112.
15. R. I. Gelb, L. M. Schwartz, R. F. Johnson, and D. A. Laufer, *J. Am. Chem. Soc.*, **101**, 1869 (1979).
16. S.-F. Lin and K. A. Connors, *J. Pharm. Sci.*, **72**, 1333 (1983).
17. I. Sanemasa, T. Mizoguchi, and T. Deguchi, *Bull. Chem. Soc. Jpn.*, **57**, 1358 (1984).
18. T. K. Korpela and J. P. Himanen, *J. Chromatogr.*, **290**, 351 (1984).
19. S. F. Sun, S. W. Kuo, and R. A. Nash, *J. Chromatogr.*, **288**, 377 (1984).
20. M. DeVylder and D. DeKeukeleire, *Bull. Soc. Chim. Belg.*, **87**, 497 (1978).
21. I. M. Klotz, F. M. Walker, and R. B. Pivan, *J. Am. Chem. Soc.*, **68**, 1486 (1946).
22. A. N. Glazer, *J. Biol. Chem.*, **242**, 4528 (1967).
23. K. A. Connors, S.-F. Lin, and A. B. Wong, *J. Pharm. Sci.*, **71**, 217 (1982).

APPENDIX **A**

THE RECTANGULAR HYPERBOLA IN CHEMISTRY

The binding isotherm for $1:1$ stoichiometry, and for some other special cases, has the form of Eq. (A.1),

$$y = \frac{dx}{f + ex} \tag{A.1}$$

where d, e, f are parameters. Equation (A.2) is a more general expression of the same functional form.

$$y = \frac{c + dx}{f + ex} \tag{A.2}$$

Because of the widespread occurrence of this function in the field of binding, its properties are investigated here. We shall find that Eqs. (A.1) and (A.2) are the equations of a rectangular hyperbola, as many authors have pointed out.

The Standard Form

The hyperbola is the locus of a point that moves in a plane such that the difference of its distances from two fixed points is a constant. The general equation resulting from this definition is

$$\frac{x^2}{a^2} - \frac{y^2}{b^2} = 1$$

Setting $y = 0$ shows that the curve intersects the x axis at $x = \pm a$; these points are the *vertices* of the hyperbola. We also find that there is no intersection with the y axis. Moreover, replacement of x by $-x$ and y by $-y$ yields the same equation, showing that the hyperbola is symmetrical about both axes.

When $a = b$ we get the special case, Eq. (A.3), called a *rectangular hyperbola*.

$$\frac{x^2}{a^2} - \frac{y^2}{a^2} = 1 \qquad (A.3)$$

Figure A.1 is a plot of Eq. (A.3), taking $a = 3$ units. The hyperbola

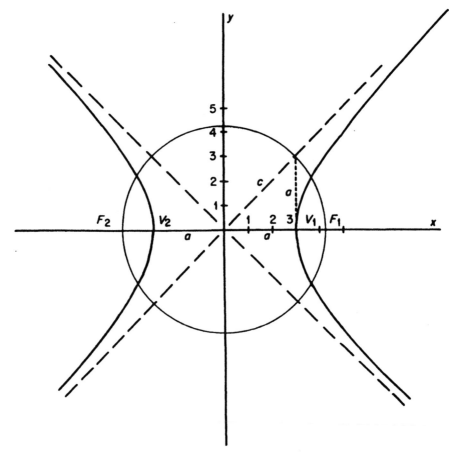

Figure A.1. A rectangular hyperbola having $a = 3$. F_1, F_2 are the foci, and V_1, V_2 are the vertices. The asymptotes are shown as dashed lines; the circle has a radius $c = \sqrt{2}\,a$.

consists of two branches. The x axis, upon which the vertices lie, is the transverse axis, and the y axis is the conjugate axis. (The hyperbola with the equation $y^2/a^2 - x^2/a^2 = 1$ has the y axis as its transverse axis.)

The quantity c is defined $c^2 = a^2 + b^2$, or in the case of the rectangular hyperbola, $c^2 = 2a^2$. The *foci* of the hyperbola are located on the x axis at distances $\pm c$ from the origin; these relationships are shown in Figure A.1. Note the triangle defined by the sides a, a, c; because this is an isosceles right triangle, the rectangular hyperbola is sometimes called an equilateral hyperbola.

The hyperbola has straight-line asymptotes whose equations are $y = \pm bx/a$, or, for the rectangular hyperbola, $y = \pm x$. Thus the asymptotes of the rectangular hyperbola are normal to each other (hence the name) and have slopes of ± 1. Note that the statements $y = \pm x$ and slope $= \pm 1$ are completely general for the asymptotes of the rectangular hyperbola, but their graphical representation as perpendicular lines requires that the x and y axes be marked off with the same scales.

Transformation to the Standard Form

It is not evident that Eqs. (A.1) and (A.2) are rectangular hyperbolae, and we next show how these equations may be placed in the standard form of Eq. (A.3). Since Eq. (A.2) includes Eq. (A.1) as a special case, we work with (A.2) and later derive the special case as needed.

The general quadratic equation may be written

$$Ax^2 + Bxy + Cy^2 + Dx + Ey + F = 0 \qquad \text{(A.4)}$$

Rearrangement of Eq. (A.2) gives $exy - dx + fy - c = 0$. Comparison shows that the coefficients in Eq. (A.4) have the values given in Eq. (4.5). Note that in all chemically interesting cases c, d, e, f are positive numbers or zero.

$$A = 0$$
$$B = e$$
$$C = 0 \qquad \text{(A.5)}$$
$$D = -d$$
$$E = f$$
$$F = -c$$

The quantity $B^2 - 4AC$ is called the discriminant, because its value establishes the nature of the function. If this quantity is greater than zero, the curve is a hyperbola. For our equation, we get $B^2 - 4AC = e^2$, a positive number, hence we know that Eq. (A.2) is a hyperbola. Next we find its equation.

We perform a rotation about the origin in the x, y coordinate frame in order to eliminate the cross-product term (the B term). This is always possible; the condition for achieving it is to rotate through the angle α in a counterclockwise manner, where α is defined by

$$\cot 2\alpha = \frac{A - C}{B}$$

Thus we desire $\cot 2\alpha = 0$. This condition is satisfied by rotations of $\alpha = 45°$, $135°$, $225°$, $315°$, ... Any one of these will result in elimination of the cross-product, and for reasons that will shortly become evident we choose a rotation of $+315°$, which is equivalent to $-45°$.

Rotation of axes through the angle α transforms Eq. (A.4) into Eq. (A.6), where the coefficients are related by Eqs. (A.7), the general transformation equations for rotation of axes.

$$A'x'^2 + B'x'y' + C'y'^2 + D'x' + E'y' + F' = 0 \qquad \text{(A.6)}$$

$$A' = A \cos^2 \alpha + B \cos \alpha \sin \alpha + C \sin^2 \alpha \qquad \text{(A.7a)}$$

$$B' = B(\cos^2 \alpha - \sin^2 \alpha) + 2(C - A) \sin \alpha \cos \alpha \qquad \text{(A.7b)}$$

$$C' = A \sin^2 \alpha - B \sin \alpha \cos \alpha + C \cos^2 \alpha \qquad \text{(A.7c)}$$

$$D' = D \cos \alpha + E \sin \alpha \qquad \text{(A.7d)}$$

$$E' = -D \sin \alpha + E \cos \alpha \qquad \text{(A.7e)}$$

$$F' = F \qquad \text{(A.7f)}$$

[Equations (A.7) are obtained by combining Eq. (A.4) with the rotation of axes equations, $x = x' \cos \alpha - y' \sin \alpha$ and $y = x' \sin \alpha + y' \cos \alpha$.]

For a rotation of $-45°$ we have

$$\sin \alpha = -\frac{1}{\sqrt{2}}, \qquad \cos \alpha = \frac{1}{\sqrt{2}}$$

Putting these values and Eq. (A.5) into Eq. (A.7) yields

$$A' = -\frac{e}{2}$$

$$B' = 0$$

$$C' = \frac{e}{2}$$

$$D' = -\frac{f+d}{\sqrt{2}} \qquad (A.8)$$

$$E' = \frac{f-d}{\sqrt{2}}$$

$$F' = -c$$

Substituting Eq. (A.8) into Eq. (A.6) gives

$$\left[\frac{e}{\sqrt{2}} x'^2 + (f+d)x'\right] - \left[\frac{e}{\sqrt{2}} y'^2 + (f-d)t'\right] = -\sqrt{2}\, c \quad (A.9)$$

We complete the squares of the terms in brackets according to this identity:

$$px^2 \pm qx = p\left(x \pm \frac{q}{2p}\right)^2 - \frac{q^2}{4p}$$

The result, after rearrangement, is

$$\frac{[x' + (f+d)/\sqrt{2}e]^2}{2(df-ce)/e^2} - \frac{[y' + (f-d)/\sqrt{2}e]^2}{2(df-ce)/e^2} = 1 \qquad (A.10)$$

Equation (A.10) may be written

$$\frac{X^2}{a^2} - \frac{Y^2}{a^2} = 1$$

where

$$X = x' + \frac{(f+d)}{\sqrt{2}e} \qquad (A.11a)$$

$$Y = y' + \frac{(f-d)}{\sqrt{2}e} \qquad (A.11b)$$

$$a^2 = \frac{2(df-ce)}{e^2} \qquad (A.11c)$$

Equation (A.10) is thus the equation of a rectangular hyperbola. The procedure in going from Eq. (A.2) to Eq. (A.10) consists, graphically, of a rotation of $-45°$ about the origin in the x, y frame to give the x', y' frame, followed by a translation of axes to give the X, Y frame, in which the hyperbola has the standard form.

Properties of the Curve

We shall explore this curve in the context of the experimental situation, which requires that the parameters c, d, e, f and the variables x, y all be zero or positive. We wish to understand the relationship of the observed quantities to the mathematical function. It is helpful for this purpose to display graphically a curve calculated for a hypothetical system. We shall use Eq. (A.1) for this calculation.

Figure A.2 is a plot of Eq. (A.1) calculated with $c = 0$, $d = 1$, $f = 2$, $e = 4$; the figure explicitly shows the axes in the x, y, the x', y', and the X, Y coordinate systems. It will now be evident why we performed a rotation of $+315° = -45°$, for this operation transformed (ultimately, after translation) the x axis into X and the y axis into Y. (A rotation of $+45°$ would have taken x into Y and y into $-X$; or would have generated a hyperbola with foci on the conjugate axis.)

Now since only positive values of x and y are physically meaningful, we see that only that portion of the hyperbola lying in the first quadrant of the x, y frame is experimentally accessible. This portion therefore constitutes a segment of one branch of the rectangular hyperbola.

Before we analyze the curve further, one important point must be made. Distances in the rotated frames (x', y' and X, Y) are measured in a different scale than distances in the original x, y frame; they are related by the equations for rotation of axes. As a consequence, the appearance of the curve depends on the relative scales of the two axes. The most dramatic result is that the coordinates of what appear to be the vertices change with the relative plotting scale. We shall shortly see, in fact, that the vertex (of the positive X branch) of the hyperbola plotted in Figure A.2 actually appears in the third quadrant of the x, y plane, despite its apparent location in the first quadrant of Figure A.2, which is plotted with different scales on the x and y axes.

Now we find the x, y coordinates corresponding to the origin in the X, Y frame. We use Eq. (A.11), setting $X = 0$, $Y = 0$, and solve for x', y', getting

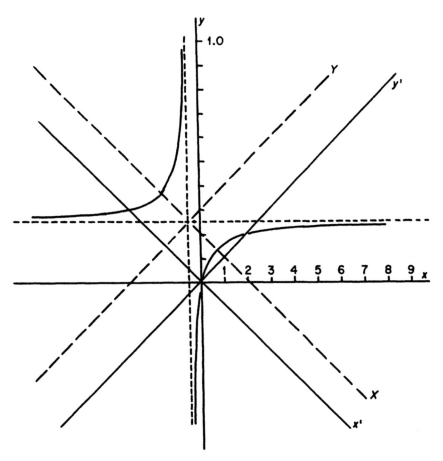

Figure A.2. Equation (A.1), with $c = 0$, $d = 1$, $e = 4$, $f = 2$, plotted in the x, y frame. The figure also shows the rotated x', y' frame, and the rotated, translated X, Y frame. The asymptotes are also drawn.

$$x'_0 = -\frac{f + d}{\sqrt{2}e}$$

$$y'_0 = -\frac{f - d}{\sqrt{2}e}$$

(A.12)

For the rotation of $-45°$, the x, y and x', y' systems are related by

$$x = \frac{x' + y'}{\sqrt{2}}$$

$$y = \frac{y' - x'}{\sqrt{2}}$$

(A.13)

Solving (A.12) and (A.13) for x and y gives

$$x_0 = -\frac{f}{e}$$

$$y_0 = \frac{d}{e}$$

(A.14)

Equations (A.14) also allow the asymptotes to be drawn, for they must pass through the point x_0, y_0 (corresponding to $X = 0$, $Y = 0$) and are parallel to the x, y axes (because they have slopes of ± 1 in the X, Y frame, which is itself rotated 45° with respect to the x, y frame).

Next we find the x, y coordinates of the vertex at $X = +a$. Again we use Eqs. (A.11), with a given by Eq. (A.11c). Solving for x_a', y_a', then combining with Eq. (A.13), we get

$$x_v = \frac{w - f}{e}$$

$$y_v = \frac{d - w}{e}$$

(A.15)

where $w = (df - ce)^{1/2}$.

Figure A.3 is a plot of the same calculated curve shown in Figure A.2, but in Figure A.3 the scales are identical on both axes, so they change in the same way upon rotation. Equations (A.14) give the origin in the X, Y frame in both figures, but only Figure A.3 represents the vertex correctly as it is given in Eqs. (A.15).

For Eq. (A.1), Eq. (A.15) becomes

$$x_v = \frac{\sqrt{df} - f}{e}$$

$$y_v = \frac{d - \sqrt{df}}{e}$$

(A.16)

Now, if $f > d$, then the vertex will appear in the third quadrant (x_v, y_v

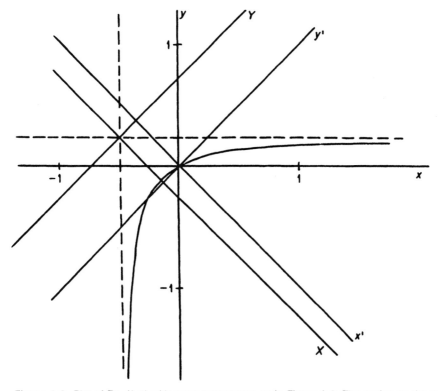

Figure A.3. Plot of Eq. (A.1) with same parameters as in Figure A.2. The scales on the x and y axes are equal; compare with Figure A.2.

both negative). If $f < d$, the vertex will be in the first quadrant. If $f = d$, the vertex coincides with the x, y origin.

Let us next find the slope of the hyperbola in the X, Y frame. This is

$$\frac{dY}{dX} = \frac{X}{(X^2 - a^2)^{1/2}} \tag{A.17}$$

At the vertex, $X = \pm a$. Inserting this into Eq. (A.17) shows that $dY/dX = \infty$ at the vertices. That is, the tangent to the curve at the vertex is parallel with the Y axis. But the Y axis is itself at the angle of 45° with respect to the y axis; hence at the $+a$ vertex the slope is at 45° to the y axis. Therefore in the x, y plane, the slope at the vertex is $+1$.

Now consider the behavior of the curve in the x, y frame in the vicinity of the origin. Inspection of Eq. (A.1) shows that if x is made so small that $ex \lll f$, Eq. (A.1) approaches $y = (d/f)x$; that is, y becomes a linear function of x for very small x. This is also seen from the derivative of (A.1):

$$\frac{dy}{dx} = \frac{df}{(f + ex)^2} \tag{A.18}$$

It should not be inferred from this, however, that the curve is "most linear" at the origin; in fact, we will find that it is most curved at the origin. We saw that when $f = d$, the vertex coincides with the origin in the x, y frame. We also saw that the apparent location of the vertex depends on the scales employed in laying out the x and y axes. Let us consider the experimental manifestations of Eq. (A.1) in these terms. Equations (A.19) to (A.22) show some of these equations.

$$f_{11} = \frac{K_{11}[L]}{1 + K_{11}[L]} \tag{A.19}$$

$$F_{HA} = \frac{[H^+]}{K_a + [H^+]} \tag{A.20}$$

$$\frac{\Delta A}{b} = \frac{K_{11} S_t \, \Delta\epsilon \, [L]}{1 + K_{11}[L]} \tag{A.21}$$

$$v = \frac{V_m[S]}{K_m + [S]} \tag{A.22}$$

These are, respectively, the 1:1 binding isotherm, the fraction of weak acid HA in the conjugate acid form, the spectrophotometric measure of 1:1 binding, and the Michaelis–Menten equation of enzyme kinetics. Evidently, in the conventional manner of defining x and y for plotting, there is no meaning in the requirement that these quantities be plotted on the same scale, for they have different dimensions. However, all of these equations can be rewritten in the reduced form of the rectangular hyperbola, Eq. (A.23), by defining the variables as in Table A.1.

$$y_r = \frac{x_r}{1 + x_r} \tag{A.23}$$

Since y_r and x_r are dimensionless, they can be plotted on the same scale. Note that y_r is normalized, that is, constrained within the limits 0 and 1.

Table A.1. Reduced Forms of Some Experimental Curves

Usual form	Reduced form	y_r	x_r
Eq. (A.19)	$f_{11} = \dfrac{K_{11}[L]}{1 + K_{11}[L]}$	f_{11}	$K_{11}[L]$
Eq. (A.20)	$F_{HA} = \dfrac{(H^+)/K_a}{1 + (H^+)/K_a}$	F_{HA}	$\dfrac{(H^+)}{K_a}$
Eq. (A.21)	$\dfrac{\Delta A}{S_t b\, \Delta \epsilon} = \dfrac{K_{11}(L)}{1 + K_{11}(L)}$	$\dfrac{\Delta A}{S_t b\, \Delta \epsilon}$	$K_{11}(L)$
Eq. (A.22)	$\dfrac{v}{V_m} = \dfrac{[S]/K_m}{1 + [S]/K_m}$	$\dfrac{v}{V_m}$	$\dfrac{[S]}{K_m}$

Moreover, in the reduced form $f = d$, so the origin in the x_r, y_r frame coincides with the vertex of Eq. (A.23). The experimental results all lie in the first quadrant; by symmetry, reflection in the rotated ($\alpha = -45°$) axis gives the other half of the branch, which is in the third quadrant.

It is not possible actually to present raw data in this reduced form, because x_r and y_r require, for their expression as in Table A.1, quantities that are not yet known, being in fact the goal of the study. However, even though we cannot plot the data in the reduced form, the accessible range is not diminished by this inability, and we conclude that the origin coincides with the vertex for any curve that can be placed in reduced form.

Table A.2. Calculated Points for the Normalized Rectangular Hyperbola, Eq. (A.23)

x_r	y_r
0.0	0.000
0.3	0.231
0.6	0.375
0.9	0.474
1.2	0.545
1.5	0.600
1.8	0.643
2.1	0.677
2.4	0.706
2.7	0.730
3.0	0.750

It is graphically obvious, and can be shown analytically, that the radius of curvature of a hyperbola is smallest at its vertex; that is, the curvature is greatest at this point. Since we have found that the experimental origin coincides with the vertex, the experimental curve is most curved at the origin.

In Chapter 2, Section 2.5, the reduced rectangular hyperbola is used to demonstrate graphical methods of presentation. Calculated points are given in Table A.2.

APPENDIX B

GENERAL LEAST-SQUARES ANALYSIS

It is necessary to make a distinction between two kinds of functions. If a function can be written in the form of

$$f(x_i) = a_0 + a_1 x_1 + a_2 x_2 + \cdots + a_n x_n \tag{B.1}$$

it is said to be *linear*. The terminology means that it is linear *in the parameters* a_i. The quantities x_i may or may not be linear; for example, we might define $x_i = x^i$; then Eq. (B.1) would be a polynomial in x, but it would remain a linear function in the present sense. If a function is not linear in its parameters, it is a *nonlinear* function. If a nonlinear function can be transformed into a linear function, it is said to be intrinsically linear. Other functions cannot be so transformed; these are called intrinsically nonlinear functions.

The Least-Squares Principle

Suppose we have a set of paired experimental data points (X_i, Y_i) that we wish to fit to a function of empirical or theoretical basis. Write the function $y = f(x, \alpha, \beta, \ldots)$, where $\alpha, \beta, \ldots$ represent the parameters of the model function. At the conclusion of the least-squares analysis we will have $y = f(x, a, b, \ldots)$, where $a, b, \ldots$ are the least-squares estimates of $\alpha, \beta, \ldots$ and x, y are the variables defining the calculated regression line. We define the residuals e_{yi} by

$$e_{yi} = Y_i - y_i \qquad (B.2)$$

That is, e_{yi} is the difference between the observed and calculated values of the dependent variable at the value x_i of the independent variable. We evidently wish to minimize these residuals. Their sum is $\Sigma_i e_{yi} = 0$; however, if we square the residuals we obtain positive numbers, and the sum of the squares of the residuals will be a positive number, because of the inevitable presence of experimental error. This important quantity, the sum of the squares of the residuals, is denoted S.

$$S = \sum_i e_{yi}^2 \qquad (B.3)$$

Now we choose the estimates $a, b, \ldots$ subject to the condition that S be minimized. This is the principle of least squares. (In fact, we refine this by minimizing the sum of the squares of the weighted residuals; weighting is explained later.)

The method is to differentiate S with respect to each of the parameters, setting these derivatives equal to zero and solving for the parameters. For the function being considered we write

$$\frac{\partial S}{\partial \alpha} = 2 \sum e_{yi} \frac{\partial f(\alpha)}{\partial \alpha} = 0 \qquad (B.4a)$$

$$\frac{\partial S}{\partial \beta} = 2 \sum e_{yi} \frac{\partial f(\beta)}{\partial \beta} = 0 \qquad (B.4b)$$

and so on. Now if the function f is linear in the parameters, the parameters do not appear in the derivatives of f in Eqs. (B.4), which are then readily solved for $a, b, \ldots$. If, however, f is nonlinear, then its derivatives include the parameters, and the resulting equations cannot in general be solved for the parameters. This is the basic problem in nonlinear regression.

Several methods have been devised to solve this problem (1–5). All involve approximations. A characteristic feature of nonlinear regression is that preliminary estimates of the parameters are needed. The numerical procedure consists of subjecting these estimates to the least-squares constraints (in their necessarily approximate forms) to obtain improved parameter estimates. This process may be repeated many times until no significant improvement is seen. In some instances the iterations do not converge, and the nonlinear analysis may fail. For the kinds of systems we shall consider, however, rapid convergence is usually seen.

We shall follow Deming's (1, 2) approach, with the equations being written out in more detail than is usual in statistical writing in order to make transparently clear the concepts and the mathematical procedure. Although in principle any number of variables and parameters may be included in the function to be fitted, we shall choose to work with two variables and three parameters. This case will adequately represent many chemical systems, and its extension will be obvious.

The Condition Equations

Let the observed quantities (variables) be $X_1, X_2, \ldots, X_n$ and $Y_1, Y_2, \ldots, Y_n$, where X_i, Y_i represent a paired observation i; we may consider the X_i to be independent variables and Y_i dependent variables. Further let $x_1, x_2, \ldots, x_n; y_1, y_2, \ldots, y_n$ be the calculated values of the variables corresponding to the experimental values. Deming calls the x_i, y_i the "adjusted values"; in current terminology they are the values calculated from the regression equation. Let the model function have three parameters α, β, γ; the least-squares estimates of these parameters are a, b, c.

We can write the model function as $\eta = f(\xi, \alpha, \beta, \gamma)$ or, transposing terms, as $F(\xi, \eta, \alpha, \beta, \gamma) = 0$, where ξ, η are the true values of the variables. In general the true and calculated curves will be different, and the experimental points will not lie on either. Thus in general $F(X, Y, a, b, c) \neq 0$. However, for each of the n observed points, we can write an equation as follows.

$$F^1(x_1, y_1, a, b, c) = 0$$
$$F^2(x_2, y_2, a, b, c) = 0$$
$$\vdots$$
$$F^n(x_n, y_n, a, b, c) = 0$$

(B.5)

Equations (B.5) are called the condition equations, because they establish conditions that the solution must meet. Each of these functions F^i is identically equal to zero because the x_i, y_i, a, b, c have been chosen (by the least-squares criterion) to make it so.

We can also write equations analogous to Eq. (B.5) as shown in (B.6):

$$F_0^1 = F^1(X, Y_1, a_0, b_0, c_0)$$
$$F_0^2 = F^2(X_2, Y_2, a_0, b_0, c_0)$$
$$\vdots$$
$$F_0^n = F^n(X_n, Y_n, a_0, b_0, c_0)$$

(B.6)

In Eqs. (B.6) the subscript (0) indicates an approximation; for example, F_0^1 is an approximation to F^1 because it is calculated with the observed quantities X_1, Y_1 and the initial parameter estimates a_0, b_0, c_0 rather than the final "adjusted" quantities. The F_0^i are small numbers, but not in general zero.

The Linearization Approximation

Since the model function may be nonlinear in the parameters, we linearize it by performing a Taylor's series expansion about the point $(X_i, Y_i, a_0, b_0, c_0)$ (9), and truncating the series after the linear term. (This is the approximation that allows nonlinear functions to be fitted.) The Taylor's series expansion of a function $f(x)$ about the point $x = p$ is

$$f(x) = f(p) + f'(p)\cdot(x-p) + \frac{f''(p)}{2}\cdot(x-p)^2 + \cdots \qquad (B.7)$$

where f', f'', ... are first, second, ... derivatives evaluated at $x = p$.

The partial derivatives needed are $F_{xi} = \partial F^i/\partial x_i$, $F_{yi} = \partial F^i/\partial y_i$, $F_a^i = \partial F^i/\partial a$, $F_b^i = \partial F^i/\partial b$, $F_c^i = \partial F^i/\partial c$. Residuals are defined in Eq. (B.8).

$$e_{xi} = X_i - x_i$$
$$e_{yi} = Y_i - y_i$$
$$A = a_0 - a \qquad (B.8)$$
$$B = b_0 - b$$
$$C = c_0 - c$$

Thus Eqs. (B.5), the condition equations, when linearized by the Taylor's series expansion, become

$$F^1 = F_0^1(X_1, Y_1, a_0, b_0, c_0) - F_{x1}e_{x1} - F_{y1}e_{y1}$$
$$- F_a^1 A - F_b^1 B - F_c^1 C = 0$$
$$F^2 = F_0^2(X_2, Y_2, a_0, b_0, c_0) - F_{x2}e_{x2} - F_{y2}e_{y2}$$
$$- F_a^2 A - F_b^2 B - F_c^2 C = 0 \qquad (B.9)$$
$$\vdots$$
$$F^n = F_0^n(X_n, Y_n, a_0, b_0, c_0) - F_{xn}e_{xn} - F_{yn}e_{yn}$$
$$- F_a^n A - F_b^n B - F_c^n C = 0$$

where the negative signs arise because of the definitions of the intervals in Eq. (B.8).

Eqs. (B.9) give Eqs. (B.10), which Deming (2) calls the reduced conditions. These equations are linear in the parameters as a consequence of the Taylor's series truncation.

$$F_{x1}e_{xi} + F_{y1}e_{y1} + F_a^1 A + F_b^1 B + F_c^1 C = F_0^1$$

$$F_{x2}e_{x2} + F_{y2}e_{y2} + F_a^2 A + F_b^2 B + F_c^2 C = F_0^2 \qquad \text{(B.10)}$$

$$\vdots$$

$$F_{xn}e_{xn} + F_{yn}e_{yn} + F_a^n A + F_b^n B + F_c^n C = F_0^n$$

The Normal Equations

The sum of squares of the weighted residuals is defined

$$S = \sum (w_{xi}e_{xi}^2 + w_{yi}e_{yi}^2) \qquad \text{(B.11)}$$

where residuals in both variables are included, and the weights w_{xi}, w_{yi} provide a means for assigning greater or less influence to an experimental point depending on its reliability. The least-squares principle requires that S be minimized, or

$$dS = 2\sum (w_{xi}e_{xi}\,de_{xi} + w_{yi}e_{yi}\,de_{yi}) = 0 \qquad \text{(B.12)}$$

The differentials in Eq. (B.12) cannot vary arbitrarily, since they must also satisfy the condition equations. We therefore differentiate Eqs. (B.10):

$$F_{x1}\,de_{x1} + F_{y1}\,de_{y1} + F_a^1\,dA + F_b^1\,dB + F_c^1\,dC = 0$$

$$F_{x2}\,de_{x2} + F_{y2}\,de_{y2} + F_a^2\,dA + F_b^2\,dB + F_c^2\,dC = 0 \qquad \text{(B.13)}$$

$$\vdots$$

$$F_{xn}\,de_{xn} + F_{yn}\,de_{yn} + F_a^n\,dA + F_b^n\,dB + F_c^n\,dC = 0$$

Among Eqs. (B.12) and (B.13) we have $(n + 1)$ equations, but we have $(2n + 3)$ differentials. Because of the least-squares constraint, we are able to solve this system of equations by means of the method of Lagrange multipliers. This is the principle.

Let us suppose we are interested in an extremum of $F(x, y, z)$, the conditions being $(\partial F/\partial x) = (\partial F/\partial y) = (\partial F/\partial z) = 0$. However, suppose it also happens that we have the further constraint $G(x, y, z) = 0$. Then we form the function

$$H(x, y, z, \lambda) = F(x, y, z) + \lambda G(x, y, z)$$

where λ is the *Lagrange multiplier*. Now solve the three equations

$$\left(\frac{\partial H}{\partial x}\right)_{y,z,\lambda} = \left(\frac{\partial H}{\partial y}\right)_{x,z,\lambda} = \left(\frac{\partial H}{\partial z}\right)_{x,y,\lambda} = 0$$

along with the equation of constraint.

In our case Eqs. (B.13) play the role of G and Eq. (B.12) is F. First multiply Eqs. (B.13) by $-\lambda_i$, giving

$$-\lambda_1 F_{x1}\, de_{x1} - \lambda_1 F_{y1}\, de_{y1} - \lambda_1 F_a^1\, dA - \lambda_1 F_b^1\, dB - \lambda_1 F_c^1\, dC = 0$$

$$-\lambda_2 F_{x2}\, de_{x2} - \lambda_2 F_{y2}\, de_{y2} - \lambda_2 F_a^2\, dA - \lambda_2 F_b^2\, dB - \lambda_2 F_c^2\, dC = 0$$

$$\vdots \qquad\qquad\qquad\qquad\qquad\qquad\qquad\qquad \text{(B.14)}$$

$$-\lambda_n F_{xn}\, de_{xn} - \lambda_n F_{yn}\, de_{yn} - \lambda_n F_a^n\, dA - \lambda_n F_b^n\, dB - \lambda_n F_c^n\, dC = 0$$

Now add Eq. (B.12) to Eqs. (B.14) to form the function $H(x, y, a, b, c, \lambda)$. After collecting terms we get

$$\sum_{}^{n} (w_{xi}e_{xi} - \lambda_i F_{xi})\, de_{xi} + \sum_{}^{n} (w_{yi}e_{yi} - \lambda_i F_{yi})\, de_{yi}$$

$$- \sum_{}^{n} \lambda_i F_a^i\, dA - \sum_{}^{n} \lambda_i F_b^i\, dB - \sum_{}^{n} \lambda_i F_c^i\, dC = 0 \qquad \text{(B.15)}$$

Equation (B.15) is function H. Now, there are n variables de_{xi} and n variables de_{yi}, but there are only three parameters, and the dA, dB, dC can be factored out of their sums. Thus the partial derivatives of H become

$$\left(\frac{\partial H}{\partial de_{x1}}\right)_{x_2\ldots x_n, y_1\ldots y_n, a, b, c} = w_{x1}e_{x1} - \lambda_1 F_{x1} = 0$$

and similarly with the other terms in de_{xi} and de_{yi}. On the other hand, we get

$$\frac{\partial H}{\partial dA} = \sum_{}^{n} (\lambda_i F_a^i) = 0$$

and so on. Thus we find

$$w_{xi}e_{xi} - \lambda_i F_{xi} = 0 \quad (i = 1, 2, \ldots, n) \tag{B.16a}$$

$$w_{yi}e_{yi} - \lambda_i F_{yi} = 0 \quad (i = 1, 2, \ldots, n) \tag{B.16b}$$

$$\sum_{i}^{n} \lambda_i F_a^i = 0 \tag{B.16c}$$

$$\sum_{i}^{n} \lambda_i F_b^i = 0 \tag{B.16d}$$

$$\sum_{i}^{n} \lambda_i F_c^i = 0 \tag{B.16e}$$

There are n Eqs. (B.16a), n (B.16b), but only one each of Eqs. (B.16c, 16d, 16e).

We solve Eqs. (B.16a, 16b) for the residuals,

$$e_{xi} = \frac{\lambda_i F_{xi}}{w_{xi}} \tag{B.17a}$$

$$e_{yi} = \frac{\lambda_i F_{yi}}{w_{yi}} \tag{B.17b}$$

Note that each residual is inversely proportional to the weight of the corresponding observation. Thus if the weight is very large the residual is very small.

The residuals from Eqs. (B.17) are substituted into the reduced condition equations (B.10), giving

$$\lambda_i \frac{F_{xi}F_{xi}}{w_{xi}} + \lambda_i \frac{F_{yi}F_{yi}}{w_{yi}} + F_a^i A + F_b^i B + F_c^i C = F_0^i \quad (i = 1, 2, \ldots, n) \tag{B.18}$$

Now define

$$L_i = \frac{F_{xi}F_{xi}}{w_{xi}} + \frac{F_{yi}F_{yi}}{w_{yi}} \tag{B.19}$$

Combining (B.18) and (B.19),

$$\lambda_i L_i + F_a^i A + F_b^i B + F_c^i C = F_0^i \quad (i = 1, 2, \ldots, n) \tag{B.20}$$

These n equations are solved for the λ_i:

$$\lambda_i = \frac{1}{L_i}(F_0^i - F_a^i A - F_b^i B - F_c^i C) \tag{B.21}$$

The λ_i are next eliminated by substitution of Eqs. (B.21) into Eqs. (B.16c–e), giving, after rearrangement, the *three* equations

$$\sum \frac{F_a^i F_a^i}{L_i} A + \sum \frac{F_a^i F_b^i}{L_i} B + \sum \frac{F_a^i F_c^i}{L_i} C = \sum \frac{F_a^i F_0^i}{L_i}$$

$$\sum \frac{F_b^i F_a^i}{L_i} A + \sum \frac{F_b^i F_b^i}{L_i} B + \sum \frac{F_b^i F_c^i}{L_i} C = \sum \frac{F_b^i F_0^i}{L_i} \tag{B.22}$$

$$\sum \frac{F_c^i F_a^i}{L_i} A + \sum \frac{F_c^i F_b^i}{L_i} B + \sum \frac{F_c^i F_c^i}{L_i} C = \sum \frac{F_c^i F_0^i}{L_i}$$

Equations (B.22) are the *normal equations* for this problem. They constitute three equations in the three unknowns A, B, C. The partial derivatives are defined just prior to Eq. (B.8), and they are evaluated at X_i, Y_i, a_0, b_0, c_0. The L_i are defined in Eq. (B.19). The weights w are *defined* to be inversely proportional to the variance at the point i, or

$$w_{xi} = \frac{k}{\sigma_{xi}^2}$$

$$w_{yi} = \frac{k}{\sigma_{yi}^2} \tag{B.23}$$

The quantity k is a proportionality constant, chosen for convenience, since it cancels out of the normal equations. It may be interpreted as the variance of unit weight. Further discussion of k is given in Section 3.2. To use Eq. (B.23), and thus to assign values to the L_i, requires estimates of σ_{xi}^2 and σ_{yi}^2. If these are constant, independent of i, they cancel from the equations, which consequently become unweighted least-squares normal equations. If one of the variables, say the independent variable X_i, is known with high precision relative to Y_i, then the σ_{xi}^2 are very small, the w_{xi} are very large, and the term in x drops out of the expression for L_i, leaving only the weighting of the dependent variable. However, the normal equations (B.22) are capable of expressing uncertainty in both variables through the factors L_i.

The symmetry in Eqs. (B.22) allows other cases to be written out directly without a full derivation. A model equation having p parameters will lead to p normal equations. For example, the normal equations for a four-parameter model will be [writing only the subscripts corresponding to Eqs. (B.22)].

$$
\begin{array}{llll}
aa & ab & ac & ad = a0 \\
ba & bb & bc & bd = b0 \\
ca & cb & cc & cd = c0 \\
da & db & dc & dd = d0
\end{array}
\tag{B.24}
$$

Notice that $F_h^i F_k^i = F_k^i F_n^i$, so that the array of normal equations is symmetrical about its diagonal $h = k$.

Solution of the Normal Equations

The normal equations, Eqs. (B.22) for our three-parameter model, are to be solved for A, B, C, which then through Eqs. (B.8) yield the least-squares parameter estimates of a, b, c. The entire process is now repeated, with these parameter estimates serving as a_0, b_0, c_0 in the next iteration. This procedure is carried out until the parameter estimates converge to constant values (as decided by an appropriate statistical criterion).

In principle it is possible to solve the normal equations by ordinary algebraic substitution and elimination. For a two-parameter model this is easy to do, and it is feasible though lengthy for the three-parameter case. But for $p > 3$ it is not a practical procedure, and matrix algebra is used instead. Presentation of the normal equations in the form of Eqs. (B.24) makes it very obvious how these equations can be written in matrix form with great simplicity. For readers unfamiliar with matrix algebra the minimum information required to carry out the present operations is presented in Appendix C.

To begin, we rewrite the normal equations (B.22) as Eqs. (B.25), where the meanings of the m_{hk} and q_h will be obvious from the comparison.

$$
m_{11}A + m_{12}B + m_{13}C = q_1
$$

$$
m_{21}A + m_{22}B + m_{23}C = q_2
\tag{B.25}
$$

$$
m_{31}A + m_{32}B + m_{33}C = q_3
$$

Now define these matrices:

$$M = \begin{bmatrix} m_{11} & m_{12} & m_{13} \\ m_{11} & m_{12} & m_{13} \\ m_{31} & m_{32} & m_{33} \end{bmatrix} \tag{B.26}$$

$$P = \begin{bmatrix} A \\ B \\ C \end{bmatrix} \tag{B.27}$$

$$Q = \begin{bmatrix} q_1 \\ q_2 \\ q_3 \end{bmatrix} \tag{B.28}$$

Then Eqs. (B.25) can be written as the matrix equation

$$MP = Q \tag{B.29}$$

Solving for P,

$$P = M^{-1}Q \tag{B.30}$$

where M^{-1} is the inverse matrix. Since M and Q are known matrices, the inverse M^{-1} can be found, and thus matrix P is obtained. This gives A, B, C directly from Eq. (B.27).

For the matrix solution to succeed, the inverse matrix must exist. In the calculation of M^{-1} it is necessary to divide by the determinant of M. If this quantity is very small, the matrix solution may be unstable. If the determinant of M is equal to zero, the matrix is said to be singular, and its inverse does not exist; this situation results if some of the normal equations are linear combinations of others. In algebraic terms this means that there are more unknowns than independent equations.

Note that M is a symmetrical matrix, with $m_{12} = m_{21}$, $m_{13} = m_{31}$, $m_{23} = m_{32}$. The inverse of a symmetrical matrix is itself a symmetrical matrix.

We have here used matrices merely as a means for solving the normal equations. Modern presentations of regression analysis tend to dispense entirely with the algebraic and summation formulation and to develop the subject solely through matrices.

REFERENCES

1. W. E. Deming, *Phil. Mag.*, **11**, 146 (1931); **17**, 804 (1934); **19**, 389 (1935).
2. W. E. Deming, *Statistical Adjustment of Data*, Wiley, New York, 1943; Dover, New York, 1964.
3. G. E. P. Box, W. H. Hunter, and J. S. Hunter, *Statistics for Experimenters*, Wiley, New York, 1978, pp. 483–487.
4. N. R. Draper and H. Smith, *Applied Regression Analysis*, 2nd ed., Wiley, New York, 1981, Ch. 10.
5. Y. Bard, *Nonlinear Parameter Estimation*, Academic, New York, 1974.

APPENDIX C

OPERATIONS WITH MATRICES

It is emphasized that the following treatment is sketchy in the extreme, with only enough substance to allow the reader to carry out the operations needed to solve the normal equations. Fuller descriptions are readily available (1).

A matrix is a rectangular array of elements. Two matrices are identical only if all of their elements are identical. We denote by a_{ij} the element in the ith row and jth column of the matrix; a matrix having p rows and q columns is called a $p \times q$ matrix. If $p = q$, the matrix is a square matrix. The main diagonal of a square matrix consists of those elements having $i = j$. A *unit matrix* I is a square matrix (of any order p) whose elements on the main diagonal are 1 and whose other elements are 0.

To add (or subtract) two matrices **A** and **B**, add (or subtract) corresponding elements. **A** and **B** must have the same numbers of rows and columns. If the elements of **A** and **B** are a_{ij} and b_{ij} respectively, then the elements of $\mathbf{A} \pm \mathbf{B}$ are $(a_{ij} \pm b_{ij})$. Multiplication of a matrix by a constant k is equivalent to addition of k identical matrices; thus $k\mathbf{A}$ is a matrix with elements ka_{ij}.

To multiply two matrices **A** and **B**, the matrix **A** must have the same number of columns as **B** has rows; the matrices are then said to be conformable. Suppose **A** is a $p \times q$ matrix and **B** is a $q \times r$ matrix. Then the product matrix **AB** is defined, and will be of order $p \times r$; but **BA** is not defined unless $p = r$. Thus matrix multiplication is not necessarily commutative. To find the product **AB** of conformable matrices, the

397

element in row i and column j of **AB** is equal to row i of **A** times column j of **B**.

Example C.1. Find the product **AB** where

$$\mathbf{A} = \begin{bmatrix} 3 & -2 \\ 5 & 1 \end{bmatrix} \qquad \mathbf{B} = \begin{bmatrix} 1 & 8 & 2 \\ 0 & -2 & 3 \end{bmatrix}$$

$$\mathbf{AB} = \begin{bmatrix} (3\cdot1+0\cdot-2) & (3\cdot8+-2\cdot-2) & (3\cdot2+3\cdot-2) \\ (5\cdot1+0\cdot1) & (5\cdot8+1\cdot-2) & (5\cdot2+3\cdot1) \end{bmatrix}$$

$$\mathbf{AB} = \begin{bmatrix} 3 & 28 & 0 \\ 5 & 38 & 13 \end{bmatrix}$$

Example C.2. Multiply the matrices **A** and **P** as shown.

$$\begin{bmatrix} a_{11} & a_{12} & a_{13} \\ a_{21} & a_{22} & a_{23} \\ a_{31} & a_{32} & a_{33} \end{bmatrix} \begin{bmatrix} A \\ B \\ C \end{bmatrix} = \mathbf{AP} = \begin{bmatrix} (a_{11}A + a_{12}B + a_{13}C) \\ (a_{21}A + a_{22}B + a_{23}C) \\ (a_{31}A + a_{32}B + a_{33}C) \end{bmatrix}$$

Notice the correspondence between this operation and Eqs. (B.25)–(B.28).

We define a *determinant* as a square array of elements to which can be assigned a value. Note that a matrix does not have a value, though we may calculate the value of the determinant of a square matrix. To distinguish between matrices and determinants we enclose matrices in brackets (or parentheses) and determinants in vertical bars.

As the reader probably recalls, a 2×2 determinant is evaluated according to the following formula:

$$\begin{vmatrix} a_{11} & a_{12} \\ a_{21} & a_{22} \end{vmatrix} = a_{11}a_{22} - a_{21}a_{12} \tag{C.1}$$

This formula is part of Cramer's rule for solving simultaneous equations. To find the value of a determinant of higher order than 2×2 we make use of the Laplace development to reduce the order, stepwise, until the determinant has been reduced to 2×2 determinants, which are then evaluated by Eq. (C.1).

We make these definitions (2). Suppose we have a determinant of order p. If we delete the entire row and entire column containing the element a_{ij}, then we are left with a determinant of order $p - 1$; the remaining determinant is called the *minor* of a_{ij}. We give a sign to the minor according to the formula $(-1)^{i+j}$, and this signed minor is called the *cofactor* of a_{ij}. An easy way to remember the sign of the minor is to use this alternating pattern starting with $i = j = 1$.

$$
\begin{array}{ccccccccc}
+ & - & + & - & + & \cdot & \cdot & \cdot & \cdot \\
- & + & - & + & - & \cdot & \cdot & \cdot & \cdot \\
+ & - & + & - & + & \cdot & \cdot & \cdot & \cdot \\
- & + & - & + & - & \cdot & \cdot & \cdot & \cdot \\
+ & - & + & - & + & \cdot & \cdot & \cdot & \cdot \\
\cdot & \cdot & \cdot & \cdot & \cdot & & & & \\
\cdot & \cdot & \cdot & \cdot & \cdot & & & & \\
\cdot & \cdot & \cdot & & & & & & \\
\cdot & & & & & & & &
\end{array}
$$

Now the value of a determinant is found by multiplying each element of one row (or one column) by its cofactor and adding the results. If the determinant is third-order, this process gives three second-order determinants that are evaluated by Eq. (C.1). A fourth-order determinant is first reduced to a third-order one, which then gives second-order determinants.

Example C.3. Evaluate this determinant:

$$
\begin{vmatrix}
1 & 1 & 3 \\
2 & 1 & 1 \\
3 & 4 & -2
\end{vmatrix}
$$

Let us choose row 1 for the evaluation. Then, letting Δ equal the value of the determinant,

$$
\Delta = 1\begin{vmatrix} 1 & 1 \\ 4 & -2 \end{vmatrix} - 1\begin{vmatrix} 2 & 1 \\ 3 & -2 \end{vmatrix} + 3\begin{vmatrix} 2 & 1 \\ 3 & 4 \end{vmatrix}
$$

$$
\Delta = 1(-2-4) - 1(-4-3) + 3(8-3)
$$

$$
\Delta = \quad -6 \quad + \quad 7 \quad + \quad 15 \quad = 16
$$

We define the *transpose* $\mathbf{M}'$ of matrix $\mathbf{M}$ as the matrix obtained by interchanging the rows and columns of $\mathbf{M}$.

Example C.4. Find the transposes of matrices **A** and **P** of Example C.2.

$$\mathbf{A}' = \begin{bmatrix} a_{11} & a_{21} & a_{31} \\ a_{12} & a_{22} & a_{32} \\ a_{13} & a_{23} & a_{33} \end{bmatrix}$$

$$\mathbf{P}' = \begin{bmatrix} A & B & C \end{bmatrix}$$

The *adjoint* of a matrix is found by replacing each element of the matrix by its cofactor and then transposing.

Example C.5. Find the adjoint of matrix **A** in Example C.2.

First we replace each element a_{ij} of **A** by its cofactor to obtain

$$\begin{bmatrix} (a_{22}a_{33} - a_{32}a_{23}) & -(a_{21}a_{33} - a_{31}a_{23}) & (a_{21}a_{32} - a_{31}a_{22}) \\ -(a_{12}a_{33} - a_{32}a_{13}) & (a_{11}a_{33} - a_{31}a_{13}) & -(a_{11}a_{32} - a_{31}a_{12}) \\ (a_{12}a_{23} - a_{22}a_{13}) & -(a_{11}a_{23} - a_{13}a_{21}) & (a_{11}a_{22} - a_{21}a_{12}) \end{bmatrix}$$

We now transpose to get the adjoint.

$$\mathrm{Adj}\,\mathbf{A} = \begin{bmatrix} (a_{22}a_{33} - a_{32}a_{23}) & -(a_{12}a_{33} - a_{32}a_{13}) & (a_{12}a_{23} - a_{22}a_{13}) \\ -(a_{21}a_{33} - a_{31}a_{23}) & (a_{11}a_{33} - a_{31}a_{13}) & -(a_{11}a_{23} - a_{13}a_{21}) \\ (a_{21}a_{32} - a_{31}a_{22}) & -(a_{11}a_{32} - a_{31}a_{12}) & (a_{11}a_{22} - a_{21}a_{12}) \end{bmatrix}$$

Division by a matrix **A** is defined as multiplication by the *inverse matrix* $\mathbf{A}^{-1}$. The inverse matrix is defined by Eq. (C.2), where **I** is the unit matrix.

$$\mathbf{A}\mathbf{A}^{-1} = \mathbf{A}^{-1}\mathbf{A} = \mathbf{I} \tag{C.2}$$

The inverse matrix can be evaluated by Eq. (C.3).

$$\mathbf{A}^{-1} = \frac{1}{\Delta}\,\mathrm{adj}\,\mathbf{A} \tag{C.3}$$

where $\Delta = \det \mathbf{A} =$ the determinant of **A**. Thus the inverse matrix $\mathbf{A}^{-1}$ exists only if $\det \mathbf{A} \neq 0$. The inverse matrix is also called the reciprocal matrix.

Example C.6. Find the inverse matrix $\mathbf{A}^{-1}$ if

$$A = \begin{bmatrix} 1 & 1 & 3 \\ 2 & 1 & 1 \\ 3 & 4 & -2 \end{bmatrix}$$

We could use the general formula derived in Example C.5 for the adjoint of A, but it is easier to do it by inspection. First replace each element by its cofactor, getting

$$\begin{bmatrix} -6 & 7 & 5 \\ 14 & -11 & -1 \\ -1 & 5 & -1 \end{bmatrix}$$

Next transpose to obtain the adjoint.

$$\text{adj } A = \begin{bmatrix} -6 & 14 & -2 \\ 7 & -11 & 5 \\ 5 & -1 & -1 \end{bmatrix}$$

The inverse is then given by Eq. (C.3). We have already evaluated the determinant of A in Example C.3, finding $\Delta = 16$. Thus

$$A^{-1} = \frac{1}{16} \begin{bmatrix} -6 & 14 & -2 \\ 7 & -11 & 5 \\ 5 & -1 & -1 \end{bmatrix}$$

or

$$A^{-1} = \begin{bmatrix} -6/16 & 14/16 & -2/16 \\ 7/16 & -11/16 & 5/16 \\ 5/16 & -1/16 & -1/16 \end{bmatrix}$$

We can use matrices to represent, very concisely, systems of equations, and to solve the equations. Suppose a physical problem generates the set of three linear independent equations

$$a_{11}x + a_{12}y + a_{13}z = c_1$$

$$a_{21}x + a_{22}y + a_{23}z = c_2 \qquad (C.4)$$

$$a_{31}x + a_{32}y + a_{33}z = c_3$$

where x, y, z are unknowns. Define these matrices:

$$A = \begin{bmatrix} a_{11} & a_{12} & a_{13} \\ a_{21} & a_{22} & a_{23} \\ a_{31} & a_{32} & a_{33} \end{bmatrix}$$

$$r = \begin{bmatrix} x \\ y \\ z \end{bmatrix}$$

$$c = \begin{bmatrix} c_1 \\ c_2 \\ c_3 \end{bmatrix}$$

Then the rule for the multiplication of matrices shows that the product Ar is

$$Ar = \begin{bmatrix} a_{11}x + a_{12}y + a_{13}z \\ a_{21}x + a_{22}y + a_{23}z \\ a_{31}x + a_{32}y + a_{33}z \end{bmatrix}$$

Since two matrices are identical if their corresponding elements are identical, comparison of the matrices Ar and c with the set of algebraic equations shows that

$$Ar = c \qquad \text{(C.5)}$$

That is, the matrix equation (C.5) is equivalent to Eq. (C.4).

Now let us multiply Eq. (C.5) by the inverse matrix A^{-1}.

$$A^{-1}Ar = A^{-1}c$$

But by Eq. (C.2) $A^{-1}A = I$, the unit matrix, which plays the same part in matrix algebra that 1 does in algebra. Hence the solution of the matrix equation is

$$r = A^{-1}c \qquad \text{(C.6)}$$

Example C.7. Solve the equations

$$x + 2y + 3z = 9$$

$$2x - y + 2z = 11$$

$$3x + 4y - 2z = -4$$

Define matrices

$$A = \begin{bmatrix} 1 & 2 & 3 \\ 2 & -1 & 2 \\ 3 & 4 & -2 \end{bmatrix}$$

$$r = \begin{bmatrix} x \\ y \\ z \end{bmatrix}$$

$$c = \begin{bmatrix} 9 \\ 11 \\ -4 \end{bmatrix}$$

Then the matrix equation $Ar = c$ is solved to give $r = A^{-1}c$. We therefore need the inverse matrix of A. First obtain the determinant of A:

$$\begin{aligned} \det A &= 1(2-8) - 2(-4-6) + 3(8+3) \\ &= \quad -6 \quad + \quad 20 \quad + \quad 33 \quad = 47 \end{aligned}$$

Next find the adjoint of A. Replacing each element by its cofactor gives

$$\begin{bmatrix} (2-8) & -(-4-6) & (8+3) \\ -(-4-12) & (-2-9) & -(4-6) \\ (4+3) & -(2-6) & (-1-4) \end{bmatrix} = \begin{bmatrix} -6 & 10 & 11 \\ 16 & -11 & 2 \\ 7 & 4 & -5 \end{bmatrix}$$

Now transpose:

$$\text{adj } A = \begin{bmatrix} -6 & 16 & 7 \\ 10 & -11 & 4 \\ 11 & 2 & -5 \end{bmatrix}$$

The inverse matrix is therefore

$$A^{-1} = \frac{1}{47} \begin{bmatrix} -6 & 16 & 7 \\ 10 & -11 & 4 \\ 11 & 2 & -5 \end{bmatrix}$$

Thus the solution is

$$r = A^{-1}c = \frac{1}{47} \begin{bmatrix} -6 & 16 & 7 \\ 10 & -11 & 4 \\ 11 & 2 & -5 \end{bmatrix} \begin{bmatrix} 9 \\ 11 \\ -4 \end{bmatrix}$$

$$\mathbf{r} = \frac{1}{47}\begin{bmatrix} (-54 + 176 - 28) \\ (90 - 121 - 16) \\ (99 + 22 + 20) \end{bmatrix} = \frac{1}{47}\begin{bmatrix} 94 \\ -47 \\ 141 \end{bmatrix}$$

$$\mathbf{r} = \begin{bmatrix} 2 \\ -1 \\ 3 \end{bmatrix} = \begin{bmatrix} x \\ y \\ z \end{bmatrix}$$

thus $x = 2$, $y = -1$, $z = 3$.

REFERENCES

1. P. J. Davis, *The Mathematics of Matrices*, 2nd ed., Xerox College Publishing, Lexington, MA, 1973.
2. M. L. Boas, *Mathematical Methods in the Physical Sciences*, Wiley, New York, 1966, Ch. 3.

INDEX

Printed in the United States
149546LV00002BA/25/A